The Packaging User's Handbook

The Packaging User's Handbook

Edited by

F. A. PAINE

B.Sc., C.Chem., F.R.S.C., F.I.F.S.T., F.Inst.Pkg., F.I.D.
Secretary General
International Association of Packaging Research Institutes
and
Adjunct Professor
School of Packaging
Michigan State University

Published with the authority of
The Institute of Packaging

BLACKIE ACADEMIC & PROFESSIONAL
An Imprint of Chapman & Hall
London · Glasgow · Weinheim · New York · Tokyo · Melbourne · Madras

Published by
Blackie Academic & Professional, an imprint of Chapman & Hall,
Wester Cleddens Road, Bishopbriggs, Glasgow G64 2NZ, UK

Chapman & Hall, 2-6 Boundary Row, London SE1 8HN, UK

Blackie Academic & Professional, Wester Cleddens Road, Bishopbriggs, Glasgow G64 2NZ, UK

Chapman & Hall GmbH, Pappelallee 3, 69469 Weinheim, Germany

Chapman & Hall USA, One Penn Plaza, 41st Floor, New York, NY 10119, USA

Chapman & Hall Japan, ITP-Japan, Kyowa Building, 3F, 2-2-1 Hirakawacho, Chiyoda-ku, Tokyo 102, Japan

DA Book (Aust.) Pty Ltd, 648 Whitehorse Road, Mitcham 3132, Victoria, Australia

Chapman & Hall India, R. Seshadri, 32 Second Main Road, CIT East, Madras 600 035, India

Distributed in Canada by Nelson Canada, 1120 Birchmount Road, Scarborough, Ontario, M1K 5G4, Canada

First edition 1991
Reprinted 1993, 1994, 1995

© 1991 Chapman & Hall

Typeset in 10/12pt Times by Advanced Filmsetters Ltd, Glasgow
Printed in Great Britain by TJ Press, Padstow, Cornwall

ISBN   0 7514 0151 X

A Catalogue record for this book is available from the British Library

Library of Congress Cataloging-in-Publication Data
The Packaging user's handbook/edited by Frank A. Paine.
    p.   cm.
Includes bibliographical references.
ISBN 0-422-30283-5
1. Packaging-Materials. 2. Containers-Materials. 3. Packing
for shipment. I. Paine, Frank Albert.
TS198.2.P33 1990
688.8-dc20   98-29397
                                                 CIP

∞ Printed on acid-free text paper, manufactured in accordance with
ANSI/NISO Z39.48-1992 (Permanence of Paper).

Preface

The first version of this book, *Packaging Materials and Containers* was published in 1967 and was revised extensively ten years later under the title *The Packaging Media*. Some thirty or so authors were involved in producing the initial texts for these books, and I must acknowledge their material, much of which is still valid. It is now thirteen years since *The Packaging Media*—high time to take stock and incorporate the considerable advances in materials, forms, techniques and machinery that have taken place. In 1977, wherever possible, we asked the original authors to carry out the revisions, but retirements and job changes have now eliminated over twenty of the original authors.

We have therefore appointed an Editorial Board to advise on this more extensive revision, and I wish to thank them for their detailed and helpful assistance: Dr C. J. Mackson and Professor Y. Dagel for general comments and guidance on the overall plan and, in particular, the Introduction (chapter 1); Graham Gordon and Harri Mostyn for assistance with much of Part D on Distribution Packages, and Dennis Hine and Susan Selke for their work in respect of paperboard and plastics retail packaging, respectively. A major contribution was made by the seventh member of the Editorial Board, David Osborne, who advised in the area of glass.

Additionally, my thanks are due to Ken Booth of Swift Adhesives, Eric Corner of E. R. Corner & Associates, and Mike Ford of Scanpac for their comments in respect of adhesives, corrugated packaging and the use of building boards, respectively.

Our thanks are also due to John Wiley and Sons, Inc., of New York for permission to reproduce several figures from the *Wiley Encyclopedia of Packaging*, and to other organisations for continued use of their material. All are acknowledged at the specific point of use.

There are many others, from whom we have obtained assistance—for example, all those listed in the Bibliography. This new Handbook is as accurate as we could make it and we hope that it provides not just facts, but interesting reading for both the student and the more experienced practitioner.

F. A. Paine

Editorial board

Dr Susan Selke,
Associate Professor,
School of Packaging, Michigan State University,
1541 Snyder Road,
East Lansing, Michigan MI49923,
U.S.A.

List of contributors to *The Packaging Media*:

Philip Adcock, Ing.Chem.(Lausanne); D. C. Allen, C.Eng., M.I.Mech.E., A.F.R.Ae.S.; W. G. Atkins; A. D. Brazier, B.Sc., C.Chem., F.R.S.C., M.Inst.Pkg.; J. H. Briston, B.Sc., C.Chem., M.R.S.C., F.P.R.I., F.Inst.Pkg.; A. O. D. Davies, M.Inst.Pkg.; R. F. D'Lemos; D. J. Flatman, F.Inst.Pkg.; M. J. Ford, B.Sc.; R. R. Goddard, M.Inst.Pkg.; Joan U. Gooch, M.I.Inf.Sci.; J. R. Green, L.I.M., M.Inst.Pkg.; D. J. Hine, M.Sc., M.Inst.Pkg., F.Inst.Pkg.; B. Lindop, M.Inst.Pkg., M.Inst.M.; R. M. C. Logan; A. R. Lott, B.Tech., M.Inst.Pkg.; J. M. Montresor, M.A., M.Inst.Pkg.; H. Mostyn, B.Sc., M.Inst.Pkg.; C. R. Oswin, M.A., D.Sc., C.Chem., F.R.S.C.; G. J. P. Porteous; D. Price, M.I.Mech.E.; J. Radcliffe, M.Inst.Pkg.; M. Rawson, M.Inst.Pkg.; P. Robinson, B.A.; A. Simpson, F.Inst.Pkg.; C. Weeden, B.Sc.(Econ.), F.S.G.T.

Contents

PART A
Introduction

1
Introduction

Package functions

The basic function of all packaging is to identify the product and carry it safely through the distribution system to the final user. Packaging which is designed and constructed solely for this purpose adds little or nothing to the value of the product. It merely preserves farm or processor fresh quality, or prevents physical damage. Cost effectiveness is the sole criterion for success. If, however, the packaging facilitates the use of the product, is re-usable or has an after-use, some extra value is added which can justify extra cost. Such developments must be judged on their merits.

Efficient packaging is a necessity for almost every type of product whether it is mined, grown, hunted, extracted or manufactured. It is an essential link between the product maker and his customers. Unless the packaging operation is performed correctly, the reputation of the product will suffer and the goodwill of the customer will be lost. All the skill, quality and reliability built into the product during development and production will be wasted, unless care is taken to see that it reaches the ultimate user in the correct condition. Properly designed packaging is the main way of ensuring safe delivery to the final user in good condition at an economic cost.

The packaging components come together with the products either in complete or component form, and progress through the packaging engineering operations. The packaged product is then protected throughout the distribution system until, when the product has been delivered, the packaging is ready for disposal or recycling. As is shown in Figure 1.1, even the materials used for protective purposes should be recovered and recycled in some way to achieve the best economic result.

Thus the packaging functions require specialized knowledge and skills, in addition to specific machinery and facilities to produce a package which will provide most, if not all, of a number of basic requirements. Of these, the most important are containment, protection and presentation, communication, machineability (i.e. suitability for the packaging line) and convenience in shape, size and weight for handling and storage. The package must be adapted for the use of the product it contains and must be environmentally friendly in respect of manufacture, use and disposal.

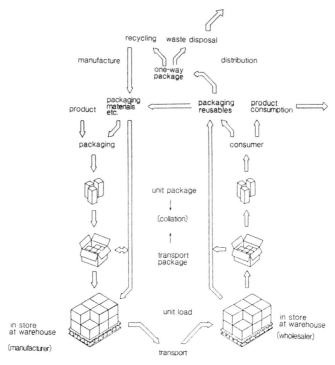

Figure 1.1   Life cycle of single trip and returnable packaging through production, distribution and retail/consumer sectors.

Moreover, these basics must be provided at one or all of the three levels of packaging usually employed: the primary pack; the secondary package or shipping container; and the unit load. Let us consider a number of these basic needs in a little more detail.

1. *Containment:* Obviously, the packaging must hold the contents and keep them secure during the period between the end of the packaging line and the time when the last of the contents have been utilized.
2. *Protection and preservation:* The packaging must protect the product from the mechanical and climatic hazards of the environment encountered during distribution and use.
3. *Communication:* All retail packages must communicate, for not only do they have to identify the contents, but they must also assist in selling. The unit load and/or the shipping container must inform the carrier about the destination, provide any instruction about the handling and stowage of the goods, and perhaps inform the user as to the method of opening the package and assembling the contents.
4. *Machineability:* The great majority of modern retail packages and many transport packages are today erected, filled, closed and collated on machinery operating at speeds of 1000 units or more per minute.

The packaging machinery must therefore perform without too many stoppages or the process will be wasteful of material and uneconomic. Even when the numbers concerned are small and items specialized, the need for a good performance in the filling and closing operations is important.

5. *Convenience and use:* The most common impression of convenience in terms of retail packaging is that of providing easy opening, dispensing and/or after-use. But convenience requires much more to be considered than this. Both the shipping container and the retail package must provide convenience at all stages, from the packaging line, through warehousing and distribution, to satisfying the needs of the consumer.

The need for packaging may therefore be summarized in the following common definitions.

1. Packaging is a coordinated system of preparing goods for transport, distribution, storage, retail sale and end use.
2. Packaging is a means of achieving safe delivery in sound condition to the final user at a minimum overall cost.
3. Packaging is a techno-economic function for optimizing the costs of delivering goods whilst maximizing sales and profits.

Packaging has also been described as a complex, dynamic, scientific, artistic and controversial segment of business. It is certainly dynamic and is constantly changing. New materials need new methods, new methods demand new machinery, new machinery results in better quality; this in turn opens up new market opportunities, new markets require changes, and the cycle starts again.

Packaging material usage

The main packaging materials and their value share of the total packaging used are shown in Table 1.1. These figures relate to Europe in 1986, but

Table 1.1 Main packaging materials: their value and tonnage (Europe 1986).

Material	% of total value	% of total tonnage
Paper and paperboard	39	38
Metal (principally steel and aluminium)	21	9
Plastics, mouldings and film, including cellulose film	26	13
Glass	10	31
Wood	4	9
Miscellaneous	17	

similar proportions of the market will be applicable in almost all the sophisticated countries in the Western world.

As Table 1.1 shows, the principal materials used in packaging today are paper, glass, steel, aluminium and plastics. Wood and textiles are still used to some extent, but they are steadily being replaced. The vast range of products available in packaged form requires sophisticated materials capable of containing, protecting, transporting, preserving and dispensing each product according to its particular characteristics and purpose. In many instances, various materials are combined to produce the most suitable container. The advantages of the most commonly used materials are as follows.

1. *Paper:*
   - can be produced in many grades and converted into many forms
   - can be readily combined with other materials to form coated or laminated products
   - can be made to varying degrees of opacity
   - is particularly suitable in box or carton form for transport
   - can be converted to rigid packaging
   - maintains its characteristics over a wide temperature range
   - is recyclable
2. *Glass:*
   - is transparent but can be coloured to suit a particular packaging application
   - is chemically inert and impervious to contents and to outside contamination
   - has a high vertical compression strength
   - is rigid and suitable for high speed packaging lines
   - can be made into a variety of shapes and sizes
   - provides many resealable packages
   - is both reusable and recyclable
3. *Steel:*
   - provides strength for transport over long distances
   - is easily fabricated
   - can be electronically welded
   - can be hermetically sealed
   - can accept tin or organic coatings
   - permits thermal processing of contents and is not affected by heat or cold
   - can be decorated
   - is impenetrable to light, moisture and odours
   - can be extracted from solid waste
4. *Aluminium:*
   - has a good light weight/high strength ratio

- forms an impenetrable barrier to light, moisture and odours
- has a high quality surface for decorating by printing
- is flexible or rigid (according to thickness)
- can be combined with paper or plastics to form laminates
- is compatible with a wide range of heat-sealing resins and coatings for various closure systems
- has value as scrap

5. *Plastics:* No two plastic materials are identical in their combination of properties, Most plastics, however:
   - are light weight
   - are resistant to passage of moisture and gases
   - resist fungi and bacteria
   - provide good heat insulation

In addition, certain plastics exhibit such properties as transparency, flexibility, and high tensile and impact strength. To provide an even greater range of properties, plastics are often combined with paper or aluminium foil or with other plastic materials to form a laminate. The major plastic materials used in packaging applications are polyethylene, polyvinyl chloride (PVC), polystyrene, polypropylene and polyethylene terephthalate (PET).

Table 1.2 compares the consumption of packaging materials in Europe, United States and Japan (the leading industrialized areas) with the figures for the world as a whole. The differences are very striking. A further comparison for the same areas on a per capita basis is given in Table 1.3. An estimate of the anticipated changes by the year 2000 in the world's consumption of packaging materials is given in Table 1.4 and illustrated in Figure 1.2.

Table 1.2 Consumption of packaging materials and turnover of packaging in Europe, United States, Japan and the world.

	Approx. consumption (kt)			
Material	Europe (351)[a]	United States (234)	Japan (121)	World (5000)
Wood	4000	5000	1600	20000
Paper and board	14200	24500	10000	64000
Metal	3800	4800	1800	15000
Glass	12500	14000	2200	35000
Plastics	5200	5000	2400	16000
Totals (kt)	40000	54000	18000	150000
Approx. value in billions Stg £	32	40	21	120

[a]Figures in parantheses are the population in millions.

Table 1.3 Per capita comparison for the areas of Table 1.2 (kg/cap per year).

Material	Europe	United States	Japan	World
Wood	11.4 (9.5)[a]	21.4 (9.3)	13.8 (8.9)	4.0 (13.3)
Paper and board	44.0 (36.7)	104.7 (45.5)	84 (54.2)	12.8 (42.7)
Metal	10.2 (8.5)	20.5 (8.9)	15.2 (9.8)	3.0 (10.0)
Glass	37.5 (31.2)	53.8 (23.4)	18.2 (11.7)	7.0 (23.3)
Plastics	15.1 (12.6)	23.4 (10.2)	22.9 (14.8)	3.2 (10.7)
Totals	120	230	155	30

[a]Figures in parentheses are the percentages of the totals for each of the five packaging media.

Table 1.4 World consumption of packaging materials in the world in 1985 and estimates for the year 2000 (in million tonnes).

	Year		Increase over 15 years (%)
Packaging material	1985	2000	
Paper and board	64	107	67
Glass	35	48	37
Wood	20	21.5	7.5
Plastics	16	53	232
Metal	15	17.5	16
Total	150	247.0	64.5

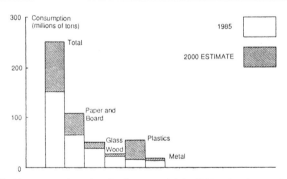

Figure 1.2 World consumption of packaging materials in 1985 and estimates for the year 2000. Source: IFEC.

Of course, estimates such as those given in Table 1.4 and Figure 1.2 often fail to materialize because of unpredicted major events in world history. Predictions made in the early 1920s were nullified by the world depression in 1929–1931. The revised versions were altered by World War II. The oil price rises of 1973/1974 precipitated a major upheaval in the costs of many materials and now the current political developments in the planned economy countries of Eastern Europe and China will have an equally unpredicatable effect.

Table 1.5 Estimate of the per capita consumption of packaging materials according to the per capita income in the year 2000 (kg/capita per year).

	Class of country				Mean consumption of packaging materials at world level	
Packaging material	Advanced industralized country (pop: 500)[a]	Less advanced industrialized country (pop: 740)	Developing country (pop: 2800)	Under-developed country (pop: 2164)	1985	2000
	10,000[b]	5000[b]	1000[b]	100[b]		
Metal	8	4	3	1	3	2.8
Glass	45	10	5	2	7	7.8
Paper and board	70	45	10	5	12.8	17.3
Wood	10	5	3	2	4	3.3
Plastics	30	20	6	3	3.2	8.6
Total	163	84	27	13	–	–

[a]Population figure in millions.
[b]Income per capita.

Nevertheless, it is instructive to examine the predictions for the year 2000, not for specific countries but for the various areas of the world classified according to the degree of development they are experiencing (see Table 1.5).

The forms in which the main packaging materials are employed are shown in Table 1.6, which also gives some idea of the wide variety available.

Packaging costs

Meaningful figures for packaging expenditure in specific industries are difficult to obtain. Prices for particular primary packages are misleading without a knowledge of several other factors, such as the cost of the shipping unit, the amortization cost of packaging machinery, the labour requirement of the packaging operation, and so on.

In addition, while at any point in time, the cost of any specific packaging will be fixed in relation to all its alternatives, the cost in relation to the product selling price will depend on the value of that product. For example the cost of a 0.5-litre glass container complete with all accessories might be 20p. When filled with mineral water of less than 2p in value, the ratio of cost of packaging/cost of product is 10. When filled with milk selling at about 60p/litre, the packaging/product cost ratio is reduced to 20/30 or about 0.7. Fill it with a liqueur whisky which retails in a 'duty-free' shop at £10.00, and the packaging/product ratio becomes fractional at 20/1000 or 0.02.

Table 1.6   The forms in which the main packaging materials are employed.

Material	Package types
Paper and board	Wrapping papers, bags and carrier bags Boxes and cartons Fitments in board Tubes, spiral and straight-wound Fibreboard cases and fittings Multiwall sacks Fibre drums
Metal	Cans and boxes (including aerosols) Aluminium foil, laminates and labels Collapsible tubes Closures Metal strapping and banding Barrels, kegs and drums Crates or boxes
Glass	Bottles, jars Vials and ampoules
Plastics (including cellulose and rubber)	Films, laminates and sheets Bags, pouches, sachets, etc. Sacks (film and woven tape) Moulded bottles, jars, pots, etc. Thermoformed trays, pots, blisters and fitments, etc. Cushioning materials and fittings Caps and closures Drums Crates Boxes
Timber (including plywood)	Boxes and crates Casks and kegs Pallets and containers Baskets and punnets Wood wool
Textiles	Sacks and bags Baling materials

However, if we had set up returnable bottle systems for the mineral water or the milk and had achieved 20 trips on average for each bottle, the bottle cost alone of 1p per trip would look much more reasonable. We must take into account the extra costs of returning the bottle to the filling point and cleaning it, which may raise the cost per trip to 2p, giving packaging/product ratios of 1.0 for the mineral water and 2/30 or 0.07 for the milk.

Again the bottle cost, as with all packaging, will depend on the size of the order. A unit price of 20p for an order below 10 000 bottles could reduce to 15p for 25 000, and might fall to 12p for 1 million. Table 1.7 gives some typical examples of costs in relation to quantity supplied for milk packagings. The costs are in German marks for the year 1984.

Table 1.7 Cost prices for different quantities of various packaging.

Packaging	Volume of package (litres)	Cost for x thousand packs in DM/100							
		x = 2.5	6	10	12	15	25	50	100
Glass bottle[a]	0.33	20.5	18.5	–	17.7	16.8	14.5	14.1	13.9
Euro-glas bottle[a]	0.55	20	18	–	16.3	15.2	13.5	13.3	12.9
Light weight glass	0.33	10	9.35	–	8.75	8.40	8	7.85	7.7
Tinplate can	0.35	–	–	22	–	–	–	21	20.5
Aluminium can	0.35	–	–	22	–	–	–	21	20.5
Plastic tub	0.25	–	–	7.3	–	–	–	7	6.7
Plastic tub	0.4	–	12	11	–	–	–	9.9	9.3
Polypak[b]	1.0	–	–	2.7	–	–	–	2.6	2.5
Tetrabrik	1.0	–	–	6.7	–	–	–	6.4	6.15
Blocpak	1.0	–	–	7.8	–	–	–	7.5	7.35

[a]Multi-trip returnables—remainder all one-trip.
[b]Polyethylene pouch.

Leonard has produced an excellent example of the complexity of comparing the costs of packaging materials (Tables 1.8 and 1.9), where the costs of packaging an instant beverage powder in six possible ways are set out. The result indicates that packaging in glass is the most expensive and cartons with barrier liners are the least costly. If we consider the costs of the filling, closing and packing operations and include the machinery write-off rates (amortization) and the labour content of the operation, the final cost comparison could change.

Shipping containers

Shipping containers are principally made from wood, fibreboard and metal. There is a limited usage of glass for corrosive liquids, such as acids, etchants and other chemicals, but the quantities are small, and the glass must always be protected from impact. Wood is generally used for large packages (e.g. machines), or even for small packages if the product is of high density

Table 1.8 Product X, basis for cost calculations (US dollars, 1983).

Product: a dry, moisture-sensitive granular substance
Density: 0.4798 g/cm^3
Net package weight: 454 g (1 lb)
Package size: 454/0.4798 = 946.2 cm^3 (1 qt)
Primary package options: glass jar   composite can
                         plastic jar   paperboard carton
                         metal can   flexible-foil bag
Type of shipping container: corrugated box
Primary packages/shipping container: 12
Shipping containers required for 1000 packages: 83.3
Other requirements: re-closability of primary package

Table 1.9   Product X, comparative package costs (US dollars×1000).

	Primary package	Closure/ re-closure	Label	Corrugated box	Total
Glass jars	250.00	35.00	9.50	a	294.50
Plastic jars	175.00	35.00	8.70	28.00	246.70
Metal cans	175.00	25.00	9.65	16.80	226.45
Composite cans	157.00	25.00	b	16.80	198.80
Paperboard cartons	53.10	12.75	c	14.60	80.45
Flexible-foil bags	70.00	d	e	11.70	81.70

[a]Glass prices normally include the corrugated box unless specifically excluded by the buyer.
[b]Labels are part of spiral-wound composite cans as delivered.
[c]Paperboard cartons are preprinted.
[d]Foil bags do not require a separate closure. They are folded at the top after filling and heat-sealed or glue-sealed.
[e]Foil bags are preprinted.
Note: Costs are estimates based on requirements of 10–20 million (10^6) packages. Actual costs depend on volumes ordered, location of delivery within the continental United States and specification details.

(e.g. minted coins and other metal items). Timber cases and crates are used extensively for weights above 100 kg (2 cwt) while below this weight, fibreboard, both solid and corrugated, is the favoured material. Timber is also used for casks for wine, beer, etc., but is becoming replaced by metal (stainless steel or aluminium), either alone or with inner plastic liners.

Plastic materials are also used for shipping containers but, since the basic material cost is high, the main use is for returnable containers. Plastic crates have long been established in the dairy industry, and also in the more exacting use for bottled beers, minerals and soft drinks. Beer crates are usually stacked higher and for longer periods than milk crates; polypropylene is therefore preferred for beer, while high density polyethylene has proved adequate for milk crates. Polyethylene is also used in carboys for many liquids, thus avoiding the protection required for their glass counterparts. Polyethylene casks or barrels are also used.

Expanded polystyrene is employed in shipping containers for tomatoes and grapes, and also for both cured and wet fish. In the last two uses, the heat insulation properties are useful in keeping the product cold with the minimum of solid coolant.

Corrugated fibreboard cases are probably the most widely used shipping containers. They combine convenience with economy and hygiene. The most common type of case is the one-piece (or regular) slotted container, although open-tray and wrap-around styles are used extensively. The normal range of weight which corrugated fibreboard cases carry lies between 5 and 20 kg without any special fittings being used. Specially reinforced containers can be produced, capable of carrying loads of powdered or granular material of 500 kg or more.

In the last decade, a revolution in the movement of goods has taken place.

Palletization, modular packaging, IBCs (intermediate bulk containers) and freight containers have arrived. In some instances this has led to the pallet superseding the wooden case or crate, and in others the almost complete elimination of the conventional shipping container, since the goods are loaded straight into a freight container and secured within. This is particularly applicable to heavy machinery.

The main purposes of a shipping container can be listed as follows:

1. It must contain the product efficiently throughout the journey.
2. It must provide protection against external climatic conditions and contaminants.
3. It must be compatible with the product.
4. It must be easily and efficiently filled and closed.
5. It must be easily handled by the appropriate mechanical or other means.
6. It must remain securely closed in transit, but open easily and, when required (as for customs inspection), be capable of efficient and secure reclosure.
7. It must communicate to the customer, the carrier, the wholesaler and the manufacturer, all that is necessary to know about the product and its destination, as well as how to handle and open the package.
8. Where the product is dangerous or potentially harmful (as for chemicals and acids), the package must be virtually unbreakable.
9. The package must be inexpensive and cost no more than is absolutely necessary to do its job.
10. The package must be readily disposable, or be re-usable, or have an after-use for some other purpose.

Very few containers will need to satisfy all these requirements. The emphasis as to which particular ones predominate will be very much dependent upon the product packed.

Retail packaging

Successful selling of any product involves both producing a good quality article and then ensuring that it reaches the ultimate user in as near perfect condition as possible. Product design or formulation controls the first, and packaging for a specific distribution system takes care of 90% of the second.

Since packaging is an activity practised by almost every industry and is employed universally to deliver every type of product to the ultimate user, it is useful to recognize that there are at least four kinds of users of packaging:

1. Makers and packers of products
2. Distributors and warehouse staff
3. Retailers of all kinds
4. The final customer

Each of these users has a set of requirements for the packaging, and all the requirements need to be taken care of so that the final retail package reaching the ultimate consumer has done its job.

Four major considerations are involved in retail package design: the product, its market, the package production problem, and the economics of the packaging operation.

Under product considerations come such questions as susceptibility to damage or deterioration, the need for protection against temperature changes, moisture changes, oxygen, light, mould, or insects, and whether the product will interact with the potential container.

Market considerations include the characteristics of the customer, such as age group and income level, the unit quantities required, aspects of the retail and wholesale operations, and probable methods of handling, warehousing and distribution. The graphics considered include brand name, whether the product is part of a range, whether there are colours that may not, or must be used, any pertinent government regulations, and the economic effect of all these factors. Recyclability is now an increasingly important marketing factor.

Approximately two-thirds of all packaging expenditure is on retail packages. Metal cans and boxes account for about 32%, paper boxes and cartons for 20%, plastic packages for 20%, glass for 12%, with paper bags and foil accounting for about 6% each.

Retail packages are sometimes criticized, and it is suggested that a good retail package should comply with the following guidelines.

1. The package must comply with all legal requirements.
2. The package should ADEQUATELY contain, protect and preserve the product.
3. The packaging must NOT MISLEAD the customer about the nature, quality or quantity of the contents.
4. Packagings for hazardous products must contain them SAFELY, incorporate SUITABLE warnings and be fitted with APPROPRIATE closures.
5. The packaging must be made from materials that have no ADVERSE effect on the product.
6. Packaging should be CONVENIENT and SAFE in NORMAL use and have regard to the nature and purpose of both pack and contents.
7. The package should contain CONVENIENT quantities of product.
8. The packaging should PROPERLY identify the product and provide instructions for use, where APPROPRIATE.
9. The packaging should be designed with regard to its re-use and recycling and ultimate disposal.

Apart from the legal requirements, all the guidelines have words (in capital letters) which are open to interpretation. This tends to reduce them to trite, though well-intentioned statements. The UK Code (see Table 1.10) attempts to be more specific.

All retail packages must contain, protect and preserve, communicate, and

Table 1.10 Code for the packaging of consumer goods.

The prime function of packaging is to enable consumers to receive products in good condition at the lowest reasonable price. Any manufacturer, distributor or retailer concerned with design or use of packaging has a responsibility to ensure that there is a regular review of packaging having regard to the economies of the total manufacturing/distribution chain and to considerations of re-use, and disposal. Marketing and commercial considerations should be reconciled as far as possible with economy in the use of materials and energy, and the environment.

(1) Packaging must comply with all legal requirements

(2) In containing a product the package must be designed to use materials as economically as practicable while at the same time having due regard to protection, preservation and the presentation of the product

(3) Packaging must adequately protect the contents under the normal foreseeable conditions of distribution and retailing and also in the home

(4) The package must be constructed of materials which have no adverse effects on the contents

(5) The package must not contain any unnecessary void volume, nor mislead as to the amount, character or nature of the product it contains

(6) The package should be convenient for the consumer to handle and use; opening (and reclosure where required) should either be obvious or indicated and convenient and appropriate for the particular product and its use

(7) All relevant information about the product should be presented concisely and clearly on the package

(8) The package should be designed with due regard to its possible effect on the environment, its ultimate disposal, and to possible recycling and re-use where appropriate

be easily filled and closed appropriately. Many must also sell and provide convenience either in terms of opening, closure and/or reclosure, or by providing useful dispensing methods or simplifying the use of the product in other ways. Unless such advances are real and valuable, the package rapidly ceases to be used.

A good retail package can sell a poor product once, but a badly designed or unduly expensive package will much more easily ruin the market for an excellent product. The consumer certainly understands *caveat emptor* far more than many anti-packaging pressure groups believe.

Packaging and the environment

During the last forty years, public awareness of the use of materials in the distribution of goods of all kinds has increased considerably. In the last fifteen years, awareness of the utilization of fuels of all types for the production of energy has moved from an attitude of almost complete indifference to one of concern. The solutions to questions of material resources, energy usage and conservation are all dependent on the type of society in which we desire to live.

The so-called throw-away society has developed in response to a desire to save labour and time, both of which have in the past been priced at a level above that of most materials and of energy. Consequently, packaging designers have produced packages tailored to carry out all their functions effectively at the lowest overall cost in money terms. Packaging has therefore become a techno-economic function in the distribution of products, with the objective of minimizing overall costs whilst maximizing sales and profits. Society determines its needs, and the package-making industries, the producers of goods, the transport organizations and the retail outlets try to meet them as efficiently as possible.

Criticisms have been levelled at the whole concept of packaging by some ecologists and environmental pressure groups, that packaging makes excessive use of resources in a wasteful way, adds to the burden of waste disposal and uses large amounts of energy, including fossil oil as feedstock for plastics. To these specific criticisms, a number of others have been added at different times, including some from consumers that the packaging adds unnecessarily to the cost of the product, that it limits consumer choice, may be used in a deceptive manner, and is a major constituent of litter.

Packaging is, in fact, a relatively small user of energy (Table 1.11); it is one of the major forces for preserving and protecting goods and for keeping waste to a minimum; and continually strives to use less material to carry out its very specific goals. All concerned with making and using packaging continually seek to understand better why they operate as they do and, where inefficiences exist, to rectify them. Because packaging materials and techniques are continually being improved, 100% efficiency is never likely to be possible and, no doubt, at any given time, about 25–30% of all packaging could be improved. Let us examine the main criticisms in more detail.

Table 1.11   Energy used[a] in container production (tonnes of oil equivalent per tonne).

	Aluminium (%)	Plastics (%)	Paper (%)	Tinplate (%)	Glass (%)
Raw material production	6.00(95.3)	2.30(78.7)	1.45(91.2)	1.00(86.3)	0.35(92.5)
Conversion to containers	0.20(3.2)	0.40(13.7)	0.05(3.2)	0.10(8.6)	
Heating and lighting factories	0.08(1.2)	0.16(5.5)	0.07(4.4)	0.04(3.4)	0.02(5.0)
Transport to user	0.02(0.3)	0.06(2.1)	0.02(1.2)	0.02(1.7)	0.01(2.5)
Total	6.30	2.92	1.59	1.16	0.38

[a]Note that most of the energy is used in producing the material, (metal, glass, plastics or paper) and the actual conversion into packaging, even with plastics, consumes less than ⅕ of the total.

Energy considerations

In principle, the estimation of the energy required for any purpose should be straightforward, once the operations to be included have been decided. In practice, it is just these decisions that cause difficulty, together with the fact that few industries at present monitor the energy requirements of individual machines.

For example, the energy used to produce a can for beer must obviously include the operations of mining the ore, extracting the metal, rolling it into sheets, converting the sheets to bodies and ends, and assembling the final can. But should we include the energy needed to make the machinery used in all these operations? Should we include the energy used in transporting ore to the smelter, sheets to the can-maker and cans to the packer? Is the energy used in making labels, printing inks and varnishes to be included in our calculations?

These questions may be answered as shown in Table 1.12, which gives the operations to be accounted for in the case of aluminium metal. Thirteen operations were selected, and for each the energy required was estimated. Table 1.13 reports the energy estimates for some common packaging materials (note that because the figures are from a different source, they differ from those in Table 1.11).

Table 1.12 Energy to produce 1 kg of aluminium metal in the United Kingdom from its ore.

Operation	Energy needed (MJ)[a]
1. Mining bauxite	3.1
2. Bayer process	44.6
3. Transport	1.0
4. Smelting	190.1
5. Mining fluorspar	0.3
6. Producing hydrogen fluoride	1.3
7. Conversion to aluminium fluoride	1.1
8. Brine extraction	0.3
9. Mining limestone and calcining it	0.9
10. Brine electrolysis	0.4
11. Synthesizing cryolite	0.8
12. Petroleum coke production	42.0
13. Pitch production	7.2
Total	293.1 or 6.8 tonnes oil equivalent (TOE/per tonne Al)[b]

[a]The conversion factors are 1 kg oil = 43.2 MJ and MJ per kg×430.4 = BTUs per lb.
[b]Notice that although this figure is close to that given in Table 1.11, it is not the same, because the figures were not obtained using the same derivation. It has been found that the energy required is dependent on the location of the ore and the place of production, as well as the manufacturing process. For example, if in the case of aluminium we took account of all the steps in the processing of all the aluminium plants in the world in relation to the total global quantity of aluminium produced, we would obtain a figure for the energy requirement of about 6.0 TOE/per tonne of aluminium, but variations between 5 and 8 (and possibly more) will be found in specific instances.

Table 1.13　Typical energy requirements to make 1 kg of various materials from raw materials in the ground.

Packaging material	Total energy	
	MJ/kg	TOE
Aluminium	293	6.8
Cellulose film	192	4.4
Shrink wrap film	187	4.3
Polyethylene terephthalate resin	183	4.2
Can sealing compound	180	4.17
Polypropylene film	173	4.0
Low density polyethylene resin	104	2.4
Paperboard	99	2.3
Kraft paper	82	1.9
Tinplate	50	1.2
Glass containers	22	0.5

Notice the difference in Table 1.13 between the energy required by the polyethylene resin and the polypropylene film (a similar hydrocarbon). Could this be due to the energy used in extruding the film from the resin? Compare the figures with those in Table 1.11 derived from a different source.

This discussion indicates the difficulties and emphasizes that if values are to be compared, they must be the values for the actual packagings involved. Generalizations can only mislead.

Finally, the values in Table 1.13 are derived on a weight basis (energy per kg of material). Packaging materials are used on an area or volume basis, and the area of any film material obtained from 1 kg will be much greater than that from 1 kg of tinplate. Likewise the volume of product packed by 1 kg of glass will depend on the weight of each container. Thus, although glass and tinplate use the least energy per unit weight, they have the highest densities and this must also be taken into account. After examining these figures, it could be said that a little oil can go a long way (see Table 1.14).

The energy values directly related to the problem are the ones to use whenever possible, and some idea of their accuracy (i.e. probable range of values) is also required. To simplify matters, however, relatively rough estimates may be used for materials which contribute very little to the total energy needs.

Similar problems exist with transport and the heating of production areas (depending on geographic conditions). Energy may be supplied as electricity, or as solid, liquid or gas fuels, and the thermal energy of the specific fuels used are the values to be taken into account. The actual transport distances should be used and estimates for the energy requirements of the modes of transport actually employed. Finally the measured values for heating production spaces should be used in estimating the energy consumption.

Table 1.14 Oil equivalent required as feedstock and/or energy for manufacture of various packages.

Packaging[a]	Oil as feedstock and/or as energy (tonnes)
One million m² of film	
Polypropylene	110
Cellulose	155
One million fertilizer sacks	
Polythylene film sacks	460
Three-ply sacks	700
One million 0.33 litre single trip containers	
Polyvinyl chloride bottles	81
Glass bottles	90
Aluminium cans	142
Tinplate cans	75

[a]Oil is required as the feedstock for the plastic material and this has been added to the energy requirement in tonnes of oil equivalent.

Providing the complete process is studied, it is not necessary to consider any internal recycling of materials, e.g. scrap fed back into the system, but if post-consumer packaging (e.g. glass from a bottle bank collection) is involved, then this must be accounted for in the estimate.

Generally, the following should be presented in any energy study:

1. The content and the boundaries of the energy system
2. The inputs of energy that are included
3. The combustion values of the actual fuels employed
4. The efficiency quotas of any energy transformations
5. The estimates of energy requirements of any materials, where measured values are not available

The recommended method for energy calculation is based upon a division of the total energy system into subsystems. These subsystems can be used as building blocks that can easily be put together for various purposes, provided that they represent mutually independent and directly additive quantities.

An energy-consuming system may be divided into five main subsystems, each of which consists of three basic building blocks. The five main subsystems are:

1. Procurement of basic raw materials
2. Manufacture of packaging materials
3. Conversion of packaging materials to packages
4. Use of packages
5. Handling of used packages

Each of the five main subsystems consists of the following three basic building blocks:

1. One process part
2. One transport part
3. One process space part

The energy-consuming system may be presented as an energy matrix, where each row contains one of the five main subsytems. The columns, therefore, represent each one of the basic building blocks which can be repeated for each geographic area considered in a particular energy study. A total mapping of one energy-consuming system results in one or more matrices, each one containing at least 15 subunits (a 5×3 matrix).

An energy study may, of course, be limited to only one or a small number of these basic building blocks. It is still a good idea to use the blocks defined here to retain the simple additive properties of the method. The model used is an input-output model, where the cell or block boundaries are so chosen that all material and energy inputs are either known, can be calculated, or can be estimated. The same is obviously true for the corresponding outputs.

Material resources

Table 1.15 indicates the materials that are required to deal with the situation in today's complex society. Figures in parentheses indicate the usage of those materials that are concerned with packaging. In no instance can packaging be said to be utilizing 'scarce' materials. It is also a fact that apart from an increase in growth of the plastics used in packaging, most other packaging media are either static in terms of tonnage used, or their use is decreasing. This is very largely due to the fact that over the past 50 years, a

Table 1.15 Consumption of some raw materials used in modern society based on figures from the United Kingdom

	Tonnage (million of tonnes per annum)[a]
Sand, gravel and stone	200
Coal	154
Oil and gas	95
Bricks	25
Iron and steel	24 (0.85)
Cement	16
Food	15
Wood	7.7 (1.0?)
Paper and board	7.3 (3.5)
Glass	2.5 (1.65)
Plastics	1.4 (0.35)
Copper	0.7
Aluminium	0.6 (0.1)
Rubber (natural and synthetic)	0.5

[a]Figures in parentheses are estimates of packaging usage.

continual process of light-weighting of containers of all kinds—metal, glass, paperboard and plastics—has been in progress. Glass milk bottles, for example, now weigh nearly one-third of what they did 40 years ago. Tinplate containers for many products are down to thicknesses weighing between one-half and two-thirds of their weight before World War II. Paperboard materials particularly for corrugated cases have decreased in weight by a third or more. In addition to decreases in weight in the actual materials used, in many instances less weight to carry has been achieved by switching from one medium to another. Thus many goods, previously packed in wooden cases, are today packed in paperboard packages and goods originally packed in paperboard are now packed in shrinkwraps and film. Hence, although the area of packaging material used has increased, the tonnage of materials used remains either constant or is decreasing.

Recycling of packaging materials

Recycling has been a feature of the packaging industries for decades— mainly in paper, paperboard and glass-making. In these industries, not only has scrap been fed back into the process, but used and waste materials from industry and domestic sources have been re-used. Waste paper and other fibres have been recycled to produce materials for building purposes and chipboards for packaging applications, while glass containers have been re-used after cleaning or employed with 'cullet' to improve the flow and heating characteristics of the charge for the furnace.

Most paper and board is not made from tropical hardwoods and therefore does not contribute to the destruction of tropical rain forests in South America or elsewhere. In fact, papermaking indirectly increases the world's softwood forest reserves. In Sweden, for example, new tree plantings exceed 500 million every year, a figure greater than the number felled. One company expects to plant three seedlings for every single tree cut down. Swedish forest reserves are today greater than they were a decade ago.

Glass, because of its chemical composition, is virtually unchanged by repeated heating and so does not degrade on recycling as do paper and plastics. Traditionally, glass works have always used at least 15–20% of broken glass or 'cullet' in the charge for the furnace, and this has been increasing in recent years, both to save energy (because the broken glass melts more readily than the raw materials) and to conserve those materials. In Switzerland, over 1.5 million tonnes of glass per annum are collected in a bottle bank scheme, and this is used to produce 75% of all new glass bottles, saving some 20 million Swiss francs annually. All glass containers in West Germany and the Netherlands are reported to contain some 50% recycled material, and in Japan a figure of 55% has been claimed. In the United Kingdom, recycled material has reached some 35–40%.

Metal packaging is also recycled extensively. Aluminium and tinplate scrap arising from container-making is recycled in toto and advances have been and are continuing to be made in the recycling of used containers. Aluminium has a ready market from such sources and markets for greater amounts of recovered tinplate are being explored. Plastic process scrap is also largely recycled but so far plastics in domestic waste have been used only as fuel in incineration plants. Some alternative markets are under development.

To summarize, the main functions of packaging are to prevent damage and hence reduce waste—not only of goods and materials but also of energy and labour. When the package has done its job, it can be used for something else and, when it finally becomes a waste material to be disposed of, it often has already reduced the waste disposal problem by reducing the amount of trim (husks, bone, etc.) appearing in urban waste. Even where the primary function of packaging is not that of conservation—it is designed either to improve sales or to provide convenience—both will lead to better profits and will contribute to improvements in the quality of life.

Packaging will continue in the future to serve mankind by containing, protecting, preserving and identifying the products that are needed. It will contribute to improvements in the quality of life in all countries, from the developing to the most sophisticated, by getting goods in the right quantity, at the right time and in prime condition, to the people who need them, at the minimum overall cost. It will continue to do this, although the manner in which it does so may well change in a world where the costs of raw materials and of energy are increasing disproportionately.

There are not likely to be any fundamental changes in the way packaging performs its tasks, and the three trends that have been taking place over the last forty years will probably continue into the 21st century. These are:

1. Packaging designers will continue to develop lighter weight, more economical and more convenient packages for the purposes for which they are needed. Disposal and recycling properties will become increasingly more critical.
2. The functions of containment, protection and preservation, together with identification, will be necessary for all types of packaging. The amount that will be spent on the selling function and the convenience and service aspects of packaging may, however, change in the course of the next decade. In general, the overall cost of the operation of delivering products to consumers will still be a major criterion in the decision on how such changes will occur.
3. Whenever packaging can contribute to a reduction in labour costs, particularly when these are high, then the interaction between packaging materials and filling, forming, and closing machines will play a big part in determining which packaging material will be used for any particular purpose. This will be particularly true where flexible materials are concerned in the packaging of food products and other household goods.

PART B
Packaging Materials

2

Wood and wood-based materials

From tree to merchant

Trees are felled and the branches trimmed off, then the trunks are cross cut into suitable lengths for hauling to the sawmill. At the mill the logs are washed, further cross cut if necessary, and then sawn into timbers or boards. The boards are sorted, graded and dried. After drying and inspection, boards are further dressed as required into finished lumber.

The nature of wood

The 'grain' of wood is a result of its fibrous structural formation. Since these fibres are arranged parallel to the long axis of the tree trunk and in concentric rings, the mechanical properties of a piece of wood are related to the orientation of the grain direction, and also to the pattern of sawing. If a log were to be sawn into boards, the centre plank would have its grain more or less perpendicular to the face, and in the outer planks the grain would be more or less parallel. By quarter sawing, these grain differences are minimized.

Wood resists damage by crushing, stretching, bending and twisting, whether these forces are applied slowly or rapidly. It is also resilient enough to yield temporarily to a load and then return to its former shape. With regard to compression and tension, wood is strongest along the grain and weakest at right angles to the grain. In shear, wood is strongest at right angles to the grain. A beam will best resist bending if the load is applied perpendicular to the grain.

The fibrous nature and other strength properties of wood give it a unique property compared with metals or other amorphous materials; that is, it can be split along planes parallel to the grain leaving a rather smooth surface.

Woods for box-making

Wood has long been used for packaging. There are many kinds of wood—some unsuited for packaging and some suitable but too expensive. Wood is easily shaped and joined, and the suitable varieties have the correct strength properties. The main forest products are logs, roundwood products, sawn wood products, plywood and other laminated products, particle boards, fibreboards, wood pulp, paper and paperboards.

In certain parts of the world, roundwood is still used for baskets and boxes, but apart from wood used in the manufacture of pulp for papers, paperboard and regenerated cellulose, and some usage of building boards and plywoods, the majority of wooden packaging is derived from sawn wood. Particle boards are not used to any great extent for packaging.

The principal species used (generally softwoods) are listed in Table 2.1, grouped according to similarity of characteristics. Table 2.2 gives an indication of the properties of woods typically used for box-making. In general, the order of preference decreases from Group 1 to Group 4.

Timber for wooden packaging should be well seasoned, either by air drying or kiln drying, and have somewhere around 15–20% moisture

Table 2.1   Woods for box-making.

Group 1	Group 2	Group 3	Group 4
Alpine fir	Agba	Black ash	Beech
Aspen	Douglas fir	Black gum	Birch
Basswood	Hemlock	Maple	Hickory
Cedar	Larch and various pines	Red gum	Oak
Chestnut	Gaboon	Sycamore	Rock elm
Cottonwood	Lime	White elm	White ash
Jack pine	Obeche		
Lodgepole pine			
Norway pine			
Spruce			
Sugar pine			
White fir			
White pine			
Willow			
Yellow poplar			

Table 2.2   Properties of typical box-making woods.

Species	Density $(g/m^3)^a$	Volume shrinkage %	Static elastic bending modulus $(kg/cm^2)^b$
Oak (Quercus alba)	0.71	15	125 000
Beech (Fagus sylvatica)	0.68	18	160 000
Birch (Betula verra cosa)	0.61	14	165 000
Pine (Pinus sylvestris)	0.49	12	120 000
Spruce (Picea abies)	0.43	12	110 000

[a]On oven dried wood.
[b]Measured on small clear specimens at 12% moisture.

content. Drying is an essential preparation of all kinds of wood for further use, because it reduces the magnitude of dimensional changes due to shrinkage and swelling, protects the wood from microorganisms, reduces weight, prepares the wood better for most finishing and preservation methods and increases its strength. The object of air drying is to reduce the moisture content of the wood to the lowest value obtainable under the weather conditions in the shortest possible time without producing defects. Naturally, the level to which the moisture can be reduced will depend upon the temperature and the relative humidity. Wind will reduce the time required, but direct precipitation (rain or snow) wets the wood and hinders the progress of drying.

An air drying yard must have reasonable air movement, unobstructed by tall trees or buildings, and a ground surface which is free from debris and vegetation. Gangways must be provided for the working areas and to permit air movement, and the first row of any timber being dried is kept about 40 cm above the ground to allow space for air circulation. As layers are added, spaces are also left for this purpose. Some sort of roof, usually of low-grade timber or panelled material, will be placed on top of each pile, and the time required to air dry from green to 20% moisture will vary from about 20 to 300 days for 2.5-cm thick material, depending upon the species, the place and the time of year.

The other method of drying, kiln drying, is conducted in a closed chamber under artificially induced and controlled conditions of temperature, relative humidity and air circulation. This permits much more rapid reduction of moisture content to levels required, independent of weather conditions. A reduction of moisture from green to 16% will be accomplished in 2–50 days for 2.5-cm thick material. Heating is usually by steam circulation in pipe coils, and some control of the relative humidity by allowing steam to enter the chamber is necessary in order to regulate the exit of moisture, to avoid warping, and so on.

All the species of woods shown in Table 2.1 may be used for packaging, including nailed wooden boxes, battened plywood boxes and framed wooden cases, subject to the following reservations. Group 4 woods should not be used in large plywood sheathed wooden cases, and their use in large framed wooden cases should be limited to rubbing strips, skids, sills and intermediate members forming the framing where the length of the piece does not exceed 2 m (7 ft). In nailed wooden cases, use of Group 4 woods should be limited to ends and battens, if the buyer agrees. Group 3 woods may be used unless the buyer specifically prohibits it. Oak, sweet chestnut and other acid woods should never be used in boxes with metal parts liable to corrosion, and for the same reason unseasoned or high moisture content woods should also be excluded.

Wood for box-making is obtained by resawing from standard thicknesses and nominal (not actual) thickness should be specified for reasons of economy. (Actual thicknesses are about 2 mm (1/16–1/8 in) less than nominal as they allow for loss in sawing standard boards.)

Newly cut or green timber will contain between 30% and 60% moisture determined on initial weight (i.e. 40–200% on dry weight) depending on the species, the time of felling and conditions of transport and storage. Seasoning is carried out by air or kiln drying and should result in a moisture content around 15–16% on dry weight (about 20% as received). Wet woods will dry out with time and boxes made from them will suffer shrinking, warping, loosening of nails, splitting and gapping of joints. All of these problems are avoided by the use of seasoned timber.

Defects in wood

For economic reasons, woods for packaging can rarely be completely free from defects. Those defects allowed, however, should not materially weaken the case or interfere with nailing. Cross grain and knots are the most serious defects to be avoided. Wood must be reasonably sound and not have any bad cross graining. Knots or knotholes should never reach the edges of a board and should not exceed one-third of its width. Cross grain increases the possibility of failure under bending stresses, and knots are most weakening when in the middle third of a supported board. Wane (the original surface of the tree) should not be accepted unless it is not likely to reduce strength or complicate asssembly.

Splits, shakes and checks. A shake is a partial or entire separation of the wood between annual rings. A check is a split which runs radially across the rings. Both are frequently due to unequal shrinkage during seasoning. Splits are due to rough handling or internal stresses and are easily confused with shakes and checks. A method of preventing splits, shakes or checks from increasing is to drive corrugated fasteners across the line of development. The corrugated fasteners should be driven only when the board can be firmly supported opposite the point of driving, as otherwise the attempted remedy may prove to be detrimental.

Finally, it should be mentioned that wood is one of the primary packaging materials and is (or was) most commonly used first in packaging developments in any particular country. In most developing countries it is going to remain one of the most important packaging materials for a long time to come. The most obvious reasons for this are:

1. Wood is often readily available as an indigenous raw material.
2. The construction and assembly of wooden crates and boxes is a labour intensive operation and most developing countries have plenty of cheap labour.
3. Wooden packages have a comparatively favourable ratio of cost/strength, especially for shipment by sea freight.
4. Wooden crates and boxes are easily tailor-made in small quantities, whereas

other shipping packs, e.g. corrugated boxes, require longer runs to be economical.

There are, however, a number of negative factors to consider in connection with the use of wooden packages:

1. The customers in industrialized target markets are less and less prepared to receive, handle, open and dispose of wooden packages.
2. Wooden packages have an unfavourable ratio of cost/weight and shipping space required.
3. The effective use of wooden packages requires good technical know-how concerning methods of constructing and assembling wooden boxes and crates, the right material specifications to use, etc.

A very important aspect is frequently overlooked by exporters in developing countries. Most business is done on an f.o.b. (free on board) basis and not enough attention is paid to the effect of packaging weight and volume on the final landed cost of the products sold, unless the buyer specifies the exact method of packing. The result is often excessive freight costs, with subsequent negative effects on the competitive position. Therefore, much more attention should be paid to the structural design of wooden packaging in general.

Being a replenishable product, good husbandry in silviculture will result in a continuing supply to meet all our needs.

Plywood

Plywood is a panel product manufactured by gluing together one or more veneers to both sides of a veneer or a solid wood core (Figures 2.1 and 2.2).

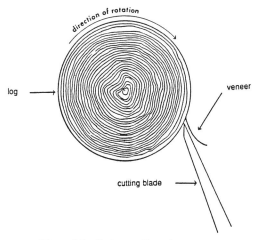

Figure 2.1   Rotary cutting of veneers.

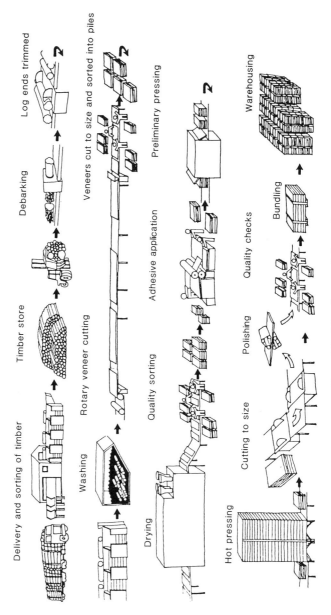

Figure 2.2 Manufacture of plywood.

The grain of alternate veneers is crossed (in general at right angles); the species, the thickness and the grain direction of each layer are matched with those of its opposite number on the other side of the core. Usually the total number of layers is odd (3, 5 or 7) and after assembly, the panels are brought to presses for cold or hot pressing. The glues employed for pressing at room temperature (cold pressing) are either the natural or synthetic resins, while for hot pressing, synthetic resins only are used.

Plywood has several advantages over natural wood, namely greater dimensional stability and greater uniformity of strength. It is, however, somewhat more expensive than ordinary timber.

Plywood should be made from the woods shown in Table 2.3, but preferably from birch, and the battens from the woods listed in Groups 1 and 2 of Table 2.3, with the exception of cottonwood. The plywood and battens need not be of the same species or from the same group.

For large plywood sheathed cases where the plywood is subject to severe conditions, one of the better grades should be used. However, in order to reserve exterior grade plywood for those cases exposed to severe weather conditions, it is recommended that water-resistant plywood of good commercial box or sheathing grade should be used for cases not subject to harmful exposure. An example of such boxes would be transit cases not used for overseas shipment.

Fibre building boards

Hardboard and softboard—known generally as *fibre building boards*—have been relatively unexploited as packaging materials until recently. These wood-based sheet materials are relatively cheap (occupying a position on the cost scale between timber and plywood on the one hand and corrugated board on the other), yet offer a useful range of properties. They are appropriate for use in situations calling for strength and protection, but where economy is an essential factor.

Table 2.3 Woods suitable for making plywood.

Group 1	Group 2	Group 3
Cottonwood	Hemlock	Beech
Gaboon	Hoop pine	Birch
Parana pine	Douglas fir (Oregon pine)	
	Gum	
	Alder	

Board types and manufacture

Fibre building boards, defined in BS 1142, are a family of sheet materials manufactured from wood fibres, the basic strength and cohesion of the boards being derived from the felting together of the fibres themselves and from their own inherent adhesive properties. (Bonding, impregnating or other agents may, however, be added during or after manufacture to modify particular characteristics.) Thus fibre building board is wood which has been reconstituted from its fibres, as opposed to the reassembly of timber by the bonding together of veneers or particles. This process offers the significant advantages of overall consistency, absence of surface defects and a higher resistance to fungal attack and timber pest infestation (since most of the sugars and starches are extracted from the fibres during manufacture). Furthermore laboratory tests in which samples of standard hardboard have been placed in contact with milk powder show that food is not tainted by odours from the hardboard. These test results are supported by the practical experience of tea shippers who often use hardboard tea chests to package their product.

The manufacturing process is as follows. Forest thinnings and sawmill waste are cut into chips. The chips are softened by steam under pressure, then reduced to fibres, either by grinding between disc knives or by ejection from a steam-heated pressure vessel. The fibres are added to water to form an aqueous pulp, and additives such as rosin size and aluminium sulphate may be added at this stage. The pulp is laid on to a slowly moving Fourdrinier-type wire mesh and, as water is removed by drainage, suction and thickening rollers, the fibres interlock or 'felt' before approximate sheet lengths of this 'wet lap' are cut by a rotary knife. The wet lap is fed on to racks through a series of drying ovens, when the remaining water evaporates to leave air cells within the board, thus producing *low density softboard*. Because it is chiefly used in the construction industry for its thermal insulating properties, this material is commonly called *insulating board*. It has a density normally about 200 kg/m^3 and is commonly produced in 13 or 19 mm thicknesses. Thicker boards (up to 50 mm) can be produced, although this may involve the lamination of separate sheets. Common sheet sizes are 1200×2400 or 2745 mm. Softboards, by virtue of their low density, tend to be absorbent and have a low bending strength. It should be noted, however, that they have a high resistance to compression in the plane of the board.

Hardboards and medium boards are much denser, and hence considerably stronger. During manufacture, the cut lengths of 'wet lap' are transferred to wire mesh carriers, which are then loaded into a multidaylight press (which may be as long as 7200 mm and as wide as 2135 mm). A heat and pressure cycle forces the remaining water out through the wire mesh (thus impressing the familiar 'screen' pattern on one face of the board), while the polished surface plates produce the very smooth glossy face. Further treatment in drying ovens follows, and the final stage before trimming and cutting involves humidity treatment to re-establish an appropriate moisture content in the board. Variations in the nature and length of the fibres used, and in the variables of the press cycle, significantly affect the physical properties of the finished product.

Thus, hardboard from one manufacturer may be better suited to applications calling for high strength characteristics or industrial painting, while that from another producer may be better suited to die-stamping or moulding techniques.

Standard hardboard is normally in the density range 800–960 kg/m^3, with a colour ranging from straw to dark brown. It is widely available in thicknesses from 2 to 6.4 mm and in sheet sizes 1200×2240 or 2745 mm. Cut sizes—even of quite small pieces—can be obtained in large quantities at costs which compare reasonably with those for full sheets. Standard hardboards have high bending strength and are particularly strong in their own plane, the denser boards having the highest strength levels and the most consistent machineability. The smooth face has a higher resistance to water absorption.

Tempered hardboard has high strength characteristics and greater resistance to water and abrasion. It is produced by impregnating the board, after pressing, with drying oils followed by further heat treatment. Thicknesses range from 3.2 to 8 mm and sheet sizes are generally as for standard hardboard.

Medium board comes in two density ranges. The lower type (Type LM) lies in the range 350–560 kg/m^3, while the higher (Type HM, or 'panelboard') ranges from 600 to 800 kg/m^3. Although less dense than standard hardboard, it is normally thicker, being available in thicknesses from 6 to 12 mm.

Processing and use techniques

Cutting. Hardboard and softboard are relatively easy to work by normal wood-working processes. For low density softboards, sharp knives are recommended for clean cut edges. Hardboard and medium boards can be cut with most common panel-sawing equipment, jig saws and routers. Softboards and some hardboards can be guillotined, whilst irregular shapes or curved outlines in hardboard can be obtained by die-stamping or punching.

Most of these techniques produce cuts at right angles to the surface of the board. Many useful effects can, however, be produced by cutting hardboard with bevelled edges. A simple version of this, to produce mitred corners, consists of using a saw blade canted at an angle of 45°. More sophisticated mitred edges can be produced by the use of V-grooving techniques. In this process, a V-shaped notch is cut to a carefully controlled depth into the board by using a specially shaped cutter block. This block removes material part of the way through the board, leaving a very thin layer of fibres (or in some cases a laminated layer of another material such as PVC film) on the side remote from the cutter. This final layer maintains the integrity of the board and acts as a hinge when the board is folded along the line of the V-groove. Before folding, glue is applied to the groove to produce a permanent rigid joint (Figure 2.3). The creation of such a change of plane gives a significant improvement in stiffness without the need for a supporting framework.

Bending and moulding. Hardboard can be bent through quite small radii by a number of simple techniques. The simplest consists of bending dry board by pressure alone, but this cannot produce radii much below 300 mm (for 3.2 mm thick board), and the curved board requires fixing to a former. The addition of moisture effects an improvement. Soaking hardboard in water for several hours facilitates the formation of bends with radii of curvature down to about 125 mm (3.2 mm thickness).

Even better is the application of heat and pressure during the bending of a soaked board; this enables permanent right-angled rigid bends to be made. Using a heated platten press and suitable moulds, the bending technique can be extended to produce complex shapes such as inner door panels for the motor industry. The same moulding process can be used for the production of domestic and industrial trays.

Creasing and folding. During 1974/1975, it was appreciated that some grades of standard hardboard could be creased to form a flexible hinge which can be folded. The process is similar to that used for many years to crease solid fibreboard. The fact that a rigid material such as hardboard can be creased, yet retain considerable strength, is of considerable importance for packaging applications.

Jointing. When panel products are used to make containers, the method of fastening the sheets together is a major consideration. The simplest methods for hardboard are the use of frameworks and mechanical fixings. Nailing or

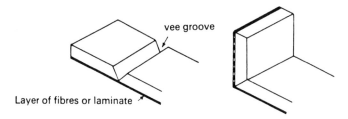

Figure 2.3 Controlled machining by a V-shaped cutter block leaves a constant thickness of fibres or flexible laminate, which acts as a hinge while the panel is glued and folded.

stapling of hardboard to softwood structures is common, as also are methods
of riveting hardboard sheets to preformed angle strips of materials such as
tinplate. A further technique being increasingly used for produce trays and
boxes is the use of angled corner stitching in which a wire staple is applied to
two panel sections held at right angles to each other—the stitching being
possible either with or without a timber corner support (Figure 2.4). All
these methods use through fastenings and do not rely on the inherent
strength of the bond between individual fibres in the hardboard. However,
the perforation of the sheet by mechanical fasteners may weaken the board.
Furthermore it is advisable that the equilibrium moisture content of both the
hardboard and framework be taken into consideration.

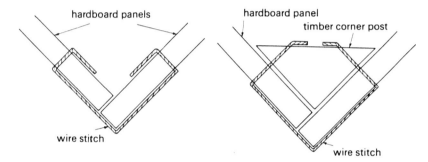

Figure 2.4   Hardboard panels fixed with a corner stitch, with and without a corner-post
attachment.

Surface fastenings make use of the strength of hardboard without
producing such perforation weaknesses. In addition, if the fastening is itself
flexible, a fold-flat property can be designed into the finished article. Surface
fastenings employ adhesives in conjunction with metal angle strips,
fibreboard angle strips, or tapes of reinforced paper, textile or plastic. The
metal strips are usually rigid, whilst the fibreboard and the tapes are flexible.

3
Pulps and papers

Materials

It was not until the middle of the nineteenth century that the paper industry started to use wood pulp for making paper. Before this, timber had been used for many other purposes—building ships, houses, and for making furniture and boxes, but the beginning of the paper-making industry using wood pulp caused great changes in the practice of cutting trees.

In the early days, trees were cut and used locally, but shortages of timber led to a search for new sources. For example, the timber requirements of the British navy for ships' masts led to the building of ships specially designed to carry them across the North Atlantic. One particular tree, the Weymouth pine, is named after a Captain John Weymouth who was one of the early timber transporters. In 1608 Captain John Smith carried eight Poles and a Dutchman for the purpose of erecting sawmills in Jamestown, Virginia, and not long after that America began to export many goods, including timber.

Increasingly after this, the demand for timber rose, and by 1900 supplies were falling. This does not mean that people had not thought about forest conservation—as early as 1682 William Penn decreed that, in Pennsylvania, for every five acres of forest cut, one acre must be replanted. However, these were small contributions, and adequate re–afforestation schemes did not begin until much later.

Many suggestions have been made about how to maintain a credit balance in forests. The problem is basically very simple to diagnose. All we have to do is speed up forest growth and reduce timber consumption, cut out forest fires and cure diseases, as well as removing the insects which live on wood and woodpeckers which also damage it. But it is quite another thing to do something about re-afforestation. Trees do not breed like rabbits. Time is essential. Tree farming means very much more than just planting the trees and then standing back to await the final crop. It involves raising the seedlings, planting them out, and spraying them against tree illnesses and insects, and then thinning at the appropriate time. While all this is going on, steps must be taken to prevent damage by fire. Nowadays the paper industry replaces the trees that it uses at a rate which is adequate for our needs at present.

In addition to wood pulp, other fibrous materials such as cotton, flax and esparto, bamboo and straw are also used for making paper. Finally, waste paper is a secondary source of material and is used for certain grades of paper and board. Trees are, however, the most important primary source for cellulose fibres.

Non-fibrous additives

It must also be borne in mind that several non-fibrous materials are used in making paper such as minerals for loading and coating, dyes for colouring and materials for sizing and water proofing. There are four main classes of non–fibrous additives (see Table 3.1):

1. *Fillers or loading agents* are white mineral pigments such as china clay. They increase the opacity and brightness of the paper, and improve the surface smoothness and printability, but reduce the strength properties if present in large amounts.
2. *Binders* may be starches, vegetable gums, synthetic resins and rubber lattices. Such additions improve the strength properties—tensile, tear and burst.
3. *Sizing agents* provide resistance to the penetration of aqueous liquids, such as water and writing ink. Blotting papers are unsized; writing papers are usually sized and there are various degrees of sizing from 'soft' to 'hard'. Resin and wax emulsions are the main agents, but some special synthetic resins are also used.
4. *Miscellaneous additives.* A number of other materials are incorporated either to impart specific properties to the paper or to assist in manufacture, e.g.

   - Optical brightening agents for increased whiteness
   - Coloured pigments and dyes for coloured papers
   - Various gums to assist in formation on the paper machine
   - Anti-foaming agents
   - Filler-retaining agents
   - Wet-strengthening agents

   Additives are inserted at several stages before the fibre suspension reaches the sheet-forming stage, i.e. in the pulping, beating or refining operations—where exactly depends on the additive.

Table 3.1 Non-fibrous additives to paper.

Agents	Effects
Fillers or loading	Improve opacity and brightness
Binders	Increase strength
Sizes	Reduce penetration of water and writing inks
For specific properties	Whiteness, colour, formation
For manufacturing aids	Filler retention, anti-foaming

What is wood?

Underneath the outer and inner layers of bark in a tree, we have a layer which contains the plant foodstuffs and it is through this layer that the tree itself grows. Then we have the main woody part which consists of bundles of cellulose fibres running vertically up the trunk, held together by a material which is called lignin (Figure 3.1). Essentially, the wood part of the tree consists of about 50% cellulose fibres, 30% lignin, 16% carbohydrates and some 4% of other materials such as proteins, resins and fats. It is principally the cellulose which eventually becomes paper. It is composed of individual fibres which are finer than human hairs and are a few millimetres in length at the most. These fibres are about 100 times as long as they are thick. Lignin is a chemically complex substance which holds the fibres together. It is best, perhaps, to think of it as the glue holding the tree in one piece.

Pulping processes

The primary intermediate product used to make paper is wood pulp. The properties of an individual paper or paperboard are extremely dependent on the properties of the pulps used. Pulp preparation from deciduous (hardwood) or conifer (softwood) species may be done by *mechanical*, *chemical* or *hybrid processes*. These hardwood or softwood pulps may be used unbleached, or they can be bleached to varying degrees by a diversity of techniques.

Mechanical pulps produce papers that are characterized by relatively high bulk, low strength, and moderate to low cost. Their use in packaging is very limited.

Figure 3.1   The structure of a tree.

For each process, first of all the bark is stripped away from logs which have been cut to a suitable length at the appropriate stage in their growth. In the mechanical pulping process, the logs may be used directly in 1.2 m (4 ft) lengths or, alternatively they may be chipped, i.e. converted into pieces of uniform size about 15–20 mm long. Two methods of mechanical pulping are employed.

In one, the logs (Figure 3.2) are pressed against the surface of a large revolving grindstone, kept wet by a stream of water which additionally removes the fibres. In the other system, the wood chips are passed between the two plates of a disc refiner with specially treated surfaces which are very close together and which are rotating at high speed. Thus, the wood chips are reduced to individual fibres. In mechanical pulping the water soluble impurities only are removed and most of the lignin still remains. In addition, by mechanical pulping, many fibre bundles and some damaged fibres are left in the pulp. Much of the grinder and disc-refined wood pulp is used for newsprint, although substantial quantities are employed mixed with chemical pulp for making certain kinds of board. Mechanical pulp is normally made from softwood, typically spruce.

Chemical pulping starts from chips, but removes all materials other than the cellulose fibres by chemical action and solution, the chemicals converting the lignin to a soluble form that can be removed by washing (Figure 3.3). This type of pulping produces cellulose fibres of a higher purity than those produced by the mechanical processes. They are generally much less damaged and, in addition, the fibre bundles are fewer. Several different chemical pulping processes exist, and the quality of the pulp depends upon the process, as well as the kind of wood fibre used. For packaging purposes, three chemical processes are of major importance. These are the 'kraft

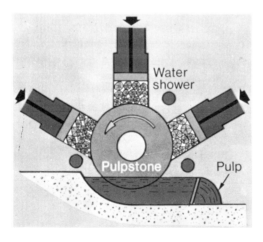

Figure 3.2   Mechanical pulping.

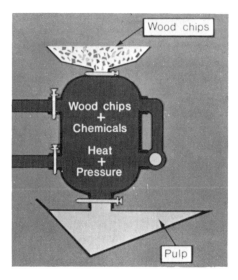

Figure 3.3   Chemical pulping.

process' which retains most strength in the fibres, the so-called 'sulphite process' which is less strong and the 'semi-chemical process'.

The kraft (sulphate) pulping process, introduced about 100 years ago, dominates the chemical pulping industry; yields are higher, pulps are stronger and process chemicals are more completely and economically recovered than with any other process. Unbleached pulps are generally stronger, stiffer and more coarse than their bleached counterparts, but many papers made of white, conformable fibres are used in many more applications.

In the kraft or sulphate process, the wood chips are digested in a solution of caustic soda and sodium sulphate for some hours, but originally only caustic soda was used. Most of this, which dissolves out the lignin, can be recovered and re-used. In the recovery process, the presence of sodium sulphate leads to the production of sodium sulphide as an end-product. When put back into the digester and heated again under pressure, the sodium sulphide produces the necessary caustic soda for dissolving out the lignin, while avoiding excess. This is known as the *kraft process*, the name being derived from the Swedish word *kraft* which means strong. In the early days, kraft paper was always associated first with a brown colour and second, with long fibres and what was called a 'wild look through'. This means that because the fibres were long, they did not form a very uniform sheet and when held up to the light, the sheet looked a little blotchy.

The *sulphite process* uses sulphur dioxide and calcium bisulphite. This is mixed with the chips in aqueous solution and heated to about 140°C. Once again the lignin is dissolved out, leaving the fibres, and after digestion the mass is washed with water and then bleached with another chemical, such as calcium hypochlorite, before pressing into pulp sheets. This process gives a very pure cellulose fibre, although the resulting pulp is not as strong as that from the kraft process.

In the *semi-chemical process*, the wood chips, which are usually from beech or birch trees, are partly treated chemically and partly mechanically, to reduce them to fibres, hence the name semi-chemical. Semi-chemical pulps are used for the manufacture of fluting medium for corrugated board.

Beating

Once pulp has been produced, it may have to be bleached to make it white or coloured, or treated in other ways. One of the most important processes in the pre-preparation of fibres for paper making is the so-called *beating process*. The object of most beating processes is to rub and brush the individual fibres, and cause them to split down their length in such a way as to produce a mass of thin fibrils which will enable them to hold together in the matted paper far more strongly. This process is called *fibrillation*. The greater the degree of fibrillation that can be induced, the higher the strength of the paper. Different pulps respond differently to this treatment. Softwood fibres will fibrillate to a greater extent than hardwood or straw fibres. Hence softwoods are potentially able to produce better and stronger papers. Beating also cuts the fibres by the action of the bars in the beater. This is undesirable to some extent, although some cutting may be necessary for good running on the machine. Furthermore, the process breaks up any fibre clumps and refines them into individual cellulose fibres. The art of beating is to maintain a high proportion of fibrillation and a low proportion of cutting in the pulp, to give the desired properties in finished paper. Figure 3.4 gives an idea of the effect of beating on paper strength characteristics in respect of burst, tensile and tear strength.

Paper-making machines

Paper was first made in the United Kingdom by John Tait in 1496. He used a hand-making process producing a slurry of pulp fibres in water, inserting a wire 'mould', lifting it and 'couching' the matted fibre on to a sheet of felt. The sheets were stacked in a press where the fibre mat compacted and water was removed, and finally the paper sheets were hung over ropes to dry. All this was an extremely slow process. The first paper machine was invented in 1799 in France by Nicholas Robert but, because of the unsettled conditions in France, the invention was soon moved to England where, in 1807, it was

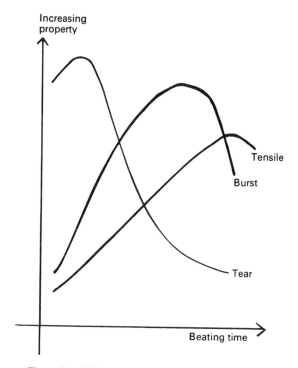

Figure 3.4   Effect of beating on strength properties.

taken up by two London stationers named Fourdrinier. They were relatively unsuccessful commercially and finally went bankrupt, but the type of machine has been given their name. The modern Fourdrinier machine has eight basic parts, four comprising what is called the wet end, and four the dry end. They are:

    The wet end
        Stuff box or chest
        Head box
        Slice
        Fourdrinier wire

    The dry end
        Presses
        Dryers
        MG dryer
        Calender stacks

Let us briefly describe the purposes of each in turn.

Stuff chest

The stuff chest holds a slurry of pulp containing about 97 parts of water to every three parts of fibre. This is much too heavy to be formed into a paper sheet, and is diluted in a mixing chamber with more water to a consistency of about 0.5%. That is 99.5% of the 'stuff' is water.

Head box

The diluted pulp passes from the stuff chest to the head box, which is both a means of agitating the slurry and a turbulence reducer. Essentially it is designed to produce a uniform fibre suspension (the stock) and to feed it to the slice.

Slice

The slice is part of the head box and consists of a narrow slot in its front face through which the stock flows on to the wire. Adjustments can be made in the opening to the slice, and it can be raised and lowered to adjust the flow.

Fourdrinier wire

The Fourdrinier wire carries the stock from the slice up to the place where the sheet is removed at the so-called 'couch' roll. The function of the wire is to allow the initial drainage of water to take place, helped by rollers underneath it and by suction boxes to remove the water, to get the mat of paper fibres into a sufficiently strong sheet to transfer to a felt which will move into the drying section. At this stage the mat of fibre consists of about 75–80% water by weight, and the first operation is to pass it through presses to remove more water. The felts on which the sheet travels are made of wool; essentially they help to blot water from the sheet (Figure 3.5).

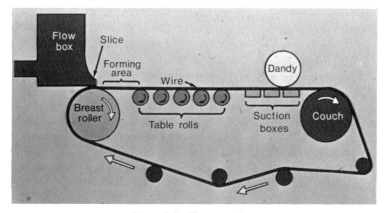

Figure 3.5 The wet end.

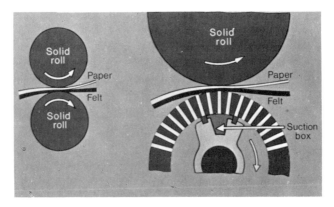

Figure 3.6   Plain (left) and suction (right) presses.

Presses

The presses (Figure 3.6) may be either plain, which just catch the water in trays, or fitted with suction to help to remove the water by pulling it through thousands of tiny holes in a suction roll. After the presses the sheet still contains some 60% water as it passes into the dryers.

Dryers

The dryers consist of a long train of heated cylinders which come into intimate contact with the paper and dry it by heat.

MG dryer

In certain instances the paper may require to be *machine glazed,* and MG papers are given their gloss on massive cylinders which have a highly polished surface. This is a similar sort of process to the glazing process used with glossy photographic prints.

Calender stacks

Machine finished (MF) papers are smoothed in what is called a *calender stack*. This consists of a series of rolls which iron the sheet by slippage. This occurs because the bottom roll in the stack is the only one which is driven, the others are turned by friction with the paper. The degree of machine finish is controlled by the number of rolls through which the paper passes and it may be increased by applying a certain amount of surface water (from water boxes) to increase it. Finally, the paper passes to the reeler and winder, and is wound up into a giant roll.

Main packaging papers

The furnish, the degree of beating, the amount of filler, binders, sizes, etc., together with the operating variables of the particular paper-making machine and specific finishing processes, can be varied to produce many types of paper. The main packaging papers, their origin and uses are outlined in Table 3.2.

Unconverted papers

We can now discuss those applications where paper-based materials are used in an unconverted form, i.e. as flat material. The largest use of paper in this way is for wrapping, in circumstances ranging from counter reels in shops to high speed automatic parcelling machines in factories. Other applications of packaging papers include interleaving and separation of materials, stiffening and supporting products, as well as in space-filling and shock absorption. Packaging papers are also used for the production of envelopes, sachets, bags and sacks, etc.

Wrapping papers

The wrapping operation may be described as the covering of an object or a collation of objects of a similar or different nature into a handleable unit. Paper in sheet or reel form may be used for this purpose. As in every area of packaging, the wrapping must perform certain clearly identifiable functions. The basic functions are to contain and protect, with subsidiary functions of communicating and providing convenience. Wrappers may be required to perform any or even all of these functions.

Containment. The wrapping of parcels of all types can involve contents which may be a single item, an assortment of different items, or a collation of identical items. The function of the wrapper for a single item may be to hold it compressed, e.g. bulky clothing; to present a smooth exterior surface, e.g. an engineering component; to obscure the markings on a transport pack to prevent confusion or to reduce the possibilities of pilferage.

In parcels for an assortment of articles, the function of the wrapper is primarily to secure these together in a single handleable unit. The more similar the articles are to one another in shape, the more easily this will be accomplished. For example, a collection of books of the same basic size but different thicknesses are relatively easy to parcel. If the various items differ widely in size, shape or density, then a simple wrap may not be suitable without first forming them into some other unit, e.g. by tying with string, or overwrapping them with special materials to bring them to a more

Table 3.2 Main packaging papers.

Basic material	How made?	Weight range		Tensile strength[a]		Properties and uses
		lb/1000 ft²	kg/1000 m²	lb/in width	kN/m	
Kraft papers	From sulphate pulp on softwoods (e.g. spruce)	14–60	70–300	MD 14–65 CD 7–30	2.4 –11.3 1.2 – 5.2	Heavy-duty paper, bleached, natural or coloured; may be wet-strengthened or made water repellent; used for bags, multi-wall sacks and liners for corrugated board; bleached varieties for food packaging where strength required
Sulphite papers	Usually bleached and generally made from mixture of softwood and hardwood	7–60	35–300	Very variable		Clean bright paper of excellent printing nature used for smaller bags, pouches, envelopes, waxed papers, labels and for foil laminating, etc.
Greaseproof papers	From heavily beaten pulp	14–30	70–150	MD 10–25 CD 5–12	1.7 – 4.4 0.85– 2.1	Grease-resistant for baked goods, industrial parts protected by greases, and fatty foods
Glassine	Similar to greaseproof but super-calendered	8–30	40–150	MD 8–30 CD 5–16	1.4 – 5.2 0.85– 2.8	Oil and grease-resistant, odour barrier for lining bags, boxes, etc., for soaps, bandages and greasy foods.
Vegetable parchment	Treatment of unsized paper with concentrated sulphuric acid	12–75	60–370	12–80	2.1 –14.0	Non-toxic, high wet strength, grease- and oil-resistant for wet and greasy food, e.g. butter, fats, fish, meat, etc.
Tissue	Lightweight paper from most pulps	4–10	20– 50	Low strength		Lightweight, soft wrapping for silverware, jewellery, flowers, hosiery, etc.

[a]MD, machine direction; CD, cross direction.

acceptable shape. Almost inevitably the strength of the wrapper required in the second instance will be greater than the first.

Protection. Protection is needed when the product, or its immediate packaging, is sensitive to some external condition or circumstance, or where the product is noxious or otherwise unacceptable when handled in an unwrapped state. The external factors likely to affect the conditions of the product are: dust and dirt; moisture; light; heat; contact with other surfaces likely to cause punctures, scuffing or abrasion; and possibly oil or grease picked up during handling in works conditions. There may also be a need to prevent a component of a product from escaping out of the wrapper, causing either loss of product itself, or damage to both the pack and adjacent packs during transit. Examples of this include engineering items, which are frequently covered with oil or grease as a protective, blood from meat and fatty materials from cooked foods.

Communication. Wrapping materials can be printed to identify the contents and the manufacturer, and also to provide a smooth opaque uniform background on which other information or instructions can be applied. Methods vary from handwriting through simple forms of printing to the attachment of labels and the provision of printed wrappers, e.g. Christmas gift paper.

Convenience. Many of the less obvious aspects of packaging are covered in this area. Wrappers can conceal the nature of the contents, either to avoid embarrassment or to deter thieves by preventing recognition of valuable contents. They may also prove that the item has been legitimately purchased in large department stores where theft can be easy without some security measures.

Interleaving

In this application the paper functions only as a separator to protect a product in flat sheet form. Typical products are sheets of window glass and metal, art prints and similar materials where the surfaces of the sheets are sensitive either to physical damage or to chemical attack. Sometimes, however, the prime function of an interleaving paper will be purely to facilitate the separation. An example of this is sliced cooked meat separated by thin greaseproof sheets.

Stiffening and supporting

Generally only the thicker grades of packaging material are used for these purposes, but even thinner materials wrapped around a fragile item, or part of an item, can provide sufficient support.

Shock absorption and space filling

Sheet packaging materials are used in some circumstances simply to fill irregular spaces in packs where voids occur because the product does not completely fill the box. Where this is so, it is preferable to use special sheet materials which will not readily compact, and which will produce the maximum degree of bulk. Examples of such materials are single-faced corrugated board, and 'poppled' or foam-backed paper. Special laminated and structured papers are produced for furniture packaging, the various layers providing a smooth surface to prevent abrasion, cushioning to reduce the effect of knocks and continuity to prevent dust settling.

Papers for conversion

Sachets, envelopes and bags consist simply of a prefabricated wrapper which only needs the product and the closure to complete the pack. Many packaging papers of various kinds, plain, printed and coated, as well as with multi-plies and laminates, are used for this purpose. Such prefabricated wrappers have always been useful where handfilling operations are involved, and nowadays automatic bagging and enveloping machines have raised the potential considerably. Almost all types of paper can be used for these purposes, and the functions of such prefabricated wrappers are very similar to those already mentioned. When assessing the performance of such materials, the quality of any seals and joins must also be taken into account.

Performance requirements

Depending on the functions required, a wide range of properties may need to be specified. These can be classified conveniently into physical and chemical properties, and they must always be within a specific price range.

Physical properties

Description. The materials must be described in a recognizable and measurable manner. and frequently such a description is the only specification. This will cover the type of paper, and its thickness or grammage. These criteria will frequently give some performance criteria within a particular range of papers. They are not critical criteria, except in respect of yield, but because the strength of a particular type of paper tends to increase as its weight and thickness increases, they do give a measure of strength. Weighing, therefore, is one of the simplest methods available of checking that the material supplied is of the nature required and this, combined with a count of the number of sheets, is frequently the only check of quantity and quality that is carried out. Even for this simple test there is a correct procedure, detailed in BS 3432.

Optical properties. Colour, opacity, gloss and brightness are most important, and standard methods of testing are available. Colour in particular is a difficult attribute to specify, since it almost certainly introduces an element of subjective judgment. It would be preferable to specify colour by means of the 'tri-stimulus value' but this is seldom done. Reference to the three primary colours in this way would require sophisticated optical equipment. Such a numerical description could also be used, however, as a reference to assess the fastness of colours.

Strength properties. These are clearly related directly to the functions of containment and protection. The most commonly quoted properties are tensile strength (the force required to break a specified strip of material of known dimensions), stretch at the breaking load (a measure of the ability of the wrapper to absorb energy), initial tearing resistance (the force required to tear from an uncut edge), internal tearing resistance (the force required to continue a tear made from a clean cut in the edge of a specimen), puncture resistance (more commonly applied to the heavier materials) and bursting strength, which is the property most frequently quoted, the test being carried out either in a dry or wet condition according to whether a wet-strength paper has been specified. Other strength properties that may be specified for particular circumstances are folding endurance, abrasion resistance and ply-bond strength, especially where laminated or multi-layer materials are concerned.

Barrier properties. When the product to be packed is sensitive to moisture or oxygen, a suitable barrier performance may be needed. Wrapping materials may also be required to provide barriers to liquid water, oils and grease, and these must also be specified and tested.

Miscellaneous. Stiffness and surface friction relate to the ease of running of wrappers on automatic machines. Smoothness and air permeability, water absorptivity and dimensional stability may also be important if the wrapper is converted into a bag, envelope or pack by mechanical means, and when printing or securing with adhesives may be involved.

Chemical properties

Packaging materials which are intended to come into direct contact with sensitive products must not contain any trace materials likely to affect the product, e.g. wrappers for highly polished metal items must be neither acid

Table 3.3   Selected test methods.

	Test, etc.	Material	Reference[a]
Pre-test procedures	Sampling	Paper Board	BS 3430
	Conditioning	Paper Board	BS 3431 ISO/R 187

	See also:
	Laboratory humidity ovens
	    Injection type
	    Non-injection type
	Laboratories, controlled atmosphere
			
			BS 3718
			BS 3898
			BS 4194

Tests for mass and density	Grammage	Paper Board	BS 3432 FEFCO 2 and 10
	Mass	Wax on paper	BS 4685
Tests for strength properties	Adhesive strength	Adhesive and adherends	BS 847
	Burst strength	Paper and board Board Paper (wet)	BS 3137 FEFCO 4 BS 2922
	Folding endurance	Paper	BS 4419
	Heat-seal strength	Plastics/laminates	PIFA 2/74
	Ply-bond strength	Paper	TAPPI RC 364
	Puncture strength	Paper, board	BS 4816
	Stiffness Ring stiffness	Paper Paper	BS 3748 ASTM D-1164

	Tear strength
	    Initial
	    Internal
		
		Paper
		Paper
			
			ASTM D 827
			BS 4468

	Tensile strength and stretch	Paper	BS 4415
Tests for surface absorption and permeability properties	Abrasion (see rub-proofness)		
	Absorption of water by	Paper, board	BS 2644
	Absorption of wax by	Paper	ASTM D 688
	Friction coefficient = 1/slip	Paper	TAPPI T 503
	Moisture content	Paper	BS 3433
	Permeability (see also water vapour transmission	Paper to air BS 2925 Sheet materials to gas	BPBIF p. 13 ISO 2556 BS 2782-514A

Table 3.3 *cont'd.*

	Test, etc.	Material	Reference[a]
	Resistance to grease	Paper	ASTM D 722
	Resistance to oil	Paper	BPBIF RMT 1
	Roughness	Paper	BS 4420
	Rubproofness	Paper	BS 3110
	Water vapour transmission	Sheet materials	BS 3177
Analytical tests	Acidity/alkalinity (pH)	Paper	BS 2924
	Ash	Paper	BS 3641
	Chlorides	Paper	BS 2924
	Contraries copper, iron	Paper	BS 1820
	Sulphates, reducible	Paper	BS 1820
Dimensional tests	Curl	Paper	BPBIF RM 4.2
	Dimensional stability	Paper	BPBIF RM 53
	Thickness	Paper Board	BS 3983 BS 4817
Optical tests	Brightness Whiteness Reflectance Opacity Gloss	Paper	BS 4432
	Light fastness	Paper	BS 4321
Miscellaneous materials tests	Crease quality	Board	BS 4818
	Odour and taint	Packing materials	BS 3755
	Resistance to blocking	Paper	ASTM D-918

[a]The methods referred to are taken from:
ISO/R   International Standards Recommendations;
BS   British Standards. In some instances the number of the standard is followed by a further refeence, e.g. BS 2782.509 is Method 509 in BS 2782;
ASTM   American Society of Testing and Materials Standards;
TAPPI   Technical Association for the Pulp and Paper Industry (USA) Standard. Where TAPPI Standards have been developed by ASTM, the latter reference has been used;
FEFCO   Fédération Européenne des Fabricants de Carton Ondule Test Methods;
BPBIF   Technical Section of the British Paper and Board Industries Federation Test Methods;
PIFA   Packaging and Industrial Films Association Standards;
BPBMA   British Paper and Board Manufacturers Association.

nor alkaline, nor contain unacceptable levels of sulphate or other corrosive chemicals. Wrappers for brass or silverware must be free from reducible sulphur, and in all these instances, specified limits for acceptance of papers are involved.

Food wrappers must not contain more than minute traces of heavy metals to prevent the possibility of contamination. Vegetable parchment, which is frequently used for wrapping materials such as butter, must be virtually free from copper or iron, which catalyse oxidation of the fat and could lead to rancidity.

Types of wrapping material available

Kraft paper. Kraft paper is made by the Fourdrinier process from a pure chemical sulphate produced pulp. The term *kraft* is of Scandinavian origin and means literally *strong*. It is only applied to paper or board made entirely from paper-pulp produced by the sulphate process. It is available either *unbleached* (brown), which is used for general wrapping, or *bleached*, which is used for bag-making and for wrapping clothing, food, medical goods and stationery. The bleached material is also used for dyeing to various colours and may have decorative applications in ribbed and striped form. Various finishes can be imparted during the final stages of the paper-making process according to requirements. Unglazed paper is smooth on both sides and does not have a gloss finish. Machine-finished (MF) papers are treated on the machine to produce a smooth, but again not a glazed surface. Water-finished (WF) papers are produced by damping one or both sides on the machine, and this gives rise to a higher gloss and a smoother finish than MF. Machine-glazed (MG) papers have a high gloss finish on one side only, and super-calendered (SC) papers possess a high finish imparted by damping on the machine and subsequent treatment in a super-calender. This last variety is used extensively where attractive wrappers are required. All of these finishes may be supplied either plain or with a ribbed structure.

In addition to wrapping applications, kraft papers are used for multi-wall sacks, bags, envelopes and gummed sealing tape.

Stretchable papers. Stretchable papers come in two varieties, crêped or uncrêped. Crêped paper is produced either on a paper machine or by a secondary operation. Extensible papers which are not crêped are produced by a special process to give particular properties.

A stretchable paper will stretch under sudden strain and then exhibit a rapid rate of stress relaxation. This property of high energy absorption is responsible for a great many of its uses. Crêped paper is produced by the removal of the moist sheet from a roll by the action of a doctor blade which piles up the paper in ripples parallel to the cross machine direction. Other methods of compacting the paper use other devices, but the result is similar.

The use of elastomers in a paper can also produce stretch, but here we will consider only those produced by the mechanical means. The characteristics of crêpe papers are expressed by stretch, ratio, or texture. *Stretch* is the percentage increase in length; *ratio* is the ratio of the extended length to the average length of the crêped test strip. Stretch is used in machine crêping, ratio in secondary crêping. The term *texture* refers to the surface of the crêped paper, which shows a varying degree of fineness or coarseness. Measuring the number of hills per linear inch gives a numerical value. The texture and the ratio can be varied independently of each other within general limits, but in secondary crêping a wider range of combinations is possible. The following generalization about the properties of crêpe paper should be made:

1. The tensile strength of the uncrêped base paper is normally higher than that of the crêped paper.
2. The tear resistance is higher in a crêped sheet than in the base paper due to the greater amount of material present.
3. The softness of the paper is increased, especially in dry crêping, due to the increase in apparent density, in sheet compressibility, and a decrease in fibre bonding.

For many years the wet creping operation was the main on-machine method available for producing stretchable papers for packaging. Although such papers were satisfactory, the texture of the surface was rough and gave problems in handling. Such a texture was also coarse, and problems arose in printing. Consequently, attention was directed towards production of extensible papers which were not crêped. The best known of these is probably that produced by the Clupak process. Here the sheet is similar in appearance to ordinary wrapping kraft paper but has the high energy-absorption properties of crêped paper. This process employs a thick rubber belt instead of the crêping blade on the machine. Figure 3.7 presents the process which is located in the dryer section at the early stages on a paper machine. The sheet meets the rubber blanket when it contains about 35% moisture, and it is then pressed between the rubber blanket and the surface of the dryer roll. Under these conditions, the rubber blanket in the nip is expanded. As it leaves the nip it shrinks, compacting the sheet as it does so. The important variables in the process include the nip pressure, the dryer roll temperature and the moisture content. No additives are used to assist crêping, and the action is capable of giving a machine-direction stretch of 20%, although this is not frequently employed. In the cross direction, the stretch is increased somewhat above the usual level, but by no means to the same extent. Such papers are frequently used in producing sacks, particularly where extra strength is required. Several other types of paper of a similar nature are produced.

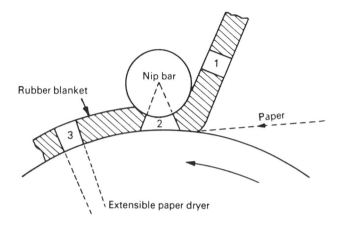

Figure 3.7   Essential part of the Clupak process. As the rubber blanket passes under the nip bar, the volume shown at 1 expands lengthwise to dimensions 2 and when the pressure is removed, reverts to its normal dimensions as at 3.

Wet-strength papers. Conventional papers are not strong when wet. Most of paper's strength is the result of hydrogen bonding between hydroxyl and carboxyl groups on adjacent fibres. The removal of water during the paper-making process generates these bonds, and the process is reversible.

A wet-strength paper is a paper that remains strong when fully saturated with water. It is easy to define this in principle, but considerably more difficult in practice. We must first of all distinguish between wet-strength papers, waterproof papers and moisture or water-vapour-resistant papers. A water-vapour-resistant paper must contain a barrier, usually a plastic or metal film, to resist the penetration of water vapour. A waterproof paper is one that sheds liquid water easily. Wet-strength papers, on the other hand, will often be saturated with water, and under these conditions they must still retain strength. By definition, wet-strength paper should retain at least 30% of its dry strength when these two strengths are measured by means of the bursting strength test.

Paper that has been parchmentized is actually stronger wet than dry, principally because of the loss of individual fibre identity. However, several more economic alternatives exist for generating wet-strength papers. In general, the chemicals used to augment the natural hydrogen bonding are cross-linked during the manufacturing process. Chemicals for producing wet-strength papers include urea, melamine, resorcinol and other phenolic or amino resins cross-linked with formaldehyde, and condensation products of polyalkylene polyamines with dicarboxylic acids cross-linked with epichlorohydrin.

The condensation of these chemicals takes place in the drying section on the paper machine, and the reaction product (being highly water-insoluble) produces an effect on the paper sheet to increase its wet strength. Wet-strength papers are particularly useful for outside packaging as they are stable in adverse atmospheric conditions. They can retain sufficient wet strength for a long time under wet conditions and are also suitable for the packaging of goods which themselves are wet. They are again frequently used for paper sacks.

Oil- and grease-resistant papers. The penetration of the surface of any paper involves both the 'wetting' of the surface and the mechanism of capillary penetration. The packaging requirement must be considered before making a selection from the several possibilities. Is there a need for minimum staining by grease or oil under very light pressure? Do we want absolute resistance for a considerable period under substantial pressure? Does the requirement lie between these extremes?

Plastic-coated paper, parchment, glassine and so-called greaseproof papers, in that order, offer decreasing protection from oils, fats and greases. The degree of protection provided is dependent on the capillarity of the paper and the wettability of the paper surface.

In selecting materials for primary packages, the economics of the various possibilites must be considered. A one-trip bag for a consumable such as French fried potatoes in a fast food outlet may only need resistance to staining for a few minutes (half an hour at most). A package for car engine oil, however, may need to provide a stain-free barrier for six months or more. The latter will almost certainly require the paper to be laminated or surface coated with a synthetic, but many of the requirements for the less demanding situations can be provided by chemical treatments. Waxing may provide moderate resistance for up to a few days for the inner surface of paper boxes for such products as cream cakes or swiss rolls. Treatment with low surface energy materials such as fluorocarbons are used for carrier bags, carry-out food packages and packages for short-life bakery products.

Vegetable parchment. The process for producing parchment paper was developed in the 1850s, making it one of the earliest special packaging papers. Parchment paper is usually produced in two stages. Pure chemical pulp stock is made into paper on a Fourdrinier machine. The paper is then passed through a bath of sulphuric acid, the strength and temperature of which is carefully controlled, as is the time of immersion. The parchmentiz-ing process is stopped gradually by passing through diluted sulphuric acid and water and, finally, the material is dried.

Soaking an absorbent paper in concentrated sulphuric acid swells the cellulose fibres tremendously and they are partially dissolved. The plasticized fibres close their pores, fill in voids in the fibre network, and thus produce extensive hydrogen bonding. Rinsing with water causes re-precipitation and consolidation, resulting in a paper that is water resistant, lint free, odour and taste free, and resistant to grease and oils.

Parchment is virtually a pure cellulose paper. The usual range of weights produced is 40–75 g/m^2, but lighter and heavier weights are possible. Most material is softened by the addition of glycerol, but unsoftened papers are sometimes used. In addition to being used for wrapping, it is also made into bags and corrugated papers and is often printed. It is extensively used to package fatty biscuits and other baked goods, butter, margarine and cooking fats, cheese, ice-cream, fish (wet, cured and fried with chips), poultry and meat. Packs for tea and coffee also often contain vegetable parchment as the immediate wrapper.

Greaseproof paper. Greaseproof paper, or imitation parchment, is made in the same way as ordinary paper, but the pulp is subjected to an intensive beating action for several hours before going to the paper machine. It is produced frequently as a substitute for vegetable parchment and is much inferior in 'whiteness', bulk, strength and greaseproof qualities. Vegetable parchment may be distinguished from greaseproof paper by boiling in water. In a relatively short period, greaseproof paper and imitation parchment will be reduced to a soft mass which is easily defibred by pulling, while vegetable parchment will stand the action of boiling water for several hours with no sign of disintegration and when it is torn there is very little evidence of a fibrous structure.

Glassine. By passing a greaseproof paper over a super-calender, the paper surface can be given a very high finish and the density of the sheet will be greatly increased. The resulting product is called a *glassine*. The close-knit structure and the hydrated condition of the cellulose in glassines enables such papers to be used for wrapping purposes, not only for greasy products but also to maintain flavours and aromas. It also, of course, will exclude any odours which might be contaminating. Untreated, however, glassines are not by any means moisture-resistant, and to achieve this quality, surface coating is applied. A smooth non-porous non-absorbent surface, such as is provided by glassine, is ideal for wax coating and for other types of coating which provide barrier properties to gases, water vapour and odours. The material therefore has considerable use in the packaging field.

Greaseproof and glassine papers are frequently plasticized to further increase their toughness. They run well on high speed packaging lines and provide an odour barrier. They can be chemically modified to enhance wet strength, adhesion and release. When waxed, they are used for food pouches to package dry cereals, potato crisps, dehydrated soups, cake mixes, bakery goods, ice-cream confections, coffee, sugar, pet food, etc. In addition to their protective functions when waxed, these papers heat-seal easily and reclose well.

Tissue papers. These specially soft thin papers, usually available in a low weight of between 17 and 30 g/m^2, are used for wrapping articles where the surface is susceptible to abrasion. All of them are generally white, but can be made available in a range of colours. There are three basic qualities available: acid-free, machine-glazed and machine-glazed mechanical tissues. Acid-free papers are invariably unglazed and bleached, and they contain no chemical impurities of any great degree. They are particularly concerned with protecting goods that are susceptible to the action of acids, e.g. silverware, and also for wrapping expensive goods such as jewellery, to enhance their appearance. The machine-glazed papers are less costly than the acid-free, although they have a very smooth texture, do not have quite the same degree of purity. They are extensively used for interleaving metals where soft paper is required. The machine-glazed mechanical tissue is coarser than machine-glazed tissue and includes some mechanical wood pulp in its furnish. This can result in damage to goods with high finishes, but it is particularly suitable for counter sales of bread and for protecting glassware where a cheap quality tissue is suitable.

Coated papers. The production of synthetic polymeric materials capable of forming self-supporting films failed to supplant the use of coated materials, even after the initial difficulties in converting had been overcome. In many instances, the thickness of the film material required to give the requisite barrier properties is well below that at which it can be handled on filling and forming machines, and the cost becomes prohibitive. Coating a more expensive film material to a thicker and less expensive substrate to achieve the desired strength and handling properties is one way to overcome this difficulty. The commonly used substrate for this purpose is paper in one of its many forms. It should be stressed that the properties of the substrate, particularly in respect of its surface smoothness and absorption character-istics, are of great importance in achieving a continuous film of the more expensive barrier material at a low rate of application. The behaviour of some substrates of different surface characteristics coated with the same barrier material is shown in Figure 3.8. The smoothest, and least absorbent, cellulose film gives the same resistance at the lowest rate of application.

The various coatings used in packaging are best considered according to the manner in which they are applied. These are:

1. From aqueous solutions
2. From solvent solutions (lacquers)
3. From aqueous dispersions
4. As hot melts
5. As extrusion coatings

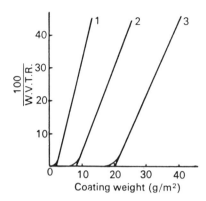

Figure 3.8 Behaviour of some substrates of different surface characteristics. Water vapour transmission rate (W.V.T.R.) in g/m^2 per 24 h at 38 C and 90% relative humidity. (1) Cellulose film; (2) glassine paper; (3) glazed imitation parchment paper.

Schematic representations of the various methods used are shown in Figure 3.9.

The main use of coatings applied from *aqueous* solution, apart from gummed tapes, is to make papers greaseproof and oil resistant. For this purpose, the water-soluble cellulose ethers are still in use, although polyvinyl alcohol has replaced them in certain instances.

In the field of *lacquer coatings*, the older materials are still being used, although many modifications both of the resins and plasticizers have taken place. This has assisted them to withstand the competition from recent additions to the resin range, and the improved methods of application developed to overcome the problems which faced converters wishing to use these newer materials.

The main attraction of *aqueous dispersions* lies in the fact that they contain little or no solvent, water being the carrier. Thus, when applied to a porous substrate such as paper, the evaporation is less than experienced with lacquers. The use of air knife doctors has removed the need for improving such dispersions in order to apply them by conventional methods. To achieve the full performance from the resin used in such dispersions it is generally necessary to apply a number of coats to paper substrates, and to subject the coated web to a high temperature, in order to fuse the material into a continuous film. Dispersions, based on polyvinylidene chloride, find outlets in the treatment of papers, where a single application of low weight produces a low cost heat-sealable paper with a fair degree of resistance to water vapour and fat-containing materials.

Hot-melt coatings comprise all the materials which can be applied in a molten state. This is probably the most attractive method of producing coated papers because no drying system is required. Providing the coating can be satisfactorily chilled after it has been applied, high production rates are possible. The coating material forms a film on the surface of the web which has properties very little different from those of the free film. Wax coatings are the oldest group of hot-melt materials in use today (Figure 3.10) and although paraffin wax is still used alone, the disadvantages which characterized such films (lack of durability, little resistance to creasing, scuffing and so on) have been considerably overcome by modification with various microcrystalline waxes, butyl rubber, polyisobutylene, polyethylene and copolymers of ethylene and vinyl acetate. Generally speaking, modified paraffin wax coatings contain up to about 15% by weight of the modifier, and these are used either singly or in combination to improve durability, crease resistance, scuff resistance, heat-seal strength and gloss.

Waxing can be performed in-line with the paper manufacturing process, in-line with printing, converting or laminating processes or as a separate process. A great many base papers are suitable for waxing, including greaseproof and glassine papers, and water-resistant papers.

The introduction of polyethylene and the design of suitable *extrusion equipment* for coating polyethylene films on to other materials in reel form, is one of the major developments in packaging in the post-war period. Extrusion coating with other plastics followed and considerable tonnages of coated paper are used for bags and for one or more of the plies in multi-wall paper sacks, particularly for hygroscopic products and certain chemicals. Much development is taking place in this area and many organizations now study coating/laminating processes using equipment such as that shown in Figure 3.11.

Non-wovens. Non-wovens are used as cloth substitutes and are made entirely or partially from cellulose fibres. They are finding increasing uses in medical, healthcare, industrial, food-processing, consumer and household product areas. Differentiated from classical paper, which is formed in water and consolidated with interfibre hydrogen bonds, non-woven manufacturing technologies include resin and thermally bonded carded web process, meltblown process and an air-laid process. Because the non-wovens industry is less than two decades old, much of the technology is proprietary.

Storage of materials

All paper-based materials are sensitive to climatic changes, especially variations in humidity, although temperature does not affect them very much. Plastic-coated papers may lose or gain moisture from one face, and as a result are particularly prone to curling. Smooth materials are prone to block if sheets are pressed together in stacks. Cast-coated materials may block causing surface damage if stacked flat in conditions of high

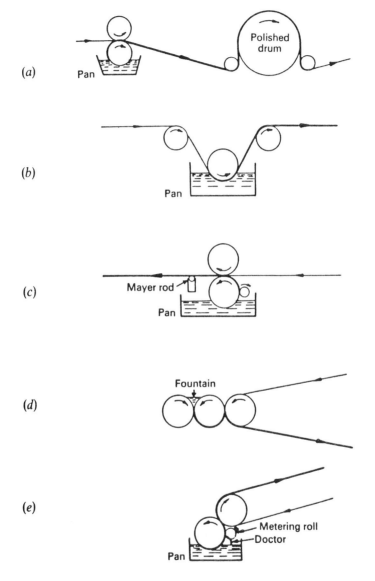

Figure 3.9 Coating methods. (a) Cast coater; (b) dip coater; (c) Mayer-type roll coater; (d) reverse roll coater; (e) roll coater.

humidity. Cellulose film materials can embrittle at freezing temperature and uncoated grades are very moisture sensitive. Adhesive-coated materials may be unusable if activated by moisture and heat.

Reeled materials should be stored either suspended on bars pushed

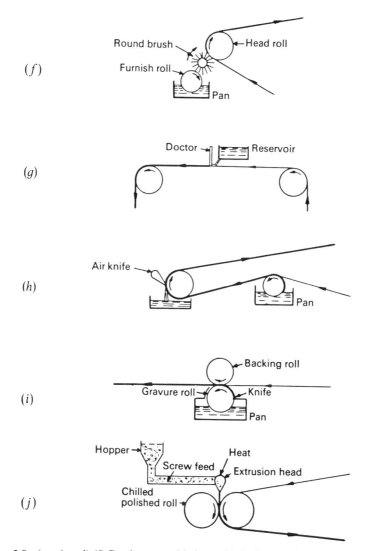

Figure 3.9 (continued) (f) Brush coater; (g) doctor blade (hot-melt) coater; (h) air brush coater; (i) gravure coater; (j) extrusion coater.

through their centre, or if overwrapped, by standing the reels on end. They should never be laid on their curved sides.

Some materials are sensitive to heat or light, e.g. latex adhesives deteriorate with age. Adequate protection should be given and stock should be rotated to minimize the storage time. Many papers, particularly treated papers (e.g. waxed), are attractive to rodents and cockroaches.

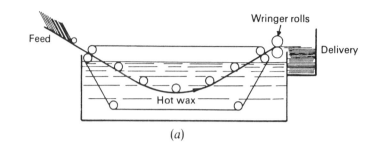

(a)

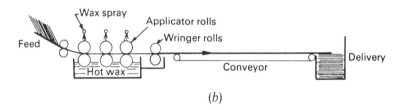

(b)

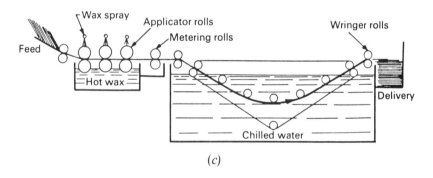

(c)

Figure 3.10 Waxing. (a) Wax saturator; (b) hot waxer; (c) cold finish waxer (wet waxing).

The optimum conditions for storing packaging materials are constant temperature of about 20°C and 50% relative humidity. If the storage area is uncontrolled and differs widely from the packaging area, then materials should be brought into the packaging area at least 24 h prior to use and not unwrapped until equilibration has occurred as far as temperature is concerned.

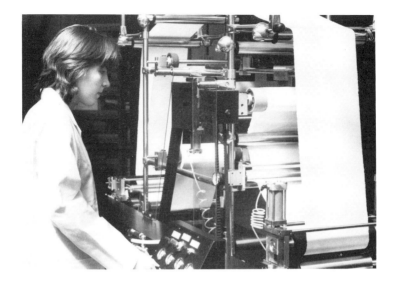

Figure 3.11 A multipurpose coater/laminator for research and development. Courtesy of Pira.

4
Paperboard

Paperboard is one of the major raw materials used in packaging. It is the generic term covering folding boxboards, chipboards, straw boards, etc. for primary packages and corrugated and solid fibreboards for transit packaging. 'Cardboard' is often erroneously used as a generic term for these materials but should be reserved to describe boards used for making cards of all kinds. There are many applications for paperboard, from simple cartons to more complex board structures for liquid packaging. It can be converted into fibre drums for bulk chemicals and combined with other materials to make containers for the transport and protection of heavy machine tools and delicate instruments. Generally, in addition to its containment and protective qualities, paperboard has at least one surface suitable for printing.

Terminology

The line of demarcation between paper and paperboard is rather loose. ISO define paper with a grammage above 250 g/m^2 as paperboard, or simply as board. However, in some parts of the world the dividing line is placed at 300 g/m^2. There are many anomalies: e.g. blotting papers are often heavier than 300 g/m^2 and fluting medium and some liner boards used for corrugated are less than 250 g/m^2.

Structure and general properties

Typical paperboards consist mainly of cellulose fibres of which the most common source is mechanical and chemical pulps derived chiefly from wood. Material recovered from used papers and boards (secondary fibres) are also widely used in the cheaper grades of material. Straw and esparto grass are also used for some boards. The structure of typical boards can vary from a single homogeneous layer or two to eight plies, each of which may or may not have the same fibre composition depending on the strength, stiffness, optical, bending or other properties required for the finished board. Typical structures for white lined folding box boards for cartons are shown in Figures 4.1–4.3. The top ply in a typical board is of bleached chemical wood pulp which will give the necessary surface strength and printability. An underliner of a lower grade (essentially white pulp) will prevent show through of the grey/brown colour of the middle plies of secondary fibre. The back may be similar to the middles but where extra strength or printability is required, a better grade of chemical/mechanical pulp mixed in the appropriate ratio, should be used. It is very important that all the plies are well bonded and this is dependent on achieving the right degree of mechanical entanglement and hydrogen bonding in the board-making process.

Some definitions

Backs: The ply that forms the under surface of the sheet.
Caliper (thickness): Also called substance, usually measured either in mils (thousandths of an inch) or microns (millionths of a metre (μm)) (1 mil = 25.4 μm).
Furnish: The fibre composition of a paper or board sheet, usually given in percentages.
Grammage: The mass per unit area of the board in g/m^2.
Liner: The ply that forms the top surface of the sheet, usually a good quality printable layer.
Machine direction (MD) and *cross direction* (CD): Sometimes called grain and cross grain. The MD is the direction along the sheet as it emerges from the board-making machine and the CD is at right angles to the MD.
Middles: Plies in the middle of the board, frequently of secondary fibres or mechanical pulps.
Ply: One of several layers of fibre that make up a multi-ply board.
Sheet size: The width of a sheet is given first. This is the dimension measured in the CD, e.g. 500×750 mm means 500 mm in the CD and 750 mm in the MD.
Underliner: A ply of lower quality than the top surface between the liner and the middles, usually to prevent the grey/brown colour of the middles showing through the top liner.

Physical characteristics

Mechanical properties

Paperboards have basically a similar structure to paper and hence a similar strength to weight relationship. The stiffness characteristics are

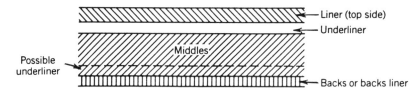

Figure 4.1 Stucture of a typical multi-ply paperboard.

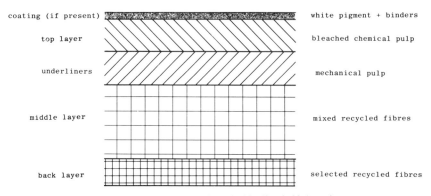

Figure 4.2 Ply construction of white lined chipboard.

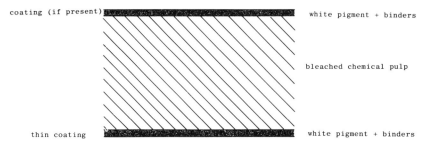

Figure 4.3 Construction of fully bleached solid board.

dependent on the furnish used in the several plies and increase with increasing use of better quality pulps in the outer plies. Paperboard has grain characteristics like paper and is stronger in the MD. The ratio of strength MD/CD varies according to how the board is made, from about 1.5/1 up to 4/1.

Surface properties

These are dependent on the furnish of the liner and backs. As with most packaging, at least one surface must be suitable for printing. Paperboard must also be glued during package manufacture and hence both surfaces must have the correct absorbency to accept an aqueous adhesive. Typically, the back of the board has a higher absorbency than the top liner and so will absorb liquid more easily during the gluing operation. When hot-melt adhesives are used, the properties of concern are the 'wettability' of the board surface by the melt and the fibre bond strength within the surface, which determines the ultimate bond strength. Like paper, paperboard suffers dimension changes when it absorbs moisture so the use of aqueous adhesives and printing by offset lithography, together with any other converting operations involving water, can cause problems such as curl of the board and print misregister. Smoothnes and friction depend on the drying and finishing processes and can be modified by surface coatings.

Optical properties

The colour (generally whiteness) and the gloss of the surfaces to be printed are important in obtaining a good printed image. This is the main reason why the use of an underliner ply immediately below the printed surface is necessary, not so much because the inks will be affected but rather because the appearance of the image is dependent very much on the degree of whiteness of the background and show-through of the middles would undoubtedly affect this.

Board making

The methods of treating fibres before the board-making process (beating, cleaning, refining, etc.) are virtually the same as those used in paper making (see Chapter 3). After the sheet has been formed, the methods of dewatering, pressing and drying, while similar in principle, differ in detail more or less depending on the method used.

Fourdrinier machines

As has already been described in the Fourdrinier process for paper (Chapter 3), the sheet is formed by pouring a slurry of fibres in water on to a moving wire sieve, when the water drains through leaving a mat which is ultimately consolidated into a sheet of paper. Attempts to produce thick materials will cause difficulty in draining off the water once a certain thickness is reached.

It is therefore only possible to make the thinner single ply paperboards by the unmodified Fourdrinier machine. Improvements have been made in flowbox design by using nozzles instead of a slice and by improving drainage methods under the wire. Modern practice replaces the table rolls by foils which generate less intense suction than table rolls and allow the length of the drainage zone to be increased. The foils are spaced closer together to provide a considerable gain in drainage capacity. When the consistency of the sheet has reached about 3–4% flat suction boxes are introduced to continue the drainage under a controlled vacuum. Other methods of forming the web from the fibres have also been developed. For the production of multi-ply boards (e.g. kraft liner board), a secondary flowbox is used. A base ply is formed first and partially de-watered before a second ply is deposited from the second flowbox. The basic idea behind these developments is to form the paperboard web more rapidly and usually between two wires so that the drainage can take place from both sides. The *twin wire former* (Figure 4.4) is, in principle, a method of letting the fibre suspension come out of the slice into a converging gap between the two wires. Water is expressed from both sides of the sheet by the tension in the wire and assisted by scraper blades pressing on the underside of the wires.

The idea dates back to the nineteenth century, but it was not until the 1950s that further development took place, taking advantage of improved equipment. The process became commercial and a viable contender in some applications. Figure 4.5. shows the basic Inverform concept, a U.K. invention and the first commercial twin wire system.

Cylinder mould or vat machines

Such methods of improving drainage can only go so far and, where we want to make thicker material in one pass, a vat or cylinder machine is used. Here the wire mesh takes the form of a cylindrical screen revolving in a vat filled with the pulp and water mixture. As the roller comes out of the mixture (Figure 4.6), it brings with it a layer of fibres, and the water drains away through the wire. These fibres form a web which is picked up by an endless belt travelling over the top of the cylinder. Some of the water is squeezed out from the web by rollers and then it passes on to another vat to join a web emerging from that one, and so on; it may be over four or five vats to build up the total thickness required (Figure 4.7). In this process, the vats are maintained at a constant level from a reservoir called a stuff chest and the pulp/water mixture is maintained at about 99% water, 1% fibre. Once the

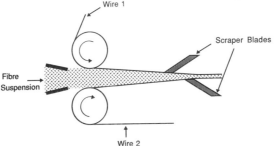

Figure 4.4   The principle of the twin-wire former.

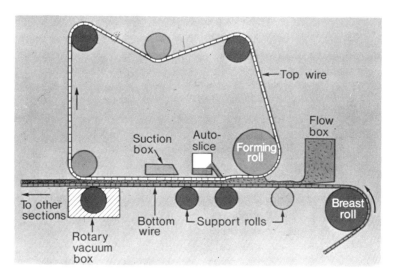

Figure 4.5 The Inverform system.

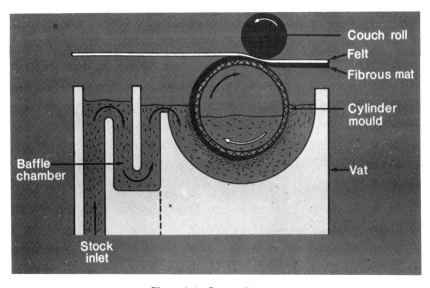

Figure 4.6 Counterflow vat.

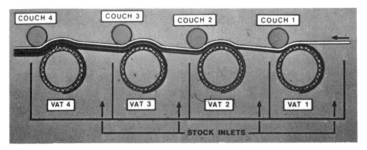

Figure 4.7   Four-vat board machine.

sheet has been formed, similar drying systems to those employed on Fourdrinier machines are used.

Roll formers

Figure 4.8 shows a typical rotary former. It consists of a cylinder-forming screen with an associated flowbox. The forming length has been considerably reduced and the drainage forces can be far higher than is the case with cylinder mould units.

The cylindrical screen can be of simple construction relying on a pressure force in the forming zone to assist drainage or it can take the form of a suction roll using a series of vacuum boxes to further assist de-watering. Compared to cylinder moulds, roll formers have three main advantages:

1. They can develop and tolerate higher levels of turbulence in the initial forming zone because the drainage zone is enclosed, whereas the initial forming zone in a cylinder mould machine starts at a free surface.
2. High pressure in the free suspension and the possibility of using suction on the underside of the forming screen permit a much higher rate of drainage.
3. The rotary former has a more uniform metering of the fibrous suspension on to the forming screen.

Multi-ply board machines

On a Fourdrinier machine, board is not built up in plies but formed as one thick layer. Drainage through this layer is very slow and different layers of several types of fibre cannot be formed. Sheet formation is, however, more uniform.

There are some Fourdrinier installations that use several flowboxes to produce a multi-ply board. A base ply is formed on the first part of the wire. Some drainage occurs before a second flowbox adds another ply. Water has now to drain through both plies. A third ply may be added in the same way (Figure 4.9).

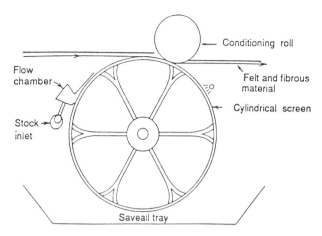

Figure 4.8 A cylinder former.

Combination machines form a base ply on a Fourdrinier wire and then combine this with several plies made on a cylinder mould machine (Figure 4.12).

Finishing processes

The surface of a paperboard web can be treated during manufacture by various means according to the characteristics required. It is not unusual to use as part of the drying process an MG (Yankee) cylinder, which imparts a smooth surface to one side of the web without too much densification of the web taking place.

We have already mentioned the processes of calendering and machine glazing as operations in producing a paper sheet. Other types of finish can also be applied. For example, surface sizing can be carried out, usually with a solution of gelatin in water containing small quantities of other chemicals to make the surface more water-resistant and to improve its printing properties. Any sizing process can be applied on the paper machine or as a separate process afterwards.

In the dryer section of the paper machine, there is often a size press where chemicals can be added to the surfaces of the web to impart certain characteristics (e.g. hard sizing or barrier properties). Paperboard machines often have coaters in-line with the operation at which one or more layers of mineral pigment such as china clay bonded with a resin can be applied. Many boards these days are clay-coated in order to improve their appearance and their printing properties. The coating suspensions are either sprayed or fed on to the web from a rotating brush soaked in the material. The coating is then distributed evenly by other brushes and the board conditioned in a steam chamber before calendering. Air-knife techniques have also been

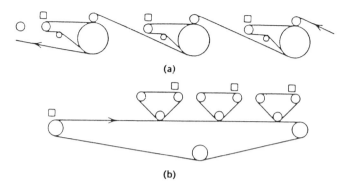

(a)

(b)

Figure 4.9 Examples of multi-ply formers. (a) Super ultra former. Examples in Japan, the United States, Canada, etc.; (b) Fourdrinier with 'on top' mini Fourdriniers. Examples in Europe and the United States.

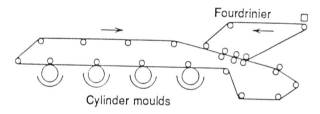

Fourdrinier

Cylinder moulds

Figure 4.10 Typical combination machine.

used. Here an excess of coating material is applied to the paper and then metered off by a 'doctor', which consists of a very fine jet of air blowing closely on to the paper surface.

Board-making machines are, with their associated stock preparation and finishing sections, large, expensive installations, capable of board production speeds of several hundred metres per minute. In order to achieve the level and consistency of board properties demanded by present day applications, extensive use is made of monitoring and feedback control, using computers. On-line measurements concentrate on caliper, grammage and moisture content of the finished board. Uniformity of these properties across the web and through the reel is controlled. Specialized sensors can be developed to monitor other properties such as stiffness.

Other surface treatments

Paperboard has the disadvantage, from a packaging point of view, of being susceptible to changes due to moisture. There are many processes of impregnating, coating and laminating board with other materials in order to improve the resistance to water (see Chapter 7). Probably the simplest and best known is waxing, and their are two main methods of carrying this out. In *dry waxing*, each sheet passes through a bath containing the wax and then through a heated section or hot nip rollers, which assist the wax to penetrate into the sheet. In this process, the wax does not form a surface film. When it is desired to have a film of wax on the top of the sheet, the process employed is called *wet waxing*. The reason for this is that immediately after the wax film has been applied to the surface of the sheet, it passes into a bath of refrigerated water, which immediately sets the wax before it has time to penetrate into the sheet. The film of wax so applied provides much greater water vapour resistance than the dry waxed sheet could ever do, but being a surface film, it is much more easily damaged. The paperboard web may also be coated with many kinds of emulsions, varnishes, lacquers and the like, and can be laminated to other materials such as plastics, either using adhesives or by direct extrusion. The objective in all these instances is usually to improve the resistance of the basic sheet to water, water vapour, gases, grease or oils.

Table 4.1 summarizes the type of construction, the characteristics and the principal uses of the main packaging boards.

Folding boxboards

The boards used for cartons are mainly made on cylinder mould or Inverform machines. Only the thinnest boards are made on single and twin wire machines.

The two main methods provide many possibilities, as all these boards have a ply structure. All plies may be re-pulped waste as in chipboard. Replacement of the outer ply by better quality pulps gives in succession cream-lined chipboards and second and first quality white-lined chipboards. Duplex boards without any waste pulp are also available. Board made entirely from chemical pulp (solid white boards) are produced for frozen food packaging and other applications where the board must be waxed. The cost of these boards obviously rises with the quality of pulp used.

Where special properties such as moisture resistance or grease resistance are required, coated and laminated boards are employed. These may be wax-laminated where the wax is the moisture barrier, glassine-lined for grease resistance or plastic-coated for special properties, including heat sealing. Paperboard cartons, even if made from plastic-coated boards, unless of special construction, cannot be expected to give much moisture protection to the contents and are therefore used with inner bags.

The external appearance and the printing quality of carton board is greatly enhanced by the use of coated board. Clay and other minerals are coated on to the lined outer surface of the board. The coating is applied either during the board-making operation or subsequently. A limitation of the boards which are coated on the board-making machine is that the coating

Table 4.1 Main packaging boards.

Type of board	Construction	Characteristics	Principal uses
Chipboard 200–800 g/m^2	100% low grade waste pulps newsprint to fibreboards	Low cost, non-printing, grey/brown, poor folders	Rigid boxes, packing pieces, fitments and tubes
Cream lined chipboard 200–800 g/m^2	Liner of mechanical and chemical, mixed pulps, backs as chipboard	Lowest cost board which can be printed and folded	Lowest quality cartons
White lined chipboard 200–800 g/m^2	Liner is 100% new pulp mainly chemical, backs chipboard	Good printability strength and foldability	Quality folding cartons for shirts, cosmetics, detergents, etc.
Duplex board 200–1000 g/m^2	Liner and back both new pulp middles mixture of both	As white lined chipboard	
Clay coated boards 200–600 g/m^2	As white lined with coated top surface	Excellent printability	Top grade cartons
Solid bleached board (food board)	100% virgin sulphate pulp	Strong board, high whiteness, odourless, good folder	All foods particularly frozen foods and baked foods
Liquid packaging boards 200–400 g/m^2	As food board but with barrier coating or lamination	As food board	Milk, creams, fruit juices, etc.
Kraft liner board 120–400 g/m^2	90% + kraft pulp	High strength	Outer component of solid board liners for corrugated board
Test liner 120–450 g/m^2	Skin of kraft, rest waste pulps	Burst strength equivalent to kraft liner grades	Liners for corrugated boards

must generally be of a somewhat thinner nature and have a less glossy surface than coatings applied off the machine as a separate operation. The latter include cast coated boards which have a very high gloss and brightness. In many instances, the extra cost of these boards limits their use to the packaging of comparatively expensive goods, such as cosmetics.

Foil-lined boards are also used for various types of cartons, not only where protection is of importance but also to give particular display effects.

Chipboard manufactured from re-pulped waste materials (newsprint, corrugated and solid fibreboard cases, etc.) accounts in its various forms for more than half the current carton board consumption in the United Kingdom today. *Plain (unlined) chipboard* is only utilized for the very simple type of carton for holding stocks of materials which do not require any great display. It is almost unprintable in terms of quality and its strength is somewhat lower than that of any other variety. Some chipboard cartons are produced, however, and are lined with printed paper wrappers.

White lined chipboard, with its improved appearance, printability and folding qualities, is used in greater quantities than any other grade. It is used for many cartons where the internal appearance is unimportant, and where no contact with foodstuff occurs. Taint is possible when foods containing flour or fats contact chipboard directly. Obviously, if a separate bag is used inside a carton, the contact between its inner surface and the food is prevented and, under these conditions, white lined chipboard may be usable.

With cartons that are opened repeatedly, the better-looking inner face of a *white lined manilla* or *duplex board* may be required, although white lined chipboard is often used for these types as well. With certain exceptions, most food must avoid direct contact with chipboard, so that either pure pulp or duplex boards are used; alternatively, glassine, sulphite or other types of paper may be laminated to the chipboard side of a white lined chipboard where foods are concerned.

Pure pulp boards are particularly popular for materials such as flour confectionery and chocolate, where staining of the carton by fats, etc. is possible. In this context also, boards lined with greaseproof paper, vegetable parchment or glassine are often used.

For a considerable number of products, protection against moisture and moisture vapour, together with grease resistance, is required and here, paraffin wax used in a number of ways can provide a low priced solution. A great amount of work has gone into producing grades and blends of paraffin wax with various other components for different applications.

Wax-laminated white-lined chipboard is widely used for soap powder cartons if moisture loss or entry is to be avoided. Such waxes are also used as the laminating adhesives for glassine and other greaseproof liners referred to above.

Wax coating on one or both sides of the board can also be carried out at any stage after printing, and waxed cartons are particularly used in the frozen-food industry, although many are employed in an unprinted state, with a printed paper or film overwrap.

More recently, polyethylene and other extruded polyolefines have been employed, both for coating boards and for laminating various types of liner to chipboards and other boards.

Developments in polymeric materials are adding continually to the range of materials available for combining with paper and paperboard in all forms. Note the increased use of such materials as polyethylene, polypropylene, polyvinylidene chloride and nylon, either in direct combination with board or as ancillary materials. By and large, rigidity, which is one of the prime factors required by all unit retail packages, can most economically be supplied by paper and paperboard, whereas barrier properties, such as moisture resistance, oxygen resistance and grease resistance, are readily supplied by comparatively thin films of plastic. The combination of the two (rigidity and resistance) in coated paperboard can result in highly protective packages for many goods.

The use of barrier materials in carton board is principally restricted by the inability of the normal types of carton closure to prevent the ingress of moisture directly without passage through the barrier. Because of this, a considerable amount of work has been directed towards the development of one-piece functional cartons, aimed initially at the frozen-food industry where the protective requirement was originally demanded. Further designs have led to the use of cartons for liquids of various kinds.

Boards for rigid boxes

It is difficult to define the principal market for rigid boxes, but they are used by the hosiery and footware trades and for small handtools, and also for the expensive perfume and cosmetic markets. The choice of board used depends very largely on the type and weight of the contents, the size of the box, the intricacies of manufacture and the degree of accuracy required.

During recent years, change has taken place within the rigid box industry, as to the type of board used. Originally, two main types were used. First, Dutch straw board was the traditional material used by the majority of box makers engaged in the mass production of machine-made boxes. It is a stiff material and makes a good rigid box. However, because of various drawbacks (which include its propensity to absorb moisture, to vary in size and to give off a musty smell when damp) box makers looked for alternative materials.

The original chipboard offered by the British board mills, although not as rigid as the Dutch straw board, offered the advantage of not shrinking to quite the same extent. It therefore was more reliable for ensuring that the correct fit was obtained between boxes and lids, and for minimizing the acute warping of shallow lids. The main disadvantage was the lack of rigidity and often a thicker board was required to obtain a similar rigidity to the Dutch straw board. This led the British board mills to produce a new grade of board known as *rigid board*, as a substitute for Dutch straw board. It was intended to combine the advantages of both Dutch straw board and home-produced chipboards. This board is now widely used in the United Kingdom and is available in thicknesses (calipers) ranging from 500 to 2900 μm.

No strict rules can be drawn for the rigid box industry as to the type of board to be used. Generally speaking, the box maker will decide on the best board for the particular job in question, and this may vary from a plain unlined rigid board, through the complete range to a high class food quality duplex board.

There are many covering papers, though generally in the rigid box industry, there are a number of standard types in use. These include enamels, flints, tints, leatherettes, and many others. The covers can, of course, be printed, embossed or gold blocked and, as with board, the type of paper used depends on what the customer requires. The main consideration for the box maker is to know the behavioural characteristics of the various types of papers and how to overcome them if they have adverse affects on production.

Fibreboards for shipping containers

Solid fibreboard

Solid fibreboard is composed of paperboard (usually chipboard) lined on one or both faces with kraft or similar paper between 0.13 and 0.30 mm thick. The total caliper of the lined board ranges from 0.80 to 2.8 mm.

Corrugated fibreboard

Corrugated fibreboard consists essentially of two flat parallel sheets (liners) of paper board, with a central fluted or corrugated sheet between them. The combined board is held together with adhesive applied to the crests of the fluted sheet. The materials used for any of the three basic components can be in a variety of weights and type, and the flute configuration can be varied.

In the United Kingdom, corrugated fibreboard is available commercially with liners of grammage of between 125 and 410 g/m^2 and fluting medium of nominal 113 or 127 g/m^2 is used. The three-component material already described is known as single-wall board. Double-wall board consists of two fluted layers, separated by a flat sheet and faced on both sides with a liner (i.e. five components). For heavy duty application triple-wall board is also made, consisting of three fluted layers and four flat sheets.

Manufacture of corrugated board

Figure 4.11 shows a diagram of a corrugator. The process of making a double face board begins just before the fluting medium enters the single facer when it passes through preheating rolls and steam showers to plasticize it and make it pliable. It then travels through the fluted rolls of the single facer where the corrugations are formed and adhesive is applied to the tips of the flutes. At the same time, the first liner (usually the inner liner) passes through a series of heated rollers and is stuck to the flute tips. This single face board then continues up into the bridge between the single facer and the double backer where it is festooned to form a temporary store.

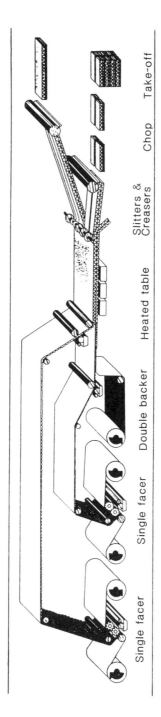

Figure 4.11   Diagram of the corrugator.

At the double backer, adhesive is applied to the exposed flute tips and a further web of preheated liner is pressed into contact with the flutes as the two webs pass between a belt conveyor and a heated 'table'. The bonding is completed under heat and pressure as the board passes down the table through a drying section until it meets the slitter/scorer and is 'chopped' into the required size blanks.

Figure 4.11 shows a corrugator with two single face units which used together with the double backer would produce double-wall board.

Single-wall board is commercially available in A, B, C and E flute construction. Double-wall board is normally available commercially in AB, CB, AA and AC fluting combinations. Triple-wall board is usually produced commercially in AAB, CCB and BAE constructions.

The corrugated medium: materials, flute height and configuration. Butcher's straw paper was the first material used for corrugating medium, the stiffness of the material being approximately what was required in those early days. A modern corrugated medium must have properties that can be classified under three headings:

1. *Combining properties.* The characteristics of the sheet must be such as to allow it to pass easily through the corrugator and to accept the flute configuration, then to be adhered to both liners at speeds in excess of 300 m/min.
2. *Conversion properties.* After combining into board with the liners, the corrugating medium must possess properties that will allow it to perform satisfactorily in respect of scoring, both at the take-off end of the corrugator and in the printer-slotter. It must also successfully resist the stresses induced during these conversion processes.
3. *Case performance properties.* The ability of a corrugated case to remain a rigid container when subjected to the normal hazards of transport is dependent, to a large extent, on the ability of the fluting medium to keep the liners apart and thus retain the stiffness of the board.

The fluting structure used for commercial corrugated fibreboard consists of four size ranges. In each size range, the material should have an approximately sinusoidal fluting form. The fluting ranges are given in Table 4.2.

Table 4.2 Fluting ranges for corrugated fibreboard.

Flute configuration	No. of flutes per metre	Flute height (mm)	Minimum flat crush (N/m^2)
A (coarse)	104–125	4.5 –4.7	140
B (fine)	150–184	2.1 –2.9	180
C (medium)	120–145	3.5 –3.7	165
E (very fine)	275–310	1.15–1.65	485

A, B and C flute corrugated fibreboards are widely used industrially for cases for the transit of goods. E flute board is widely used in display cases and similar applications, usually combined with high quality printed liners. Several types of paper are used for manufacturing commercial corrugating medium:

1. Semi-chemical papers made by treating wood chips (usually beech and/or birch) with chemicals to achieve pulp of the desired properties.
2. Straw papers made from furnishes of between 25 and 75% straw with various quantities and grades of waste pulp.
3. Kraft paper used in the United Kingdom on some grades of weather resistant board.
4. Secondary fibre, chip, or waste papers of various grades can be used in their own right after treatment or be added to any of the above materials to give a balance of properties at a moderate cost.

The liners. Four main types of liner are used:

1. *Unbleached kraft liner* is a pure pulp material with a natural brown finish which is produced from long fibre coniferous trees. This liner is most commonly used as the outer surface of corrugated fibreboard and provides a good printing surface.
2. *Multi-ply test* or *duplex test liner* consists of a layer of kraft pulp on a base manufactured from mixed waste paper or other secondary fibres. The quality and appearance of duplex test liner resembles kraft liner more closely than any other substitute liner.
3. *Single-ply test liner* consists of a single ply and is often manufactured from a pulp made from corrugated waste, e.g. recycled corrugated waste. The strength is less than that of duplex test liners but much stronger than chip. Additives are used to produce a brown kraft appearance.
4. *Chipboard liner* is manufactured entirely from recycled material. Although it has a natural grey finish, it can be dyed and glazed. It is used mainly for inner fitments such as layer pads, divisions, etc. It does not have a good printing surface.

The required liner properties for modern corrugated board can be divided into the same groups as the fluting medium and similar considerations apply. The above materials are available in various grammages as shown in Table 4.3.

Table 4.3

Material	Grammages available							
Kraft		125	150		200	250	300	400	
Duplex test					200	250	300		450
Single-ply test		125	150		200				
Chipboard	115			175	200				

5
Metal packaging

The main metals used in packaging are mild steel sheet, tinplate, terne plate, galvanized mild steel sheet, stainless steel, aluminium alloys and aluminium.

Mild steel plate, usually cold-reduced strip-mill steel, is the principal metal drum-making material. Terne plate, a mild steel coated with a tin/lead alloy and galvanized steel, with an electrolytic deposit of zinc, are also employed. For special contents, aluminium alloy sheets, commercially pure aluminium sheet and stainless steel (normally an 18-8 nickel chrome steel) are used for drums and kegs.

Tinplate is the principal material for metal boxes and cans. Tinplate is mild steel (i.e. low-carbon steel) coated on both sides with tin. The base steel plate or strip is manufactured by rolling hot steel ingots down to a strip with a thickness of 1.8 mm. The strip is then pickled continuously in a bath of hot dilute sulphuric acid and cold rolled to a finished gauge of 0.15–0.50 mm. The sheet is finally annealed and temper-rolled to impart the required hardness and surface finish.

Since the early 1960s, economies have been produced by reducing the thickness of the steel base. These economies depend on the fact that further cold reduction of the sheet produces a material with a greater intrinsic stiffness; hence a thinner sheet can be used for some applications. The plate is then known as 2CR (double cold reduced) or DCR (double reduced). Figure 5.1 indicates some of the complex processes necessary to turn steel and tin into tinplate.

Table 5.1 summarizes the wide range of mechanical properties and forming qualities that can be obtained in steel packaging materials.

The tin used for tinplate comes principally from the major tin-producing countries, i.e. Malaysia, Bolivia, Indonesia, Thailand, Nigeria, Zaire and Australia. Tinplate production is in the hands of the steel companies, who purchase their tin from the producers. Today, there are some sixty independent tinplate producers located in 34 countries worldwide.

Tin coating was originally applied by running the steel plate through a bath of molten tin (hot dipping process). Although this method is still employed, the greater bulk is now made by a continuous electroplating process (giving electrolytic tinplate). Hot dipping gives a comparatively thick coating, with a lower limit of about 22 g/m^2 (11 g/m^2 on each side). This

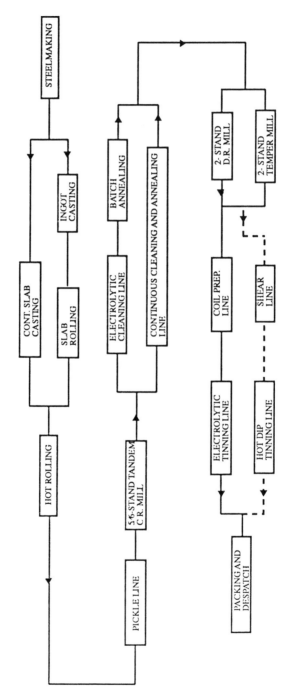

Figure 5.1 Schematic diagram of the various routes to tinplate sheet and strip.

Table 5.1 Tinplate temper designations.

Temper classification		HR30T hardness aim		Approx. UTS (N/mm^2)	Formability	Typical usage
Current	Former	Mean	Max. deviation of sample average			
T50	T1	52 Max.		330[a]	Extra deep drawing	Normally stabilized steel; deep drawn parts
T52	T2	52	+4 −4	350[a]	Deep drawing	
T55	–	55	+4 −3		General purpose	
T57	T3	57	+4 −3	370[a]	General purpose	For ends and round can bodies, non-fluting
T61	T3	61	+4 −4	415[a]	General, increased stiffness	For stiff ends and bodies, crown corks, shallow stampings
T65	T5	65	+4 −4	450[a]	Resists buckling	For stiff ends and bodies
T70	T6	70	+3 −4	530[a]	Very stiff	Beer and carbonated beverage can ends
DR550	DR8	73[a]	+3 −3	550±70[b]	Double reduced	Round can bodies and can ends
DR620	DR9	76[a]	+3 −3	620±70[b]	Doubled reduced	Round can bodies and can ends
DR660	DR9M	77[a]	+3 −3	660±70[b]	Doubled reduced	Beer and carbonated beverage can ends

[a]The UTS values for single reduced tinplate and the HR30T values for double reduce tinplate are given for guidance only. These values do not appear in any specifications.
[b]The tensile values for double reduced tinplate are the aim proof stress (0.2% non-proportional elongation).

figure corresponds to the old figure of 1 lb per basis box (1 basis box = 31 360 square inches of tinplate, i.e. 62 720 in^2 total surface area).

Hot-dipped tinplate possesses a naturally bright finish, whereas electrolytic tinning produces a rather dull coating. Electrolytic tinplate can, however, be brightened by heating momentarily either in a bath of hot oil or by electric induction. This process is known as *flow brightening*. Flow brightening not only improves appearance but also the resistance of the plate to corrosion. The dull matt finish of the as-plated sheet is due to a porous finish and the protective properties of the coating are much improved by melting the tin to give a more coherent finish. After flow brightening, electrolytic tin coatings are treated to remove any tin oxide formed during the process. They are then treated in chromic acid, dichromate or chromate/phosphate solutions to stabilize the finish. This is because tin oxide is often affected by subsequent storage or baking of the tinplate.

The high speed equipment now used for can-making necessitates the use of a film of lubricant on the tinplate surface. Cotton seed oil or synthetic oils such as di-octyl sebacate or di-butyl sebacate are normally used. The oil film is extremely thin—of the order of a few tenths of a millionth of a millimetre—and the lubricant must be compatible with any subsequent lacquer coating used.

Undoubtedly, it was the introduction of the electrolytic tinplate process which stimulated the expansion of the industry. Not only did continuous coating enable the most efficient use to be made of steel manufactured by a continuous strip process, it also brought into being a whole new range of coating grades which could not be produced by the hot dipping method previously used. Furthermore, the electrolytic process enables a different thickness of tin to be applied to the two surfaces of the steel. This 'differential tinplate' is economically beneficial to the user since it enables the most cost-effective coating to be selected to withstand the different conditions of the interior and exterior of the container.

The modern electrolytic tinplate process is capable of producing tinplate at a rate of up to 600 m/min. Thus, in one hour, a modern tinplate line can produce about 30 tons of first quality tinplate, sufficient to make more than half a million cans.

Tinplate combines the strength and formability of light gauge steel with the corrosion resistance, solderability, weldability, non-toxicity, lubricity, lacquerability and good appearance of tin. The tin coating is sufficiently adherent to the steel base to withstand, without flaking, any deformation that the steel is able to sustain. This includes the wall ironing which is a feature of one of the new can-making techniques; indeed the lubricity of the tin coating aids this process. The manufacture of the steel base and the application of the coating are entirely separate operations, so that in principle any grade of steel may be produced with any desired coating.

To the steel maker, tinplate is a finished product, but for the can maker it is only one of a number of raw materials to be used together with solders, lacquers, printing inks and sealing compounds for the fabrication of a wide range of containers.

Blackplate

Blackplate is tinplate without the tin. It does not have many applications in tin box making because the absence of tin makes it vulnerable to rust in the presence of even the slightest moisture. This can be countered by applying synthetic lacquer to the sheets before they are exposed to storage or other hazards.

Blackplate is no longer black. When new and clean, it is almost indistinguishable from tinplate to a passing observer. It is called *blackplate* because in the days when it was rolled by hand, it acquired a thin coating of black oxide which was left on the surface and not dissolved away in acid, as it would be today, before the material left the mill.

Tin-free steel

The final step in tin economy is the elimination of tin entirely, hence the name *tin-free steel*. This material is produced by electrolytically coating mild steel plate with a chromium/chromium oxide film. The process was developed in Japan and one version, known as Hi-Top, is being made in the United Kingdom. The chromium/chromium oxide layer is even thinner than the thinnest tin coatings normally used, and must be lacquered before it can be used in the manufacture of containers. It is, however, satisfactory for protecting the steel from rusting during transit and storage prior to can manufacture.

The cost of TFS is lower than that of tinplate, but is increased by the necessity for lacquer coating. Therefore, if tinplate can be used for packaging a particular product, it will normally be cheaper to use tinplate. On the other hand, if the tinplate has to be lacquered, then lacquered TFS will normally be cheaper.

Recycling

People are more and more aware that environmental interests have to be taken into account when dealing with packaging, and in several countries an infrastructure has been set up for the recycling of ferrous and steel waste material.

Significant quantities of energy can be saved by re-melting steelscrap since it requires about 50% less energy to make steel from scrap than from iron ore. The *factory scrap* of the converting industry, i.e. the can makers, is recycled almost 100%.

An attraction of tinplate lies in the simple recycling of cans collected from *household waste*. As most of the containers finally end up in the garbage stream, the normal method of recycling is to separate the cans from the rest of the waste mechanically. Fortunately, the magnetic properties of steel facilitate this separation.

Refuse is collected by local authorities and delivered to landfill sites, composing plants or incineration plants. Before incineration, composting or burial, metal may be separated magnetically. The cans are then cleaned to remove organic dirt and then treated in a de-tinning plant where tin coating is extracted and re-used. After incineration, cans can also be extracted from the ash. This material, however, contains small quantities of tin and is only acceptable in a limited way by steel makers. Ideally, extraction should be carried out before incineration.

In some countries, a recycling percentage of about 45% of the cans in household waste has already been reached. It is estimated that this percentage can grow to about 70% in the 1990s.

Aluminium

Aluminium and its alloys have long been used for the manufacture of rigid containers, although not to the same extent as tinplate. Aluminium is also used in the form of foil (both alone and laminated to other materials), and for collapsible tubes.

Aluminium is a bluish silver-white metallic element that is very malleable and ductile. Noted for its weight, good electrical and thermal conductivity, high reflectivity and resistance to oxidation, aluminium is the third most abundant element in the earth's crust, where it always occurs in combination with other elements in mineral forms such as bauxite, corundum, turquoise, spinel, kaolin, feldspar and mica. Of these, bauxite is the most economical raw material for the production of aluminium. It can contain up to 60% alumina, which is hydrated aluminium oxide. About 4 kg of bauxite will produce 1 kg of aluminium.

Alumina is converted into aluminium at a reduction plant or smelter. In the Hall-Heroult process, the alumina is first dissolved in cryolite (potassium aluminium fluoride) in steel boxes lined with carbon called pots. A carbon anode is lowered into the solution and electric current of 50 000–150 000 amps flows from the anode through the mixture to the carbon-cathode lining of the steel pot. The electric current reduces the alumina molecules into aluminium and oxygen. The oxygen combines with the anode's carbon to form carbon dioxide. The aluminum, heavier than cryolite, settles to the bottom of the pot from which it is drawn off into crucibles.

Aluminium is becoming increasingly useful to the can maker, particularly for drawn and wall-ironed (DWI) applications. Small, solid drawn cans with easy-opening ends and aluminium ends with easy-opening aids have become commonplace on beer and beverage cans. There is something to be said for having the whole of the can of the same metal, because scrap reclamation and recycling are easier. This stimulated the movement towards the all-aluminium can with an easy-opening end, which is simpler to make in aluminium than in tinplate.

Like chromium, aluminium is protected from corrosion by a thin layer of oxide which forms on the surface as soon as the metal is exposed to air. Also, like chromium-coated steel, aluminium is almost impossible to solder.

Foil

Aluminium foil is manufactured (Figure 5.2) by rolling slabs of pure aluminium (99.2–99.5% aluminium) down to a thin foil. This rolling process starts by taking an ingot of several tonnes in weight. This is rolled down

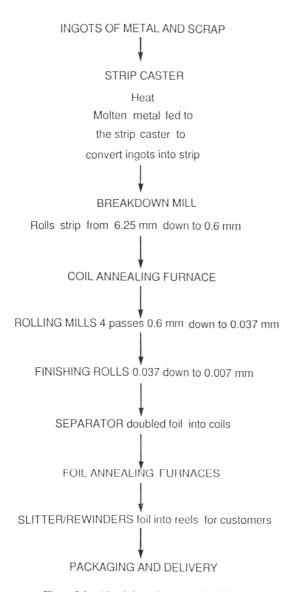

INGOTS OF METAL AND SCRAP

STRIP CASTER

Heat
Molten metal fed to
the strip caster to
convert ingots into strip

BREAKDOWN MILL

Rolls strip from 6.25 mm down to 0.6 mm

COIL ANNEALING FURNACE

ROLLING MILLS 4 passes 0.6 mm down to 0.037 mm

FINISHING ROLLS 0.037 down to 0.007 mm

SEPARATOR doubled foil into coils

FOIL ANNEALING FURNACES

SLITTER/REWINDERS foil into reels for customers

PACKAGING AND DELIVERY

Figure 5.2   Aluminium: from metal to foil.

at an elevated temperature to a slab about 60 m long. Cold rolling is then carried out by passing the slab through a line of rolling stands until the desired thickness is attained. The thinnest foil used in flexible packaging is around 0.009 mm. Thicker gauges are used for milk bottle tops or for certain semi-rigid containers. The latter may utilize foil of a thickness between 0.1 and 0.2 mm.

The high deformation caused by cold rolling makes the aluminium brittle, reduces its elongation at break, but increases its tensile strength. Aluminium foil in this form is referred to as *hard-temper foil* and is unsuitable for normal wrapping application or for the manufacture of pouches because of its tendency to break at creases. A hard-temper foil is required, however, for the so-called *push-through blister packs* for tablets, in order that the foil should break when the tablet is pushed from the plastic film side of the blister.

Corrosion of tinplate

Billions of tinplate containers are successfully used each year for packaging a wide range of products. Nevertheless, many different sorts of attack on the containers by the packaged product can occur. Many products can attack tin itself, but de-tinning is preferable, since dissolution of the tin by the product delays attack on the underlying steel. This is important, since the tin coating is a very thin one and is not 100% complete, particularly after the rigours of fabrication into a can on high speed equipment. Attack on exposed areas of steel, unchecked by preferential attack on the tin, could lead to rapid pinholing of the container. Attack on the tin is not beneficial, however, when it leads to discoloration of the product, or to off-flavours due to metal pickup. Here, the tinplate must be lacquered.

The product may also attack the lead solder where this is present. Highly alkaline products can slowly attack and dissolve lead solder, with consequent leakage at the side seam. Some organic compounds also react chemically with lead solder, resulting in blackening of the solder and the product. In aerosol shaving creams, for example, the presence of stearic acid in the formulation can result in the precipitation of lead stearate along the side seam. When the can is shaken, this precipitate may clog the aerosol valve. The problem may be reduced by lacquering the inside seam. However, perfect coverage of the seam is difficult to obtain under commercial conditions, so that high tin content solders are often used where such problems occur with lead solders. Another answer to the problem is to use a corrosion inhibitor. Testing is essential in such instances because many corrosion inhibitors work only for particular products.

Attack on the base steel normally occurs only at imperfections in the tin coating, at cut edges of the tinplate, or when the tin has been completely dissolved by the product. The use of lacquer coatings, while protecting the surfaces, cannot prevent attack at cut edges. Open-top cans do not present problems of cut edge exposure, but lever lid tins (such as paint tins) have an exposed cut edge at the ring. This is the component seamed to the top of the paint tin body to provide the seating for the push-in lid. Although the cut edge is curled under to keep it out of contact with the product, corrosion can still occur with aqueous products from water vapour in the head space.

Mechanism of corrosion

Corrosion is an electrochemical phenomenon involving a transfer or displacement of electrons. If two dissimilar metals are immersed in water or some other conducting solution (electrolyte) and then connected, a current will flow and we have a galvanic cell. The electric potential set up between the two metals provides for the transfer of electrons through the wire. (If the water was absolutely pure, only a very small current would be produced.) The two metals in such a cell are called *electrodes*. One is called the *anode*, the other the *cathode*. The flow of current is caused by the flow of electrons from the anode to the cathode.

The loss of electrons from a metal anode leaves the metal below the liquid surface as a positive ion which is soluble. The anode, therefore, slowly dissolves in the electrolyte. With a tinplate container, the two metals in contact with each other are immersed in the packaged product which acts as the electrolyte. If the tin is the anode and the steel is the cathode, then the tin is dissolved while the steel remains untouched. The tin is referred to as a 'sacrificial' anode, because by its dissolution, it protects the steel. Tin, however, can behave either as the anode or the cathode, depending on the conditions, and this provides a means of controlling the behaviour of the tin.

One of the important factors controlling the electrochemical behaviour of tin is oxygen. Normally, the concentration of oxygen in the product in a tinplate container is low. Under these conditions the tin is anodic and the steel cathodic—the desirable condition, because the tin is dissolved preferentially and thus the tinplate protects the base steel, even if the coating contains pinholes. At higher oxygen concentration, however, the tin becomes the cathode and the steel the anode, when rapid corrosion of any exposed steel occurs, usually causing perforation of the container.

Polarization

The passage of an electric current through an aqueous electrolyte leads to the formation of hydrogen gas at the negative electrode. Most of the hydrogen escapes, but some gas bubbles cling to the cathode and cause a current to flow in the opposite direction to that originally produced. The effect reduces the overall cell current. This phenomenon is known as *polarization* and in commercial galvanic cells would obviously be a drawback. In a corrosion situation, however, anything which reduces the corrosion current reduces the corrosion rate. Conversely, any factor which reduces polarization will accelerate corrosion. This gives the clue to an explanation of the effect of oxygen dissolved in the product on the corrosion of steel in contact with tin. High oxygen content in the product reacts with the hydrogen deposited at the cathode, reduces the polarization and restores the normal corrosion current.

Oxygen and moisture are the two factors necessary for corrosion of iron, and both must be present simultaneously. The corrosion product of iron (rust) is an oxide of iron. Moisture is necessary for the formation of metallic ions, which then dissolve with subsequent corrosion. Neither dry oxygen nor oxygen-free water causes appreciable corrosion of iron.

Any factor which interferes with the normal function of the electrodes (the components of the container—tin and iron) or the electrolyte (the packaged product) may slow down or even stop the corrosion process. Conversely, any factor which intensifies their function will accelerate the corrosion process.

The materials from which the electrodes are made have an important influence on the rate of corrosion, because they determine the electric potential of the corrosion cell. The common metals can be arranged in order of their tendency to go into solution and to form ions. This series is known as the *electromotive series*; metals high up the series (such as sodium) go into solution very readily, while metals at the bottom of the series, such as gold and platinum, do not go into solution at all and hence do not corrode. Two metals which are far apart produce a larger voltage when they form the electrodes in a galvanic cell than two metals close together.

Aluminium and iron are far apart and, therefore, the use of aluminium ends on tinplate containers is to be recommended only when contact of the aluminium and iron is prevented by some coating. Otherwise rapid corrosion of aluminium will occur. This holds true only if the product (electrolyte) does not change.

As might be expected, the acidity or alkalinity (pH) of the environment is an important factor in the corrosion. As a rule, acid solutions (low pH) are more corrosive than neutral or alkaline ones. However, the actual rate at which a metal corrodes in a solution of given pH depends on several other factors, including oxygen concentration and polarization, which have already been discussed. Three other factors are important from a practical standpoint.

Stress corrosion

Stress corrosion is the term applied to many patterns of corrosive attack in which stress is believed to accelerate corrosion. It often occurs when normal corrosion is almost negligible. Many theories to account for stress corrosion have been suggested, but the most acceptable is that electrochemical action occurs between stress-caused anodic areas and the more cathodic unstressed areas. Since the areas of stress are anodic, corrosion takes place here and the tensile stresses present open up crevices, thus exposing fresh metal to further attack.

The stresses can be caused by cold working of the metal or because of a too severe drawing operation in fabrication.

Presence of inhibitors

Corrosion inhibitors are often added to an environment of electrolyte to reduce corrosion attack. Such materials work by interfering with the function of the electrolyte or the electrodes and thus slow down or stop the corrosion process. Inhibitors added to a canned product are designed to reduce or stop the electric current between the tin and steel in the case of tinplate containers.

Passivity

Under certain conditions a metal electrode may cease to dissolve, although it appears unchanged. The metal is said to be in a *passive state*. The attainment of passivity depends on the electrolyte. With metals like iron it is favoured by alkalinity, but other metals such as tungsten become passive more easily in acid solutions.

Passivity can also be produced without the action of an electric current. For example, if iron is dipped into concentrated nitric acid, there is a brief reaction, which rapidly ceases and the metal becomes passive. Nitric acid is an oxidizing agent and it is significant that iron can also be rendered passive by heating it in air. The mechanism of passivity seems, therefore, to be allied in some way to oxidation and it has been found that metals in the passive state have an extremely fine oxide film on the surface. This film acts as a protective barrier and prevents further solution of the metal.

Corrosion testing

One obvious method of assessing the possibility of corrosion is to carry out storage tests on containers and the products concerned. If the conditions of storage are also equivalent to those expected in use, then this method will give good results. The only snag is the time-scale involved. If a shelf-life of two years in envisaged, then a test period of much less than 4–6 months is unlikely to give meaningful results.

Another method is electrochemical testing, commonly known as *corrosivity testing*. This is based on one of Faraday's laws of electrolysis which states that the amount of decomposition is proportional to the current and to the time for which it flows (i.e. the total quantity of electricity). Conversely, when a given weight of metal dissolves, it gives rise to an equivalent quantity of electricity which flows as an electric current. The magnitude of this electric current can be measured, as can its directional flow. This tells us which of the two metals is dissolving (or corroding) and how rapidly the corrosion is taking place.

The method is applicable to products packed in containers constructed of at least two metals (such as tinplate—one electrode is tin and the other is steel). By this method some very valuable information can be gained, even after a period of only 24 h, especially when it is evaluated in conjunction with practical experience. Thus, at the end of 24 h we know which of the two metals is anodic (and hence will corrode) and which is cathodic and being protected. We also know the magnitude of the current, and hence the magnitude of the corrosion to be expected. This method must not be used in isolation; storage tests should also be carried out in order to verify the results and to determine whether elevated temperatures can reverse the polarity or break down the product in some way.

Lacquer coatings

The basic function of an interior lacquer is protection of the product rather than protection of the container. Lacquers can be beneficial and many prolong the effective shelf-life. No lacquer is known, however, that is effective in all circumstances and the product must always be taken into consideration, as well as the likely storage conditions. Since lacquer coatings are usually applied to the sheet tinplate prior to can manufacture, the lacquer must be able to withstand the mechanical shocks associated with can-making. In addition, the lacquer must be easy to apply and cure, resistant to the product, provide the barrier required and be economical in use.

For food or pharmaceuticals, can lacquers must not give any toxic hazard and must be free from odours and flavours. For products which are paricularly susceptible to contact with metals, one additional lacquer coating is given after the can has been manufactured. This second coating is applied by spraying and serves to cover imperfections in the side seam, or those caused by mechanical damage during manufacture.

There is a wide variety of can lacquers available, including oleoresins, phenol formaldehyde, epoxy resins and vinyls. There are also specialized types, such as one containing zinc oxide. This is used for products containing sulphur-bearing proteins (e.g. processed peas). The sulphur would normally cause blackening of the can and its contents, due to iron sulphide formed by attack on the steel at lacquer imperfections. The zinc oxide in the lacquer removes the sulphur with consequent formation of zinc sulphide and prevents the blackening reaction.

Lacquers are also applied to the outside of tinplate containers to improve corrosion resistance, particularly if the cans are to be exported to tropical areas or are likely to be stored under damp conditions. With printed cans, the external decoration acts as a protection against corrosion. It should be noted, however, that external decoration does not cover the soldered side seam, so that a base lacquer may be needed in addition to the decoration in extremely corrosive conditions. Alternatively, the can may be given an external side-stripe of lacquer along the side seams.

6
Glass

The use of glass to meet some of our everyday requirements can be traced far back into history. Because of its traditional base, it is perceived by some consumers as old-fashioned and therefore incapable of adapting to modern life, whilst at the same time resisting change. The real situation is the absolute reverse. The glass industry is still vibrant and by changing and adapting modern technology to reduce costs and increase quality and reliability, it is improving its image in the market place with innovative product design, presentation and the development of new glass-based packaging systems.

Glass is manufactured from relatively simple and readily available raw materials. Sand, limestone and soda ash (the last derived from common salt) are accurately weighed in the correct proportions and then precisely mixed prior to charging into a melting furnace. The term 'batch' is used to describe mixed raw materials.

In spite of elaborate precautions taken during the extraction of these materials, the quarrying and mining techniques which are involved are bound to disturb natural habitats and the environment in the locations at which they are found. Obviously the disturbance is reduced if smaller amounts of material are removed.

Fortunately, scrap glass, or cullet as it is known in the industry, can be mixed with the batch as a raw material prior to furnace melting. The presence of cullet in the batch reduces the energy required for conversion (see Figure 6.1). Costs are reduced on two counts: first by reducing the quantity of raw materials and second, by either reducing the melting energy or at the same energy input, enabling the melting operation to be carried out at a higher throughput.

The nature of glass

As a packaging material, glass has advantages in that it will not affect the product, nor does it need additional treatment to make it impervious. It is a rigid material that does not alter its characteristics with the passage of time.

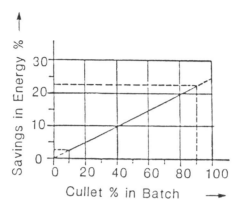

Figure 6.1 Relationship between savings in melting energy and cullet usage in glass making.

Odour, which can be of considerable concern to food packers, does not arise as a form of contamination.

There are, however, a limited number of products (blood transfusion fluids and certain drugs) which are extremely alkali-sensitive and for which special glasses or treatment are required. A process known as 'sulphating' is used to remove most of the sodium from the surface of the glass or, if the quantities are small, the containers can be made from tubing in a glass of low alkali content.

Glass compositions for container manufacture have been finely tuned over the years to reduce the percentage of soda ash (because it is the most expensive of the raw materials used for glass-making), whilst at the same time maintaining the working properties of the glass by balancing other chemical constituents. Typical chemical analyses are shown in Table 6.1.

Table 6.1 Typical glass compositions used for container manufacture.

		Flint	Amber	Autumn leaf	Emerald green
Silica	SiO_2	73.0–73.4	72.6	72.7	72.1
Soda	Na_2O	11.5–12.4	12.8	12.9	12.9
Potash	K_2O	0.44–0.49	1.01	1.04	0.87
Lime	CaO	11.2–11.3	11.1	10.9	9.8
Magnesia	MgO	1.35–2.04	0.23	0.17	1.74
Baria	BaO	0.01–0.03			
Titania	TiO_2	0.03–0.04			
Alumina	Al_2O_3	1.12–1.23	1.81	1.83	1.93
Iron Oxide	Fe_2O_3	0.040–0.068	0.34	0.28	0.37
Chromic Oxide	Cr_2O_3	–	0.002	0.005	0.17
Sulphur Trioxide	SO_3	0.17–0.19	0.08	0.05	0.09

The major constituent in soda-lime glass is silica (SiO_2), which is the major constituent in sand; sand is normally over 99% SiO_2. Depending on the impurities present in the sand deposits and the colour of the glass to be produced, the sand is purified, perhaps milled and then screened to remove particles that are too large or too small.

When available from recycling schemes, the second largest constituent in glass batches is cullet or recycled glass. One source of cullet is reject ware produced by the factory itself, but recycled glass has now become the principal source.

The next two largest constituents are soda ash (Na_2CO_3) and limestone ($CaCO_3$ or $CaCO_3.MgCO_3$). During the melting process, large amounts of carbon dioxide (nearly 200 times the volume of glass produced) are given off from these two raw materials. Alumina (Al_2O_3) is an important but not large constituent. Its major function in a soda-lime container batch is to improve the chemical durability of the glass.

Refining agents to aid melting and to remove gas from the glass are also added. Without the use of refining agents, higher temperatures and longer melting times are required to obtain glass free of bubbles. They typically contain sulphides and sulphates.

Finally, colourants such as chromic oxide (for green glass), iron, sulphur and carbon (for amber glass) and cobalt oxide (for blue glass) are added in small amounts to achieve a particular colour. Flint (clear) glass batches contain decolourizers (nickel and cobalt) to mask the colour imparted by trace amounts of impurities such as iron.

To change a furnace from one colour to another is sometimes necessary to meet planning requirements, but it is an expensive and time consuming operation. Up to 36 h can be involved before full operation is re-established. The Colorama Process attempts to overcome this problem by colouring glass in one of the forehearths which connect the furnace to the bottle-making machines. Usually only relatively small quantities of bottle glass at greater cost can be made by this method. Since a flint composition is used in the main furnace, scrap bottles from the coloured glass cannot always be returned to the recycling plant.

To maintain glass composition within predetermined specification limits and to check incoming raw materials, rapid methods of chemical analysis based on X-ray fluorescence have displaced conventional wet analysis methods, enabling a complete summary of a glass to be produced in less than a day. These laboratory measurements are supplied to the factory at regular intervals and the information is used to correct batch compositions and to reject or accept consignments of raw materials. The aim is to maintain the composition of the glass within limits of viscosity as part of the overall control of the process.

Plant overview

The production area of a typical modern glass container plant occupies more than 15 000 m^2, and the adjacent warehousing space generally exceeds 20 000 m^2. New glass container plants are frequently located near large customers and operate one to three furnaces. In a modern high productivity plant geared to the production of similar jobs, each furnace will feed at least two if not three production lines. A line consists of one feeder, one high productivity forming machine, an annealing lehr and inspection and packaging areas.

Furnaces used for the manufacture of glass containers operate continuously day and night for most of the year, and can last for up to ten years, with intermediate periods of downtime for repairs. Developments in furnace design are therefore carried out at a much slower rate. Melting energy is also a significant proportion of glass manufacturing costs (approximately 10%). Analysis of information gleaned from past records is used as a basis for improving a furnace at rebuild by modifying its design and combustion system. Over the last twenty years, fuel consumption has been reduced from an average of over 60 therms per tonne of glass melted to 45 therms per tonne, improvements which have been made possible only by understanding the technology of design and operation.

Natural gas is used widely in the European industry for melting glass, where it is available. In general, 'interruptible contracts' exist whereby glass manufacturers are forced to equip their furnaces with oil firing facilities which are brought into operation during periods of high domestic load in the winter or when the spot price of oil is low enough to justify its use in preference to natural gas.

Attempts have been made to recover waste heat from flue gases after they have passed through regenerators or recuperators. This relatively low grade energy has been used in some European factories for raising steam for space heating, generating electricity for local needs or maintaining domestic district heating schemes. In general, the benefit obtained does not adequately cover the capital expenditure required for the additional equipment and supply complications can arise when the furnace is not operational during a rebuild or intermediate repair.

Receiving raw materials

Glass-making raw materials received at the plant are unloaded under the direction of a computerized materials handling system and stored in large silos near the furnace end of the plant. The raw materials are removed from their storage silos, weighed, mixed and delivered to the furnace under the control of a computer. Water is frequently added to the batch in the mixer or in the batch charger to reduce the number of fine particles blown into the regenerators by the strong combustion gas currents present in the furnace and to reduce segregation of the materials in the batch.

Charging

The materials are typically 'charged' into the furnace either by an Archimedian screw charger or a pusher-bar charger. Charging of the furnace occurs whenever a gauge senses that the level of molten glass in the furnace has fallen below a certain minimum set point. Charging continues until the glass level reaches the maximum set point. The glass level is typically held constant to within 1 mm to aid the forming operation. Because of the intermittent nature of the charging process, the Archimedian screw charger places 'piles' of batch on top of the molten glass in the furnace. The pusher-type charger places 'logs' of batch on the molten glass. The initial decomposition of carbonates and chemical reactions between the various raw materials begin in the batch piles or logs floating on the molten glass in the furnace. In many plants, combustion quality in the furnace and the position of the piles or logs is monitored by TV cameras. As with so many other processes, furnaces are coming under the influence of computers both for ease of setting up a target and for process control.

The glass-melting furnace converts the raw materials into molten glass that is chemically homogeneous, virtually free of gaseous inclusions and of a constant viscosity. The chemical reactions occurring between the raw materials are complex, involving decomposition, solid-state reactions and liquid-solid reactions as well as melting.

Melting

After mixing, the raw materials are mechanically conveyed and charged into the melting chamber of the furnace. The furnace is rectangular in shape, holding up to 400 tons according to the number and type of machines it is feeding. The raw materials fed in at one end react, fuse and circulate until they become a homogeneous molten liquid with a viscosity similar to that of treacle. The glass passes through a throat into a zone known as the *working chamber* before it is conveyed to the forming machines via the forehearth. Most furnaces are heated by oil or gas flames sweeping across the surface of the glass, up to temperatures of 1500°C. A typical tank furnace is illustrated in Figure 6.2.

Melting by electricity, using the glass as a conductor, is technically possible but, although there are many experiments in this field, it is at present more expensive than the conventional oil- or gas-fired method. Electricity is, however, used for boosting the tonnage capability of oil and gas-fired furnaces.

The aim at this stage of the process is to produce a homogeneous glass of

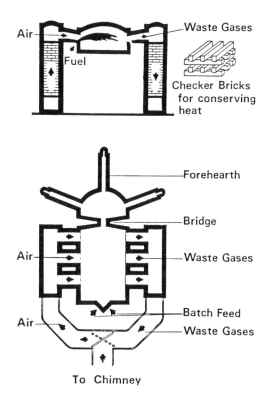

Figure 6.2 Elevation and plan view of a tank furnace.

good colour and free from seed and bubble blemish. Temperatures should remain steady and the pull on the glass should be such that it is given sufficient time to fuse and throw off the gases caused by the reaction of the materials. Otherwise the glass will contain the imperfections of insufficiently melted raw materials (known as *striae* or *cord*), small bubbles of unreleased gases (known as *seeds*), or crystalline inclusions in the glass (called *stones*).

The furnace

Glass-melting furnaces are constructed from refractory bricks. Different types of bricks are used in different parts of the furnace. For example, the bricks in contact with the molten glass are selected for high resistance to chemical attack by the glass. These bricks have a relatively high thermal conductivity and typically contain large amounts of alumina, zirconium oxide or both. Outer courses of brick are chosen mainly for their insulating properties.

There are two principal types of furnaces used for melting container glass: end-port and side-port (see Figures 6.3. and 6.4). Either type can be coal- or

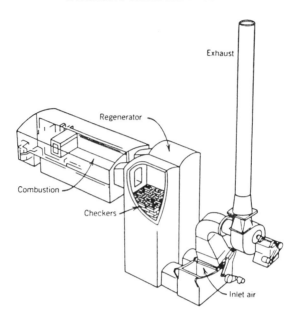

Figure 6.3 End-port furnace with forced draft stack. Reproduced with permission from John Wiley, New York.

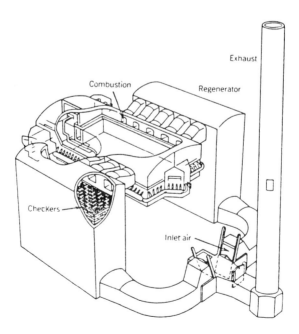

Figure 6.4 Side-port furnace with natural draft stack. Reproduced with permission from John Wiley, New York.

gas-fired, but electric boosting is frequently employed to increase the glass-melting capacity. All-electric furnaces are used in some special situations to melt glass for opal glass containers.

In an end-port furnace, there are two large openings (ports) in one end of the furnace. Air is fed through one of these while combustion gases leave the through the other. Fuel is introduced into the combustion air by various types of burner located near the mouth of the port.

Side-port furnaces have several small ports along each side of the furnace. Air and fuel are introduced along one side and the combustion gases are exhausted through the other side.

The size of a glass furnace is measured by its surface area. End-port furnaces typically contain a glass bath with a surface area less than 65 m^2. Side-port furnaces typically have surfaces larger than 55 m^2. The glass depth varies from just over 1 m for coloured glasses to almost 2 m for flint glasses. A container furnace produces between 3.5 and 8 tonnes of molten glass per hour.

7
Plastics

Plastics are comparatively new materials made from chemicals in contrast to the materials found in nature such as wood and stone, metals mined from the earth or glass which is produced from naturally occurring materials. The first plastic was celluloid and it was made in 1862 and 35 years later casein from milk was the starting point for the second plastic. This was followed at intervals by a few others but the big expansion did not occur until after World War II, when a rapid advance in thermoplastic materials took place. Originally, several plastics were made from coal tar but nowadays they are derived from crude oil.

The word plastics comes from the Greek *plastikos*, to form. The British Standard definition is: 'a group of solid composite materials that are largely organic and usually based on synthetic resins or modified polymers of natural origin which can be cast, moulded or polymerized directly into shape'. Sometimes rubbers and a few other naturally occurring products are considered to be plastic, but modern usage usually excludes them.

The ability of plastics to be shaped is a property resulting from their molecular structure. All plastics are composed of large molecules, polymers, with very long chains of repeating units derived from much shorter molecules, by one of two types of chemical reaction, addition or condensation. The formation of polyethylene in which the monomer, ethylene, $CH_2 = CH_2$, splits at the double bond and forms long chains with some branching; thus, $n(CH_2 = CH_2) = (-CH_2 - CH_2 -)_n$ is typical of an additional polymer. The formation of polyethylene terephthalate where water is eliminated between ethylene glycol (an alcohol with two $-OH$ groups) and terephthalic acid (a dibasic acid) to again form long chains is typical of condensation polymers. In general, these are formed from alcohols or amines which have at least two reactive groups in their structure ($-OH$ and $-NH_2$, respectively) with acids which have at least two $-COOH$ groups. Nylon formed by condensation of hexamethylene diamine with adipic acid is an example of an amine/carboxylic acid condensation.

All three of the polymers mentioned so far are thermoplastic, i.e. they soften on heating and harden again on cooling and this can be repeated many times. There is another class of plastics known as 'thermosets' which on heating or by treatment with chemicals, cross-link the long chain molecules with the result that they no longer soften when heated. Hence the name. Phenol formaldehyde and urea formaldehyde are examples.

Raw materials

In theory, plastics can be derived from almost any organic matter but the main raw material used at present is naphtha which is derived from crude oil and natural gas. Coal tar, once an important source, may well be used extensively in the future as the price of oil increases; waste products from the timber, pulp and paper industries are also possible sources. Wood contains three main constituents, cellulose fibres, hemicelluloses (mainly 5-carbon sugars) and lignin. Cellulose is already used as the source of cellulose acetate and regenerated cellulose film, etc. and by chemical treatment and fermentation of the resulting sugar, can be converted into ethanol which in turn can be used to produce ethylene. Currently, lignin is an undeveloped source of resins but under the right economic circumstances could be a viable possibility. Cane and beet sugar could also be used and there have been developments in this area recently. However, at present, naphtha from crude oil is the main raw material and Figure 7.1 indicates the pathways to the most common plastics currently used in packaging.

Plastics based on natural polymers

Three types of natural polymer are involved, polysaccharides, proteins and rubber.

Polysaccharides

Cellulose is a polysaccharide and was one of the earliest polymers to be studied. It occurs in cotton and wood. Cotton contains about 90% cellulose on dry weight and wood has about 40–50% cellulose, 30% lignin (another natural polymer), which holds the cellulose fibres together, some hemi-celluloses and salts. Cellulose itself is not a plastic; it does not soften on heating nor dissolve in water. Regenerated cellulose film, or cellophane as it is called in the United States, is also not a plastic in the strict sense of the term although it is frequently considered with plastic films. It is produced by treating sheets of very pure cellulose pulp with caustic soda solution for up to 4 h which dissolves it. This solution is then allowed to 'ripen' for 2 or 3 days. This reduces the length of the polymer chains a little and the sodium cellulose formed is then converted to cellulose xanthate by treatment with carbon disulphide. The xanthate is then ripened for 4–5 days more; the cellulose is then regenerated in film form by extrusion or casting into an acid bath.

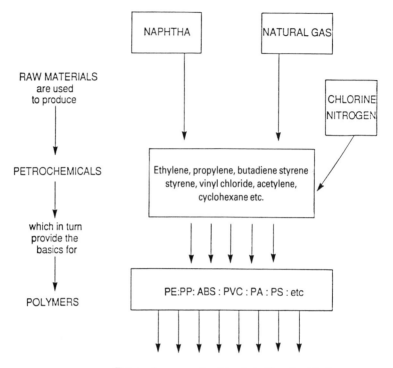

Figure 7.1   From raw materials to packaging plastic.

The cellulose polymer consists of glucose units each containing three hydroxyl groups (Figure 7.2). The process passes through the following simplified steps:

$$R(OH)_3 \rightarrow R(ONa)_m(OH)_{3-m} \rightarrow R(O-CS-S^-Na^+)_m(OH)_{3-m} \rightarrow R(OH)_3$$

cellulose   sodium cellulose   cellulose xanthate   regenerated cellulose

The value of m is about 0.5.

Cellulose esters. Organic acids will react with cellulose under certain conditions to produce esters. Acetylation, for example, is carried out in three stages to produce cellulose acetate. The 'dissolving pulp' is soaked in glacial acetic acid for about 2 h and then acetic anhydride is added; the mixture is refluxed in the presence of methylene dichloride at about 50°C. The proportions in the mix are as follows. For the first stage: 100 parts by

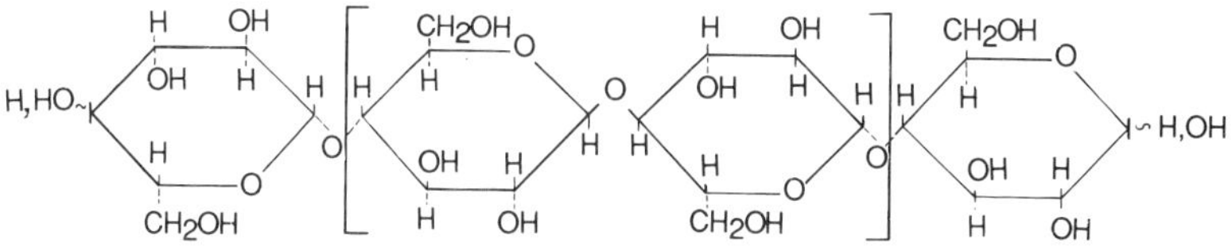

Figure 7.2   Chemical structure of cellulose.

weight cellulose; 35 parts by weight glacial acetic acid. For the second stage: 300 parts by weight acetic anhydride; 400 parts by weight methylene dichloride; 1 part by weight sulphuric acid. This converts the cellulose into the triacetate which is rarely used as a plastic but is usually partially hydrolysed by heating with water for a few days to produce the diacetate.

$$R(OH)_3 \rightarrow R(OCOCH_3)_3 \rightarrow R(OCOCH_3)_2OH$$

| cellulose | cellulose triacetate | cellulose diacetate |

While the diacetate is soluble in acetone, the triacetate dissolves only with difficulty in more powerful solvents, e.g. chloroform. Cellulose acetate is sensitive to moisture and therefore is not dimensionally stable. Its strength is similar to polystyrene but its mechanical properties are altered when it picks up moisture. It has a high clarity and is a poor barrier to the passage of water vapour. It is often used for windows in carton making.

Cellulose ethers. Ethyl cellulose and benzyl cellulose have similar properties to the acetate in respect of stiffness, clarity and chemical resistance. Other ethers such as methyl cellulose are water soluble and are used in water soluble adhesives.

Typical properties of cellulose-based packaging materials are shown in Table 7.1.

Proteins

Proteins are components of living cells which contain nitrogen and are essential to all living matter. The term was first used at the suggestion of Berzelius in 1838.

Casein is the only significant protein in packaging. It has been used for a long time as a basis for wood, textile and paper adhesives. It is precipitated from milk on acidification (e.g. using lactic, hydrochloric or sulphuric acids) in warm (35–37°C) conditions. Coagulation is accomplished at 10°C after which the precipitate is washed first in warm water and then cold water, and then dried to a powder which is soluble in dilute alkali.

Casein becomes insoluble on heat treatment at 150°C. To mould it, it is wetted with water, mixed with plasticizers, fillers and pigments as required

Table 7.1 Typical properties of cellulose based packaging materials.

Property	Cellulose acetate	Cellulose propionate	Cellulose aceto-butyrate	Cellulose nitrate	Ethyl cellulose
Density (g/cm^3)	1.28–1.32	1.17–1.24	1.15–1.22	1.35–1.40	1.09–1.17
Refractive index	1.49–1.5	1.46–1.49	1.46–1.49	1.49–1.51	1.47
Water absorption (%)					
3 mm film/24 h	2.–7.0	1.2–2.8	0.9–2.2	1.0–2.0	0.8–1.8
Effect of sunlight	Considerable	Slight	Slight	Very considerable	Slight

and cured with formaldehyde. The resulting plastic is an odourless, non-flammable solid, resistant to alcohol, ether, benzene, oils and fats but swollen on immersion in water.

Rubber

The third natural polymer, rubber, is used in packaging and has been converted into film forming materials such as rubber hydrochloride and cyclized rubber, both of which have now been superseded by more efficient and economical materials. Rubber, in particular foamed and sponge rubber, has been used for shock amelioration in the packaging of delicate articles.

Plastics based on synthetic polymers

Thermosets

There are only three thermosets used to any great extent in packaging. Phenol formaldehyde and urea formaldehyde are used mainly for bottle closures, while glass-reinforced polyesters are used for large containers.

Phenol formaldehyde (PF) resins are dark brown or black in colour and rather brittle. Various fillings such as chopped fabric or wood are added to improve the impact strength or to reduce costs. *Urea formaldehyde* (UF) resins can be obtained in white or pastel colours, but are more expensive; cellulose is used as the filler because of its whiteness.

Both materials have good chemical resistance and are mainly used for closures. They are insoluble in organic solvents, are attacked by strong acids and alkalis but are resistant to weak acids and alkalis. Urea formaldehyde resins are particularly favoured for closures by the cosmetics industry because of the wide colour range available and their resistance to oils and solvents. Phenol formaldehyde is widely used for pharmaceutical closures because it is more resistant to water. Both PF and UF closures are being displaced by thermoplastics closures, mainly polypropylene.

Glass fibre-reinforced polyesters have high strength to weight ratios and good resistance to outdoor weathering. In general, they have good chemical resistance and resistance to solvents. They are resistant to most organic and inorganic acids, except strong oxidizing agents and weak alkalis. They are, however, hydrolysed by strong alkalis. They have been extensively used for storage tanks and for large transit containers.

Thermoplastics

Six main types of thermoplastic are used in packaging.

1. *Polyethylene:* LDPE, HDPE, LLDPE, used in various grades for containers, bottles and jars, film for wrapping, bags and sacks.
2. *Polyproplyene:* PP, OPP, used for containers, bottles and jars, crisp packets, biscuit wrappers, pudding basins and boil-in-the-bag films.
3. *Polyvinylchloride:* PVC, used for squash and shampoo bottles where a clear and transparent container is needed.
4. *Polyethylene terephthalate:* PET, used for bottles for carbonated drinks.
5. *Polystyrene:* PS, HIPS, used for egg cartons and yoghurt pots.
6. *Acrylonitrile butadiene styrene:* ABS, used for tubs for margarine, vegetable salads, etc.

These and some other lesser used materials are described in the following.

Polyethylene. The properties of the several varieties of polyethylene which exist, depend on the molecular weight range, the morphology and the degree of crystallinity of the polymer and these in turn are largely determined by the method of producing it. Until Ziegler's work in 1955 on the polymerization of ethylene at low pressure using a special catalyst (alkylaluminium derivatives and titanium chloride), it had been supposed that the monomer could only be polymerized at high pressures. This discovery was one of the most significant advances in modern polymer chemistry. Karl Ziegler received a Nobel prize for the work in 1964.

Very soon after Ziegler made his discovery, Phillips Petroleum Co. and the Standard Oil Co. patented new ways of making polyethylene at medium pressures (below 7 MPa) and a temperature less than 300°C by polymerization in a hydrocarbon solvent using catalysts based on chromium(VI) oxides on alumina or silica-alumina carriers. Both low and medium pressure polymerization produce linear polymers. Both methods use a hydrocarbon solvent and require the complete absence of water and polar compounds. The reaction takes several hours to complete in both instances. Table 7.2 compares the different types of polymer.

At first, polyethylenes were differentiated into high pressure and low pressure types. High pressure material became low density polyethylene (LDPE) and the polymer made at low pressure was referred to as high density polyethylene (HDPE). With the advent of the availability of a range of polymers, other criteria such as hardness, degree of branching, etc. were

Table 7.2 Comparison of polyethylenes prepared by different methods. A, High pressure polyethylene; B, medium pressure (Phillips) polyethylene; C, Zeigler polyethylene.

Property	A	B	C
Density (g/cm^3)	0.91–0.92	0.93–0.95	0.96
No. of double bonds per 1000 carbon atoms	0.6	0.7	1.5
RCH=CH$_2$ double bonds (%)	15	43	94
RR$_*$C=CH$_2$ double bonds	68	32	1
RCH=CHR double bonds	17	25	5
No. of –CH$_3$ groups per 1000 carbon atoms	21.5	3	<1.5
No. of –CH$_3$ end groups	4.6	2	<1.5
No. of –CH$_3$ groups at branchings	2.5		
No. of –C$_2$H$_5$ groups at branchings	14.4	1	<1
Degree of crystallinity (%)			
X-ray analysis	64	87	93
NMR analysis	65	84	93
Diameter of crystallites (mm)	19	36	39

needed to classify individual polyethylenes. It is now impossible to accurately determine the boundary between the basic types; a range of properties can be made using both production methods. It would therefore appear more reasonable to use the branching of the macromolecules as the criterion. This would differentiate between the linear and branched types. Linear polyethylene (*l*PE) is still, however, more frequently called high density (HDPE) and the branched type (*b*PE), low density (LDPE).

The chemical resistance of the polyethylenes increases as the degree of crystallinity increases. At ambient temperatures they are all virtually unaffected by water, non-oxidizing acids, alkalis and salt solutions, but are attacked by oxidizing agents. They have good resistance at room conditions to polar solvents like alcohols, lower glycols and glycerol, but are swollen and attacked by non-polar solvents. At higher temperatures, some polar solvents, e.g. boiling carbon tetrachloride, benzene, toluene etc. will dissolve them to some extent. Low density polyethylene has good strength at low temperatures (brittle point about −120°C). High density articles will retain their shape at temperatures up to just above 100°C and can therefore be sterilized with boiling water, while low density polymers will soften around 80°C.

Low density polyethylene (LDPE) accounts for the greatest proportion of all the plastics used in packaging. It is very versatile, can be extruded into film, blown into bottles, injection moulded into closures and dispensers, extrusion coated on papers, cellulose films and metal foils and made into large drums by rotational casting.It is tough, semi-flexible and shock resistant, it is relatively inert chemically and almost insoluble in all solvents at ambient conditions. Some softening and swelling does occur with hydrocarbon solvents and chlorinated hydrocarbons. It is susceptible to stress cracking when exposed to surfactants. Permeability is low for water vapour but many organic vapours and essential oils will permeate rapidly. It is a poor gas barrier to oxygen and carbon dioxide and where oxidation of a product is likely, it is not suited. It can also be foamed to produce cushioning materials which are resistant to creep under load.

Linear low density polyethylene (LLDPE) is generally stronger and tougher than conventional LDPE but has similar properties.

High density polyethylene, as already mentioned, has a higher softening point than LDPE and is harder. Its barrier properties are about twice as good and for an equal wall thickness, HDPE bottles give a more rigid container. The higher molecular weight grades are used for the production of large 210 litre drums and for crates.

Polypropylene. The polymerization of propylene can lead to the atactic, syndiotactic or isotactic versions of the polymer. Atactic polypropylene is a soft material which can be used alone only with difficulty, but the isotactic polymer obtained by polymerization with the catalysts of the Ziegler–Natta type is a crystalline plastic with melting point about 170°C.

Polypropylene is similar chemically to low density polyethylene and high density polyethylene. It is harder then either, however, and has a less waxy feel. It can be injection moulded, blow moulded and extruded into film and sheet. The sheet can be thermoformed to give thin-walled trays of excellent stiffness. Polypropylene has excellent grease resistance and is also more resistant to solvents than low density polyethylene. Toluene and xylene, however, will cause swelling. Polypropylene is not subject to stress cracking and it differs in this respect from both the polyethylenes. Its softening point is higher than both polyethylenes and it is easily able to withstand steam sterilization.

One outstanding property of polypropylene is its resistance to fatigue when flexed. This means that integral hinges can be formed in an injection-moulded article, which will stand up to a very large number of flexings. Laboratory tests have, in fact, proved that even a million flexings will not cause failure of a well-designed hinge. In packaging, the obvious way of utilizing this property is to mould a box and lid in one process. The absence of subsequent assembly processes can often lead to substantial cost savings.

Although polypropylene is a rigid polymer, it is more resilient than polystyrene. The resiliency of polypropylene has been found useful in the design of screw-on closures, as it permits the moulding of slight undercuts into which can be snapped decorative inserts, ideal for the closures of cosmetic containers. In a less resilient material, such undercuts would prevent the moulding being removed from the mould without damaging the undercut. Another consequence of polypropylene's resilience is the possibility of designing a linerless closure. A thin sectioned diaphragm or fin is moulded on to the inner surface of the closure in such a position that it bears down on the upper surface of the bottle neck. A rigid material would not 'give' enough to take up inequalities in the glass surface and so would not form a good seal. On the other hand, a more flexible material such as low density polyethylene would 'give', but would not press back strongly enough to form a seal. Linerless closures not only reduce direct costs by eliminating the wad, but they can also reduce labour and inventory costs. Polypropylene has a low impact strength at low temperatures, but copolymers are available in which low temperature impact strength is improved. The improvement is sufficient to enable polypropylene copolymers to be used for the injection moulding of beer and soft drink crates.

The properties of polypropylene are summarized in Table 7.3.

Table 7.3   Properties of polypropylene.

Density	0.90–0.91 g/cm^3
Tensile strength	30 MPa
Elongation at break	700%
Elastic modulus	1500 MPa
Vicat heat resistance	30–90°C
Water absorption (7 days)	0.1%
Permanent thermal stability	100°C

Copolymers of ethylene with other monomers. Polymerization at high pressures and temperatures is used for the production of copolymers of ethylene with polar monomers, e.g. vinyl acetate and ethyl acrylate. With decreasing ethylene content, the crystallinity of the material decreases and its flexibility at room temperature increases.

Two types of ethylene copolymer are used in packaging, ethylene-vinyl acetate copolymers and copolymers of ethylene with vinyl carboxylic acids.

Ethylene-vinyl acetate copolymers (EVA) are used for food packaging, especially as shrinkable films. Blends of ethylene-vinyl acetate copolymers with low density polyethylene or with polypropylene provide intermediate properties. EVA is a polymer with the flexibility of PVC, but this flexibility is inherent and no plasticizers are necessary. It has a greater resilience than PVC and a greater flexibility than low density polyethylene. This makes it particularly suitable for snap-on caps. Permeability to water vapour and to gases is higher than for low density polyethylene and the solvent resistance is lower. Stress cracking resistance, however, is good. EVA has a high impact strength down to quite low temperatures. EVA film has a greater tendency to blocking than low density polyethylene, so that a higher percentage of anti-blocking additives is necessary. It can be heat-sealed or high-frequency welded, but for the latter, greater power is needed than with PVC.

Copolymers of ethylene with vinyl carboxylic acids, e.g. methacrylic acid, were introduced by DuPont under the generic name *Surlyn*. The term 'ionomers' was coined to describe this family of polymers in which there are ionic forces between the polymer chains, as well as the usual covalent bonds between the atoms in each chain. Although these interchain forces are strong, they are not sufficient to hold the molecules together when the polymer is heated and ionomers are still thermoplastics and not thermosets.

Surlyn A (Table 7.4) is similar in many of its properties to polyethylene, but because of the ionic interchain forces it has a high melt strength and therefore excellent drawing characteristics. Ionomers can be used in extrusion coating and very thin coatings can be obtained. Skin packaging is another field where the high melt strength of Surlyn A is useful. Chemically it is resistant to weak and strong alkalis, but suffers attack by acids. Hydrocarbons cause swelling but it resists attack by ketones and alcohols.

The presence of ionic forces between chains also modifies the crystalline structure of the material and Surlyn A is more transparent than low density polyethylene. The polar nature of the material also means that printing is easier.

Table 7.4   Properties of Surlyn A[a].

Property	Surlyn A
Density (g/cm^3)	0.94
Melt index (ASTM D 1238)	0.5–1.2
Water absorption (%)	1.4
Tensile strength (MPa)	35
Elongation at break (%)	250 MD
	540 CD

[a]Properties are dependent on molecular weight, degree of crystallinity and the nature of the metal ion.

TPX. TPX is also a polyolefin and belongs to the same family as polyethylene and polypropylene. Like them, it is resistant to acids and alkalis as well as to many solvents. It is softened by hydrocarbons and is subject to stress cracking in the same way as polyethylene. One difference between TPX and polyethylene and polypropylene is its clarity, which is almost as good as perspex. The softening points of the polyolefins range from low density polyethylene, which is below the boiling point of water, through high density polyethylene and polypropylene, with TPX as the highest. Its specific gravity is the lowest of the four materials, being only 0.83.

The impact strength of TPX is better than that of polystyrene but below that of polypropylene. Permeability of TPX to gases and water vapour is higher than that of either polyethylene or polypropylene. It can be injection moulded, blow moulded and extruded into sheet. Thermoforming of sheet is difficult, however, because TPX has a narrow melting point range.

Vinyl polymers. This important group of polymers are obtained by the polymerization of compounds with the general structure $CH_2\!=\!CH\!-\!R$. This results in polymer chains with a basic structure of

$$\ldots-CH_2-\underset{\underset{R}{|}}{CH}-CH_2-\underset{\underset{R}{|}}{CH}-CH_2-\underset{\underset{R}{|}}{CH}-CH_2-\underset{\underset{R}{|}}{CH}-\ldots$$

where R can be $OCOCH_3$ to give polyvinyl acetate, Cl to give polyvinyl chloride, OH to give polyvinyl alcohol, etc.

The most important member of the group is polyvinyl chloride (PVC). It may be processed by calendering, extruding, blowing, thermoforming and injection moulding, etc.; it can be plasticized to various degrees and has good resistance to chemicals. It can be produced by suspension, emulsion and bulk polymerization techniques, and as it is insoluble in its monomer, every method gives a powdery product. The majority is produced by suspension polymerization and therefore the product contains small quantities of auxiliary materials. Bulk polymerization, which gives a purer powder, has become established more recently. The molecular weight varies according to the production conditions and generally lies between 10 000 and 100 000. As the molecular weight increases so both the mechanical properties and thermal stability improve but it becomes more difficult to process. Polymers at the upper end of the molecular weight spectrum can be highly plasticized.

The unplasticized material is used to make rigid containers and other hard articles while the plasticized materials produce semi-rigid to highly flexible materials according to the amount and nature of the plasticizer used. Its properties provide for a wide range of packaging requirements. Unplasticized PVC is a good barrier to nitrogen, oxygen, carbon dioxide and gases in general and its WVTR is better than that of HDPE. But these barrier properties are greatly reduced by the incorporation of plasticizers which are necessary for most packaging applications.

PVC can be blow moulded into bottles but since its softening range is not far below the point at which thermal degradation occurs, the use of antioxidants and careful control of processing is essential. The bottles produced are much lighter than glass and not easily broken. It can be made into cups and caps by injection moulding. The plasticized film, used as covers, envelopes and wrappers, is very efficiently converted by high frequency welding techniques. It can also be produced as both shrink and stretch films and these are used for collating and bundling as well as protective bottle caps. The properties of PVC are summarized in Table 7.5.

Copolymers of vinyl chloride. Vinyl chloride can be copolymerized with a number of other monomers, the most important of which are probably vinylidene chloride and vinyl acetate.

Table 7.5 Properties of polyvinylchloride (PVC).

Property	Suspension polymer	Emulsion polymer		Bulk polymer
		Unmodified	Precipitated	
Density	1.40	1.39±0.01	1.40	1.40
Ash content (%)	0.01–0.1	0.5–2.0	0.05–0.5	0.03
Chlorine content (%)	56.2±0.3	55±1.0	56±0.5	56.2±0.3
Water extractables (%)	0.1–1.0	2.0–5.0	0.2–2.0	0.1–1.0

PVC/PVdC copolymers. Vinylidene chloride homopolymer is thermally unstable and therefore is very difficult to process. Copolymerizing it with vinyl chloride using thermal stabilizers, some plasticizer and other additives is used to produce films with the desired properties. The outstanding property of polyvinylidene chloride as a barrier to gases and to water vapour transmission is thus utilized, either as a film or more often as a coating on substrates providing other required properties, not the least of which is cost. Unlike ethylene vinyl alcohol (EVA) copolymers, the barrier properties of PVdC copolymers are almost unaffected by changes in moisture content (relative humidity) although the barrier characteristics are decreased at higher temperatures.

Films have good clarity, good resistance to oils and fats and to many solvents and chemicals. They are attacked by strong alkalis, esters and ketones. Coatings are applied from solution or dispersions in thin layers on materials such as regenerated cellulose film and oriented polypropylene film as well as on stretch blow-moulded polyethylene terephthalate (PET) containers to improve the barrier properties, thus extending the shelf-life, particularly of such products as oxygen sensitive beverages and sauces (e.g. wine, beer and mayonnaise). The coating provides the barrier properties and sometimes heat sealability, while the base material (usually a polyolefin, polystyrene or PET) provides strength, stiffness and other structural requirements.

PVC/PVAc. Copolymers of PVC and polyvinyl acetate are among the oldest copolymers of PVC which becomes much easier to process when its properties are modified by the acetate. The properties depend on the ratio of the components and on their molecular weight. In general, they have a lower softening point, better solubility in polar solvents and are more transparent than PVC. Copolymers of between 80 and 95% vinyl chloride are the most important and they are used as packaging films and in varnishes and coatings.

Polyvinyl alcohol. Polyvinyl alcohol is a white powder, insoluble in most organic solvents but soluble in water, particularly on warming. The monomer, vinyl alcohol, does not exist and the polymer is made by hydrolyzing polyvinyl acetate; it, therefore, usually contains a certain amount of the unhydrolyzed acetate which is sufficient to affect the properties such as solubility, which can be deliberately modified in this way. It is unusual in being water soluble and its major application is related to this property. Film sachets containing aggressive agrochemicals and dyestuffs, for example, which would be unpleasant to handle, are used to give controlled dosages in water. The unopened sachet is added to the required quantity of (usually warmed) water, the sachet dissolves and releases the contents avoiding any contact with the operator.

Polystyrene and copolymers. Styrene monomer can be made in several ways, by:

1. Alkylation of benzene and dehydrogenation
2. Pyrolysis of natural gases or crude oil
3. Condensation of acetylene with benzene
4. Oxidation of ethyl benzene

The polymerization of styrene was first accomplished in the 1830s but it was not commercially available until some 100 years later. Polymerization can also be accomplished in several ways: in solution, in suspension and as an emulsion as well as by bulk polymerization. In this last process, the material passes through a tower in which there are several zones at different temperatures from 100 to 220°C. This process gives a granular product. Solution polymerization gives a more soluble, lower molecular weight product than the rest, the suspension polymer is of the highest quality and the least energy is needed for production by emulsion polymerization.

Polystyrene is a colourless, transparent thermoplastic. It is hard with a fairly high tensile strength. It softens at about 90–95°C and is intrinsically brittle. It is resistant to strong acids and alkalis and is insoluble in aliphatic hydrocarbons and the lower alcohols, but is soluble in esters, aromatic hydrocarbons, higher alcohols, ketones and chlorinated hydrocarbons. It is a poor barrier to moisture vapour.

The brittleness of polystyrene has already been mentioned. However, toughened or high impact grades are available. The higher impact strength is achieved by blending synthetic rubbers, usually styrene butadiene or polybutadiene, with the polystyrene, either chemically or mechanically. These high impact strength grades no longer possess the clarity of basic polystyrene, but the chemical properties are almost unchanged. Toughened polystyrene is used considerably in the packaging of foods but the material must be processed with care to avoid taint problems. Thin-walled containers are made by both injection moulding and thermoforming. Newer applications are the use of multi-layer extrusions and thermoforming to make containers for aseptic food packaging.

Styrene acrylonitrile (SAN). The impact resistance of SAN lies between that of unmodified and toughened polystyrene. It has better resistance than polystyrene to acids and alkalis, aliphatic hydrocarbons and essential oils and is resistant to crazing and cracking. Colour and transparency are good but barrier properties are only marginally better than polystyrene. It can be moulded and thermoformed but applications in packaging are limited, e.g. cosmetic containers.

Acrylonitrile-butadiene-styrene (ABS). Similar in many respects to toughened polystyrene, there is some overlapping in impact strength between the very high impact grades of toughened polystyrene and the lower end of the ABS spectrum. ABS is more expensive than polystyrene. It is used for trays, pallets and tote boxes, especially for those of large area where rigidity and a minimum of warping are required.

Acrylic multi-polymer (XT polymer). This material has been suggested in the United States as a bottle blowing material suitable for food and pharmaceuticals. Impact strength is moderate and dependent on bottle shape and manufacturing conditions. Oil and grease resistance is high, as is resistance to acids, alkalis, detergents and aliphatic hydrocarbons. Resistance to aromatic and chlorinated hydrocarbons is poor, however, and products containing high concentrations of alcohol should be avoided. Gas and odour permeabilities are low but water vapour permeability is higher than that of polyethylene or PVC. Bottles made from XT polymer have good contact clarity but are somewhat hazy when empty.

Lopac. Lopac is the tradename for a material made by Monsanto Chemical Co. Ltd. and is another copolymer with an acrylic base. The main monomer is methacrylonitrile with small percentages of styrene and methylstyrene. It is still in the development stage as a possible material for packaging carbonated soft drinks. It is a hard, rather brittle polymer but has excellent barrier properties, good resistance to creep and is very clear.

Barex. Barex is the trade name for a material manufactured by Vistron Division of Standard Oil of Ohio. It consists mainly of acrylonitrile copolymerized with methylacrylate, together with a small percentage of a butadiene/acrylonitrile rubber. This is also a clear polymer with good creep and barrier properties and good impact strength. Again it was developed as a bottle blowing material for carbonated drinks.

Polyamides. Polyamides or nylons are the most important polymers where the nitrogen atom forms part of the chain. The basic unit is

$$-\!\!\!\underset{\substack{\| \\ O}}{C}\!-\!\underset{\substack{| \\ }}{N}\!-$$

Polyamides are linear condensation products characterized by repeating groups such as $-(-(CH_2)_5CONH-)_n-$. The first nylons were made by condensing a di-acid with a di-amine. They were characterized by the number of carbon atoms in the parent compounds. Thus, nylon 66 is made by condensation of hexamethylene diamine and adipic acid both of which have six carbon atoms in the molecule. The reaction between hexamethylene diamine and sebacic acid with ten carbons gives nylon 610. Later methods were developed for the manufacture of nylons by the condensation of certain amino acids which contain both the groups ($-COOH$ and NH_2), which condense to eliminate water. These nylons are characterized by the use of a single number derived from the number of carbon atoms in the parent molecule. Thus, poly(caprolactam) based on a six carbon molecule is nylon 6 and poly(aminodecanoic acid) with eleven carbons is known as nylon 11. The properties of the polyamides are summarized in Table 7.6.

Table 7.6 Properties of polyamide (PA).

Property	PA 66	PA 610	PA 6
Density	1.13	1.07	1.12
Melting temperature	250–255	210–215	215
Elastic modulus (tension) (MPa)	1700	1250	1300
Water absorption (%)	10	4	11
Max. temperature for			
Permanent use (°C)	80–100	80–100	80–100
Short-term use (°C)	150–170	140–160	140–160

Polyurethanes. Another important nitrogen-containing group of polymers used in packaging are the polyurethanes with a

$$-O-C-N-$$
$$\overset{\|}{\underset{O}{}}\ \overset{|}{}$$

grouping. These are made by the reaction between various di-isocyanates with glycols and higher alcohols. The best known are made from 1,4-butane-diol and 1,6-hexane-di-isocyanate. Both linear and cross-linked polyurethanes can be produced. They are resistant to water, dilute acids and alkalis, adhere well to metals, plastics and other materials and are resistant to abrasion. They are used in adhesives and in a foamed (expanded) form as cushioning materials.

Polyesters. This is a large group of macromolecules which includes some quite important products for the packaging field. In their early days, the polyesters were used almost entirely as the basis of coatings. They still have a significant market in this area and modified alkyd and maleic acid resins are typical.

Linear polyesters are more important in packaging and are made by direct esterification of di-carboxylic acids with di-hydric alcohols. The acids may be substituted by their anhydrides or acyl chlorides.

The polyethylene terephthalates (PET) are undoubtedly the most important of these materials. They can be used in film form for boil-in-the-bag and other applications, but must be orientated to develop the full tensile strength. They are not easily heat-sealable and are therefore often laminated to polyethylene film for bag-making purposes. Since PET was first used in the late 1970s to produce a clear lightweight shatter-resistant beverage bottle, it has probably grown in use faster than any other plastic for this use.

There are two kinds of PET extrusion and thermoforming products, an amorphous variety (APET) and the partially crystalline polyethylene terephthalate (CPET). The amorphous material consists of disorganized chains of molecules interwoven and tangled with each other. On heating to the glass transition temperature (about 70°C), the molecular chains start to 'flow' and slip past one another. On cooling, the chains remain in shape. When the crystalline CPET is heated above the glass transition temperature, the crystalline nature of the material prevents the 'flow' and the crystalline spherulites do not break down until the melting point (about 255°C) is reached. APET is a clear transparent sheet while CPET is opaque. The former is used for bottles etc. and the latter is used in trays for microwave use. There is a wide range of materials under this heading and most manufacturers have their own particular varieties.

Polyacetals. The generic name polyacetal covers polymers of formaldehyde and copolymers of formaldehyde with compounds such as ethylene oxide. They have excellent load-bearing properties and have been used extensively in light engineering applications. There have been few packaging uses. Chemically they are resistant to weak acids and alkalis but are attacked by strong ones. They have excellent solvent resistance.

The first commercial polyacetal container was marketed in the United States and was used for a hair lacquer aerosol. It was made by injection moulding two halves and joining them by spin welding—this had the advantage of giving even wall thickness but was rather expen-sive. Polyacetals can now be blow moulded to give perfectly adequate containers.

Polycarbonate. Polycarbonate has a high impact strength, a high softening point and the merits of clarity and good gloss. It is resistant to weak acids and alkalis, but is slowly attacked by strong ones. Polycarbonate is soluble in aromatic and chlorinated hydrocarbons but is insoluble in paraffins. It has good performance at high temperatures and can be used in hot-fill applications.

Table 7.7 Properties of plastic materials used in packaging.

Plastics material	Density (kg/m³)	Water absorption (24 h) (%)	Water vapour transmission rate (38°C, 90% rh) (g/25 μm/m² d)	Oxygen transmission rate (23/25°C, 50% rh) (cm³/25 μm/m² d. atmos)	Printability	Transparency	Resistance to sunlight (outdoors)
Acrylonitrile butadiene styrene	1010–1100	0.2–0.45	–	780–1100	Excellent	Poor	Poor
Acrylics (polymethyl) methacrylate	1100–1200	0.1–0.4	–	3000	Excellent	Excellent	Excellent
Cellulose acetate	1220–1340	1.7–7.0	155–630	1800–2400	Excellent	Excellent	Excellent
Cellulose acetate butyrate	1150–1220	0.9–2.2	470–630	9400–16000	Excellent	Good	Good
Polyamides	1010–1190	0.3–2.8	63–340	40–1400	Good	Fair–Good	Fair–Good
Polycarbonate	1200	0.15	172	4500	Excellent	Excellent	Good
LDPE/LLDPE	900–930	0.01	16–24	7100–7800	Good	Poor–Fair	Fair–Good
HDPE	945–965	0.01	4.7	2100–2900	Good	Poor	Poor–Fair
Polypropylene (homopolymer)	900–910	0.01–0.03	11	2400–3800	Poor	Fair	Poor

Polypropylene (copolymer)	890–910	0.03	–	–	Good	Fair–Good	Poor–Fair
Polyvinyl chloride (unplasticized)	1350–1600	0.04–0.4	14–80	80–300	Excellent	Good	Excellent
Polyvinyl chloride	1160–1400	0.15–0.75	80–500	80–9000	Excellent	Fair–Good	Fair–Good
Polystyrene (unmodified)	1040–1070	0.01–0.03	110–160	3900–5500	Excellent	Excellent	Fair–Good
Polystyrene (toughened)	1030–1070	0.05–0.07	120	2700	Excellent	Poor	Fair–Good
Styrene acrylonitrile	1060–1080	0.15–0.25	–	–	Excellent	Excellent	Fair
Polyethylene terephthalate	1340–1390	0.1–0.2	16–20	47–94	Good	Excellent	Excellent
Polyethylene vinyl alcohol copolymer	1120–1210	Very hygroscopic	24–120	0.2–1.6 (0% rh) 13–23 (100% rh)	Good	Good	Good
Polyacrylonitrile copolymer	1150	0.28	60–80	12	Good	Excellent	
Polyvinylidene chloride copolymers	1640–1740	0.1	0.3–3	0.5–9	Good	Good	Poor
Phenol formaldehyde	1240–2000	0.03–1.2	–	–	Fair	Poor	Fair
Melamine formaldehyde	1470–1520	0.1–0.8	–	–	Good	Poor	Good
Urea formaldehyde	1470–1520	0.4–0.8	–	–	Good	Poor	Good

Polytetrafluoroethylene (PTFE). PTFE is smooth and waxy to touch, has a very low coefficient of friction and excellent non-stick properties. It is a very tough plastic and can be used over a wide temperature range (-100 to $+200°C$). It is extremely inert chemically, being resistant to almost all chemicals.

Polytrifluorochloroethylene (PTFCE). A copolymer of this material is used in film form with the commercial name of Aclar. It has the lowest water vapour permeability of any polymer film and is also a good barrier to gases. It retains its flexibility down to temperatures of around $-195°C$ and has a softening point between 185 and 205°C according to grade and crystallinity.

Polyvinylfluoride (PVF). Polyvinylfluoride has excellent resistance to solvents, acids and alkalis and can even be boiled in strong acids and alkalis without losing its strength. It is unaffected by boiling in carbon tetrachloride, acetone, benzene and MEK for 2 h and is impermeable to oils and greases. It is strong, flexible and is extremely resistant to failure by flexing. The water vapour permeability of PVF film is low, as is its permeability to gases and to most organic vapours. Laminates with PVC can be vacuum-formed and PVF film has been used in packaging, giving close conformity to the shape of the product.

Table 7.7 summarizes the properties of plastic materials used in packaging.

8
Adhesives

Historical

The use of adhesives has been known for several thousand years. Prehistoric tribes placed their dead in tombs containing pieces of broken pottery which had been stuck together with a rosin. Statues over 6000 years old have been recovered from excavated Babylonian temples, on which parts had been glued together with bituminous adhesive. There is a clear early reference to the use of animal glue recorded on stone in the city of Thebes, probably about 1500 B.C. Tutankhamen's tomb provides additional historical evidence in the form of a glued wooden casket, now in a Cairo museum. Papyrus, an early non-woven has its fibres stuck together with a starch paste.

Still further evidence in history comes from Roman times, the Romans undoubtedly being knowledgeable in glues; specimens of veneering are still to be seen. They almost certainly made up adhesives which were very similar in performance to our present-day animal glues.

The discovery of British gum and dextrine is believed to have been accidental, and resulted from an observation that starch, which had been heated during a fire in a Manchester warehouse, yielded a sticky gummy solution when wet with water. In all probability the use of animal glue grew out of the fact that stews, especially those obtained from bones or skins, yielded a sticky solution which gelatinized when cooled. Violins and similar musical instruments made during the Middle Ages, especially in Italy, indicate that animal glue was used at that time. There are indications that early painters also used a glue size in preparing their canvases.

Chaucer (about 1386) writes in *Squire's Tale*:

The horse of brass that may not be remewed
I stant as it were to the ground yglewed.

Similarly in Lanfranc's *Chirurgeon* (about 1400)

as it were to bordis weren ioyned togidere with cole or with glu.

The importance of the manufacture of glue and gelatin was appreciated in Germany as a key industry. A German company, formed in 1895 with three plants, expanded until in 1912 it controlled the output of 17 plants, and also had factories in Austria, Russia, Belgium, Switzerland and France.

Further development since the turn of the century has shown an increase in animal glue and gelatin-type products, starch adhesives, dextrine and borated dextrine adhesives. In the last 40 years, polyvinyl-acetate-emulsion-based adhesives were developed, and a whole series of speciality adhesives, such as solvent-based adhesives and hot-melt adhesives have also been developed.

In packaging today, polyvinyl-acetate-based adhesives are probably the most widely used, although there is still a considerable tonnage of dextrine and starch-based adhesives. In recent years, hot-melt adhesives have become increasingly popular in packaging applications.

Definitions

Polymer and polymerization: Polymerization is the process by which a polymer –A–A–A– is made from the relevant monomer A, e.g. polyvinyl-acetate from vinyl acetate. A homopolymer –A–A–A– is made up entirely from one monomer A. A copolymer –A–B–A–B– is made up from the monomers A and B, e.g. ethylene/vinyl acetate.

Adhesion, adhesive, cohesive and adherend: Cohesive forces are the forces responsible for substances such as polyethylene or rubber having rigid shapes (i.e. the forces between the molecules within a substance) whereas adhesive forces are similar forces acting between the molecules of dissimilar substances (e.g. between paint and metal). An adhesive is the substance used to bond together two adherends (substrates).

Thermoplastic, thermosetting: A thermoplastic material softens on heating and returns to its original state on cooling, the process being indefinitely repeatable. A thermosetting material undergoes a chemical change on heating, resulting in solid material which does not revert to its original state on cooling. The change is irreversible.

Rheology: Rheology covers a multitude of physical properties, including forces such as those opposing deformation and flow. Adhesives with the same viscosity but different rheological properties will behave differently on the same machine.

Thixotropy, dilatency: A thixotropic system is one that will thin out on stirring, but is capable of reverting to its original consistency on standing. A dilatent system has the opposite properties, i.e. it will thicken on agitation.

Viscosity: The effect of internal forces within a liquid adhesive that tend to prevent the liquid from flowing.

Wetting: Wetting is the ability of an adhesive to flow out (wet) a surface by coming into intimate (molecular) contact with it.

Gel: A gel is a network of solid aggregates in which a liquid is firmly held. A gel can normally be disrupted by heat and/or mechanical forces, e.g. compression. Table jelly is a simple example.

Penetration: Penetration is the entry of the adhesive into the substrate(s). This invariably occurs to a greater extent with the substrate onto which the adhesive is first applied.

Shortness: Shortness refers to the lack of stringing, cobwebbing, and formation of threads during the separation of rollers.

Heat-set adhesive: A heat-set adhesive is one which forms a bond on the application of heat, and in which the water present is absorbed internally to form a gel—a patented process.

Heat-seal adhesive: A heat-seal adhesive is one in which a dry film is activated by heating immediately prior to bond formation.

Tack: The tack of an adhesive is the ability to form an initial bond of measurable strength immediately after the adhesive and adherend are brought into intimate contact, and generally while the adhesive is still liquid.

Blocking: Blocking is an undesirable adhesion between adjacent layers of a material, such as occurs under moderate pressure during storage, causing them to stick together.

Solids content: The solids content of an adhesive is the weight of material expressed as a percentage of the total after all solvent has evaporated (by heat).

Setting time: Setting time is the time taken to form a bond under heat, pressure, etc., by means of a chemical or physical change. It gives a handling bond (initial bond).

Open time: Open time is the time between application of the adhesive to one or both of the substrates and the bringing together of their two surfaces.

Drying time: Drying time is the time taken to form the final bond.

Plasticizer: A plasticizer is a material added to an adhesive to render the dry film of adhesive more flexible. An external plasticizer is incorporated in the adhesive as an addition after polymerization is complete, whereas an internal plasticizer is added during the polymerization process and forms an integral part of the polymer used.

Curing: Curing is a chemical reaction (cross-linking) usually brought about by increase in temperature, which results in irreversible physical change (hardening or setting).

Radio-frequency gluing (dielectric sealing): This is a technique of bonding two substances together where the glue line is heated by radio waves. It is particularly suitable for substrates that would be damaged by the application of direct heat, as in radio-frequency (RF) welding, the substrate does not become heated.

Pressure-sensitive adhesive: A pressure-sensitive adhesive adheres to a surface at room temperature by briefly applying pressure alone. Such adhesives are permanently tacky.

Peel force: Peel force is the force used to measure the adhesive strength of pressure-sensitive adhesive. Usually (180° peel) the substrate is pulled (peeled) back on itself and the force required is recorded.

Consistency: Consistency is the property of an adhesive (paste) that causes it to resist deformation.

Paste: A paste is an adhesive with the consistency of thick cream. Starch and water pastes are typical.

Retrogradation: Retrogradation is a change from low to high consistency on ageing, that occurs in starch pastes, often referred to as setting back.

Sizing: Sizing is treatment with a liquid coating performed to fill the pores present in a surface.

Easi-clean: This is a term relating to certain PVA adhesives having the property that the dried or semi-dried film (on machine parts) can be cleaned easily by a wet rag. Such adhesives tend to dry slowly and do not clog nozzle applicators.

The definitions given above are largely self-explanatory, but three in common everyday use are of such importance as to be worth further comment.

Setting time/open time/tack

The *open time* as defined is the time elapsing between applying an adhesive to one or both surfaces, and the bringing together of those two surfaces. An alternative way of defining setting time would be the time between applying an adhesive to one or both paperboard surfaces and the time after bonding at which appreciable fibre tear occurs on pulling them apart. In a carton sealing operation with aqueous adhesives, the open time will be short with respect to the setting time, i.e the time between the flaps coming together and fibre tear resulting. In the case of a hot melt, these times could be of the same order, i.e. two or three seconds only in view of the very fast setting speed of a hot melt.

Sometimes it is not possible to have a long compression time, but it may be necessary to have a fairly long open time. In such an instance the adhesive would have to have *high tack*. This could be true with water-based adhesives when the tack concerned is 'wet tack', or with a hot melt it would be 'molten tack'. The importance of tack when a long open time, short compression time system is being considered, is clear since, if the adhesive is applied to one surface, and then the two surfaces are brought together but not held firmly, it is essential that the surfaces do not separate. This can be a problem with water-based adhesives (not normally a problem with hot-melt adhesives) and therefore such water-based adhesives have to be formulated to give a high degree of wet tack in order that the flaps do not 'pop open'. It is probable that the wet-tack or molten-tack requirements of an adhesive will depend entirely on the open time/setting time/compression time character-istics of the operation in question, and it is therefore necessary to formulate a given tack requirement into an adhesive depending on the characteristics of a given packaging line. In this respect, tack is a relative rather than an absolute term.

Principles of adhesion

Mechanisms of adhesion

Even allegedly optically flat surfaces are in fact very rough when examined on a molecular scale, i.e. under high magnification. Therefore no two surfaces can ever be in 100% contact, and in fact two surfaces will seldom be in contact over more than 10% of their common area.

There are four basic theories of adhesion as follows:

Mechanical adhesion. This theory is relevant only to absorbent materials. The bond strength between such materials is achieved by the polymer molecules between the surfaces interlocking and penetrating the crevices of the surfaces to be bonded, e.g. a haystack is held together by purely mechanical forces involved in the intermingling of the straw threads. Mechanical adhesion is the prime contributing factor to bond strength in adhering paper-paper and rubber-textile systems. In bonding wood to wood, mechanical adhesion is not the major factor, as was thought until recent years. The figures in Table 8.1 illustrate this.

Table 8.1 Bond strengths in wood-to-wood adhesion.

Wood surface	Bond strength (MPa)
Planed	21.5
Sanded	16.3
Sawn	18.6
Combed	16.6

Chemical adhesion (absorption theory). This is the main theory of adhesion and involves secondary Van der Waals' surface forces, these forces being responsible for the attraction between molecules, i.e. cohesive forces in plastic materials, liquids, etc. Such forces act over only very short distances, and therefore for two substances to stick together by these forces, they must be brought into very intimate contact—the importance for good wetting out of the adherend by the adhesive becomes apparent.

The best adhesives are therefore mobile liquids that readily wet out the substrate. By the same reasoning a flexible natural rubber would be a better adhesive than the less flexible SBR rubbers, or the still less flexible styrene polymers.

Electrostatic adhesion. The theory comes into play as a bond fractures. It has been shown that 90° peel tests using PVC bonded to glass in a dry inert atmosphere are accompanied by flashes of light due to electrostatic sparks. This effect is due to an electric double layer forming as the surfaces separate. The theory has been shown to be mathematically sound, although it has limitations since non-polar materials show good adhesion and not the poor adhesion predicted if the theory were the main contributing factor.

Diffusion theory (autohesion). This is applicable only to high molecular weight materials and is due to the inter-diffusion of adhesive and adherend. This process results in no clearly defined interface after diffusion has occurred. Mutual solubilities of adherend and adhesive are an important factor, and experimental evidence from peel tests using rubber adhesives gives some support to the theory.

Having discussed the different methods by which adhesives form the final bond, let us now consider the development of bonds (Figure 8.1).

Consider a 45% solids dextrine solution bonding an absorbent board to itself. It bonds by mechanical adhesion, and the rate-determining step is the loss of water (and adhesive solution) into the board surface. When the solids content of the film between the surfaces reaches 80–90%, the film becomes coherent, develops tack strength, and hence produces fibre separation.

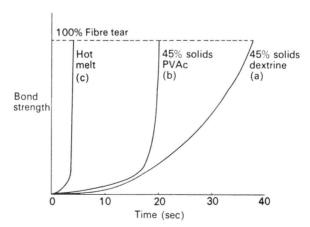

Figure 8.1   The development of bonds between absorbent boards using (a) a 45% solids dextrine solution, (b) a 45% solids PVAc emulsion and (c) a hot melt.

For an emulsion, which is a stabilized suspension of discrete particles in water, the stabilization forces are sufficiently weak so that when the particle content reaches about 65–70%, these forces are overcome, and the emulsion particles coalesce to give a continuous film. Thus an emulsion produces initial fibre tear, while 20–25% water still remains in the film. Thus a PVAc emulsion is about twice as fast in inter-bonding as a dextrine solution at the same solids/viscosity.

A hot melt forms a bond by solidification on cooling. In bonding, a little hot melt is placed on a large surface at room temperature, and hence it cools rapidly and gives fibre tear in a few seconds.

Thus hot melts form bonds fastest (by cooling), emulsions next fastest (by coalescence of the suspension), and solutions slowest (by penetration and loss of carrier).

Requirements for good bond strength

Many factors affect bond strength and can be related to both the adhesive and the substrates. The basic requirements for good bonding are as follows:

1. Good wetting-out (spreading) on the adherends by the adhesive. For this clean surfaces are essential.
2. A thin film of adhesive correctly positioned on the surfaces.
3. Increased pressure before curing (compression stage) to increase penetration of the adhesive into porous stock and/or help air trapped at the interface to dissolve into the adhesive to attain better wetting-out.
4. The thermal coefficients of expansion of the adhesive and adherends should be similar to prevent differential shrinkage on heating or cooling.

In addition to these points several factors concerning the adherends affect the final bond strength:

1. Smoothness/roughness of the surface
2. Porosity of the substrate
3. Chemical nature of the substrate
4. Temperature of the substrate at time of bonding

The smoothness of the surface will depend largely on the type of surface in question, and it is difficult to generalize. The more porous the substrate surface, the more adhesive will be required to make the bond since, for a given quantity of adhesive applied, the more highly porous surfaces will absorb more, which will therefore not be available for bonding. Ideally, the chemical nature of the surfaces should be as near as possible to that of the adhesive or, conversely, the adhesive recommended for a given bonding operation should be similar to the chemical nature of the surfaces. As a generalization, like bonds like and conversely.

A point often overlooked in commercial adhesive bonding operations is the temperature of the substrate. For ideal bonding conditions the substrate should be at the same temperature as the adhesive, i.e. in a commercial bonding operation, the substrates and the adhesive should be held for at least 24 h adjacent to the area in which the bonding operation will be carried out. This is vital if the storage conditions for any of these materials is different to that of the environment where the bonding will be carried out.

Surface energies. For an adhesive (suitably chosen chemically) to bond two surfaces, it must have the correct surface tension in order to wet out the surface.

In order that a suitable adhesive with the appropriate surface tension may wet out a surface, it is generally found that the surface energy must be 45 dyn/cm or greater for good bond formation. Between 35 and 45 dyn/cm a good bond may or may not be formed, depending on the prevailing conditions. If the surface energy is 35 dyn/cm or lower, then a satisfactory bond will not be formed.

The standard test for evaluating surface energies is given in ASTM specification D 2578-67 and uses test solutions of cellosolve (the methyl ether of ethylene glycol) and dimethylformamide in varying proportions.

Surface tension/contact angle. Mathematical relationships have been established relating the contact angle of an adhesive when applied to a surface and the surface tension of the adhesive. The mathematical equations derived will not be detailed here but, when taken in conjunction with viscosity, it is evident that for maximum bond strength and maximum rate of bond strength development an adhesive will have

1. A small contact angle to permit wetting of the surface
2. A suitably low surface tension with respect to the surfaces to be bonded to enhance the rate of wetting
3. A suitable low viscosity to ensure a maximum rate of dispersion across and into the surfaces to be bonded once the initial wetting has occurred.

The following notes are intended primarily for use by people whose activities impinge on the application or purchasing of adhesives as part of a wider job function rather than those whose work demands a more fundamental knowledge of adhesive technology.

This being the main intention, we concentrate on those adhesives that are used regularly and in significant quantities in packaging, as against attempting to provide an exhaustive survey of the entire product range now available. It should also be remembered that many adhesives are formulated to meet a specific application which, in practial terms, means that a particular property or quality has been significantly modified. The notes take account of the broad variations in properties within each adhesive type, that are possible through such formulations. Readers requiring more detailed information are recommended to contact their adhesive suppliers.

It will be seen that most packaging operations are covered, including some of the more sophisticated, such as pressure-sensitive applications (using modern 'tailor-made' acrylic lacquers) as well as the more conventional starch and dextrine types. The choice of adhesive type is never based simply on technical criteria, but always in combination with economic factors, e.g. some carton-sealing operations with non-demanding end-use factors can be quite satisfactorily performed with either dextrine or PVAc adhesives, and the choice will be made for either low-cost criteria or perhaps a preference for handling a given type of adhesive. The choice therefore depends on several factors.

Classification of adhesives

Adhesives have in the past been classified in several ways, dividing them according to their origin into natural and synthetic, according to their composition (water- or solvent-based, dispersions or emulsions, hot melts and two-component systems) and according to their physical nature (thermosets, thermoplastics, contact adhesives, pressure sensitive adhesives, hot melts, etc.).

None of these classifications is completely logical or exclusive and in the descriptions which follow we have decided to classify the major packaging adhesives under three headings:

1. Water-based systems—natural and synthetic
2. Solvent-based systems
3. Hot melts

Notes on adhesives

Water-based adhesive systems

These are the oldest and by far the largest class of adhesives used in packaging. Generally, they have advantages of easy and safe handling characteristics, lower cost than other types and good strength properties. They may be further subdivided into two categories, natural and synthetic.

Natural water-based adhesives. As one might expect the earliest adhesives for all purposes not just for packaging were all made from naturally obtained materials. In fact, until the early decades of this century there were very few adhesives that were not produced from naturally occurring materials and they still serve a large part of the market. Only since the end of World War II have some of them been replaced by synthetics with better performance.

Starch-based adhesives. Pastes are based on starch or starch derived white dextrine of medium to low solid content. They range in viscosity from a short buttery consistency (some tend to thin when agitated) to semi-fluid gelatinous and non-tacky types. Most are acid, but when up to 1% borax is added (to provide more rapid setting) they become alkaline. They generally have good resistance to humid conditions. The simpler compositions are used for seams and bottoms of paper bags, for labelling cans and glass bottles and in the winding of paper tubes.

Jelly gums possess good tack in a thin wet film and give excellent coverage. They have little tendency to crystallize when dried and hence provide good final adhesion even in humid conditions. They normally behave well in glue applicators under production conditions and are clean running. The dried film is virtually transparent. Alkaline adhesives of this type are used for labelling slightly greasy glass bottles. They have also been used in foil/paper and wet laminating.

Dextrines (without and with borating). Unborated dextrines are tacky, relatively fast drying, high solids adhesives of low or medium viscosity. They are normally brownish in colour and can be formulated to machine well in thin films on high speed applicators. They are used as re-moistenable coatings for envelopes etc. and in carton and case closing. Borated dextrines are similar to the unborated adhesives with a higher tack. They are fluid and fast setting and have a fair resistance to humid conditions. They are used for carton closures, tube winding, laminating and labelling on paperboard.

The largest single use of starch-based adhesives worldwide is probably in the manufacture of corrugated board where it is used to stick the fluting medium to the liners. The standard procedure involves a suspension of ungelatinized corn starch in a vehicle of a fluid carrier starch 'solution'. This is deposited on the tips of the flutes and as each line of adhesive hits the heated corrugating roll the heat and pressure immediately gelatinize the starch to form the bond. Various other additions can be made to the adhesive mix to promote adhesion, improve water resistance and increase the speed of set.

Protein-based adhesives. Adhesives made from animal protein (skin and bones in animal glue and casein from milk) were once very widely used but now are usually confined to relatively narrow specialized areas.

Animal glue is produced from the bones, skin and other collagen containing parts of animals by heat and hydrolysis. It is usually supplied in crystallite or pellet form and forms a thick colloidal suspension when heated in water. It is normally used hot, has a very high initial tack and in one sense might be regarded as almost the first 'hot melt' since it first 'sets' to the initial bond by cooling. Time and the availability of better synthetics means that it is used in significant quantities in two areas only: re-moistenable adhesive on gummed kraft sealing tapes and as the standard adhesive in making rigid set-up paperboard boxes.

Casein, obtained by acidifying skimmed cows' milk is usually only produced in countries like Australia, New Zealand and Argentina where surplus milk production is available. Casein glues can provide good adhesion to wood, glass, metal, papers etc. and some plastics. They are moderately water resistant and are therefore used for bottle labelling for beverages that may be cooled in ice. Casein is used as one ingredient in adhesives to laminate foils to papers being combined with styrene-butadiene latices for this purpose.

Latex adhesives. These may be based on natural rubber or synthetic latices. They can be fast setting and give good adhesion to many substrates e.g. printed, varnished and lacquered, foil and plastics. The dried films are usually water resistant and flexible. Some remain permanently tacky, and others if coated on to both substrates will provide 'self-sealing' properties since the two films can be made to adhere by pressing them together. Self-sealing envelopes are based on natural rubber latex and so-called cold seal coatings for confectionery wrappers are based on blends of natural and synthetic rubber or acrylic resins. As supplied, many latex adhesives smell strongly of ammonia and glue pots containing copper (e.g. brass or bronze) must be avoided. In addition to the uses already mentioned, these adhesives are used in carton making, as seals for polyethylene bags and for water resistant labelling of drums, cans and bottles.

Synthetic water-based adhesives. These are almost certainly the most widely used class of adhesives in packaging generally. Most are resin *emulsions* (more accurately dispersions or suspensions of very tiny solid insoluble particles in water). Most are based on polyvinyl acetate, in systems which can be designed to have a wide range of properties (viscosities, setting speeds, open time, machining characteristics, initial tack, etc.). To achieve this wide variation they also contain protective water soluble colloids like polyvinyl alcohol or 2-hydroxycellulose ether plus plasticizers, fillers, defoaming agents and preservatives, etc. When they are applied to an absorbent substrate water is released rapidly and they consequently set more rapidly than solution adhesives.

The dry films are strong and of variable flexibility depending on composition. Frequently the dry film is thermoplastic. 'Heat setting' formulations can be produced which give rapid setting with minimum steam emission when subjected to RF or other heating. In more recent times, the use of these systems has been widened by copolymerization of the vinyl acetate with ethylene or acrylic esters. This has improved the adhesion to plastics in particular and for example cross-linking acrylic-vinyl acetate emulsions have replaced polyurethane solution systems in the lamination of plastic film for packaging snacks.

They are used for practically every kind of packaging operation where fast setting and specific adhesion to the more difficult substrates are needed. Case and carton sealing, tube winding, pressure sensitive coatings, windowing, heat-seal coatings, bag making and labelling plastic bottles are all successful uses in addition to the lamination uses already mentioned. The largest areas of use are probably case and carton sealing, making the manufacturer's joint of corrugated cases and in spiral winding of tubes for composite containers.

Other synthetic water-based systems. These include the latex adhesives already mentioned and sodium silicate solutions, once widely used in many paper-based packaging applications such as the manufacture of corrugated board and case closures, but now largely confined to convolute winding of tubes for making reel cores or large fibre drums where the adhesive enhances the stiffness of the tube.

Solvent-based adhesives

Environmental considerations and the need to comply with 'clean air regulations' in many countries, problems with the so-called glue-sniffing among young people, plus factors of cost, safety (fire hazards and toxicity), and production difficulties have reduced the use of solvent-based adhesives in many areas including packaging and today this class of adhesive is the smallest; they tend to be used only when water-based or hot-melt systems are not suitable. It is expected that by the year 2000 their usage will be reduced to almost nothing.

Resins and rubbers dissolved in a suitable solvent form the basis. Those used can be almost any kind but the most usual are polyvinyl acetate, polyvinyl chloride, acrylics and synthetic rubbers. Setting times and drying speeds vary with the solvent used but are usually faster than with aqueous adhesives. They are used for pressure sensitive adhesives for tapes and labels. Solvent-based ethylene-vinyl acetate systems are being used on lidding stock for thermoformed plastic food containers for cream and jellies.

Two part systems (i.e. primarily polyurethane adhesive plus a catalyst) are possible particularly where impervious substrates are involved, e.g. lamination of cellulose and plastic films in various combinations. Some one-part polyurethane systems can be used where a lower degree of heat and fat resistance can be tolerated than that supplied by the two part systems.

Polyurethane adhesives are widely used for laminated plastic film constructions for bags, pouches and wrappers for snacks, meat and cheese packages and boil-in-bag food pouches. But movement away from solvent-based systems to solventless polyurethane (100% solid liquids) is occurring in these demanding situations. Cross-linked water-based acrylics are also in use in several areas of the snack market.

Hot melt adhesives

These have been described as 'solid solutions' which need to be heated for use, and which soften as the temperature is raised and become fluid when they reach the correct temperature for application. They contain no water or solvent. They are applied hot in the molten state and set rapidly on cooling.

The basis of any hot melt is a thermoplastic polymer. The most widely used polymer is EVA (a copolymer of ethylene and vinyl acetate which has good high temperature stability and is compatible with a large range of modifiers). The other main components of a hot melt are waxes and tackifying resins. The wax is added to reduce the viscosity and control the setting rate: paraffin, microcrystalline and synthetic waxes are all used according to the requirements. Tackifiers are drawn from lower molecular weight compounds derived from aliphatic or aromatic hydrocarbons, rosin and rosin esters, terpenes, phenol or styrene. Their function is not only that of providing tackiness; they also control wetting and assist with viscosity control. Stabilizers and antioxidants are also a necessary part of the formulation.

Low molecular weight polyethylene is the basis for a second class of hot melt for packaging applications. This is compounded with petroleum hydrocarbon tackifiers and although not so adhesive as the EVA hot melts they are used in several paper bonding uses such as case sealing and bag seaming.

A third type is based on amorphous polypropylene. Since this is formed as a by-product in the manufacture of isotactic polypropylene, it is relatively inexpensive but the adhesive power is much weaker and usage is limited to laminating paper to paper to make water resisting wraps or two-ply reinforced sealing tape.

Other hot melts are based on polyamides, polyesters, block polymers of styrene-butadiene and styrene-isoprene for specialized applications.

Hot melt adhesives are the fastest setting class of adhesives available. They set so fast, in fact, that one of the most frequent problems is that the adhesive sets before it has wet the substrate properly and this leaves a very weak bond. Environmentally they have an advantage in that no solvent is released into the atmosphere. They can be applied by roller extrusion, by jetting or by spraying.

Control of temperature and viscosity in the glue pot is most important as is the adjustment of the application speed to ensure good bonding. Hot melts have many applications: case and carton sealing, carton making, tube winding, laminating, coating and bag seaming, etc.

Adhesive applicators

Both water-based and hot-melt adhesive application systems can be divided into non-circulating or circulating systems. Non-circulating systems are less espensive and more common (Figure 8.2). An offshoot of the non-circulating system is the internally circulating hot-melt system (Figure 8.3). There is circulation between the pump and manifold but from the manifold to the gun it is a non-circulating system.

Circulating systems are used which subject the hot-melt adhesive to less thermal degradation than non-circulating systems, for example in random case sealing operations when some setup time is needed. In typical circulating installations (Figure 8.4), a number of automatic extrusion guns are connected in series with the hot-melt hose. Molten material is siphoned out of the applicator tank and pumped to the first gun in the series. The material then flows from the first gun to the second and so on until it passes through a circulation valve back to the tank. The circulation valve permits adjustment of the flow of material.

Cold adhesive systems

These systems consist mainly of applicator heads to apply adhesive either in bead, spray, or droplet patterns. The applicator heads are controlled by either an automatic pneumatic valve or a manually operated hand valve. The bead, ribbon, or spray patterns are dispensed using multiple-gun configuration systems with resin or dextrin cold adhesives. Cold adhesive droplet guns dispense cold plastisols. In bead and ribbon applicators, the tips either make contact or are in close proximity with the substrate. Spray valves deliver a mist-like pattern without touching the substrate's surface.

Hot-melt systems

Hot-melt application equipment performs three essential functions: melting the adhesive, pumping the fluid to the point of application, and dispensing the adhesive in a desired pattern.

Melting. Tank melters (Figure 8.5) are most commonly used in packaging applications. They consist of a simple open pot with a lid for loading adhesive and they can accept almost any adhesive form. Tank melters are considered the most versatile.

Dispensing. There are several methods of depositing adhesive onto a substrate once the material is in a molten state. Extrusion guns or heads (automatic or manual), slot coaters and wheel and roll dispensers. *Extrusion*

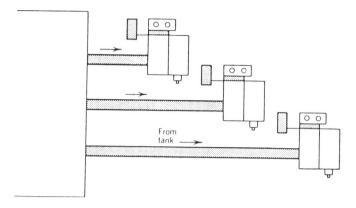

Figure 8.2   Non-circulating system with three applicators. Reproduced with permission from John Wiley, New York.

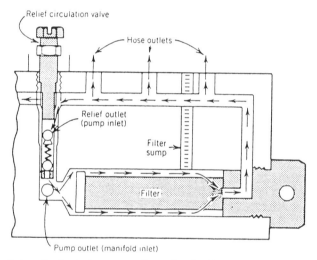

Figure 8.3   Internally circulating system. Reproduced with permission from John Wiley, New York.

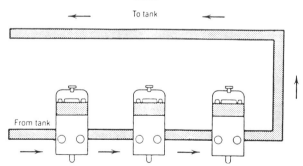

Figure 8.4   Circulating gun installation (series) system. Reproduced with permission from John Wiley, New York.

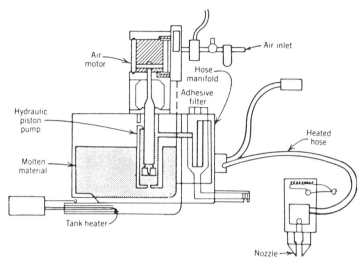

Figure 8.5   Tank-type hot-melt unit. Reproduced with permission from John Wiley, New York.

guns are used on most packaging lines. These are usually fed from the melting-pumping unit through the hose or directly from the unit itself and apply beads of hot melt. Slot coaters deposit a film of hot melt on a moving substrate and are well suited to continuous or intermittent application. They extrude an adhesive film of varying widths, patterns and thickness by adjustment of different thickness blades, by stacking shims in the slot nozzle, or by varying the adhesive supply pressure. Slot coating is well suited to applications such as labelling, tape/label, envelopes, business forms, and web lamination as in non-wovens.

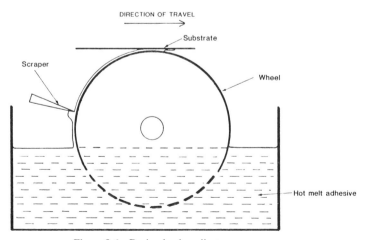

Figure 8.6   Basic wheel applicator.

In *wheel and roll* dispensers, wheels or rolls rotate in a reservoir of the molten adhesive. The wheels or rolls are etched, drilled, or engraved with desired patterns for specific pattern transfer. As the wheel rotates, it picks up the hot melt and transfers it to the moving substrate (Figure 8.6).

Since we have control over the amount of adhesive applied, we have control over the open time of the hot melt. By opening the gap between the scraper and the wheel, more adhesive can be applied and the open time of the adhesive increased. However, to achieve the best results, it is also important to adjust the height of the guide bar above the wheel. This bar is used to guide the substrate onto the wheel to pick up adhesive. On some machines a plate or roller is used in place of the bar. If the guide bar presses the substrate hard onto the wheel it will squeeze the adhesive to the sides of the wheel resulting in a 'tram line' bead of adhesives which reduces control of the open time of the adhesive. The height of the guide bar above the wheel should be adjusted so that the substrate touches the adhesive on the wheel and lifts it off as an even layer. This results in optimum control of open time and stronger bonds.

PART C
Retail Packaging

9
Packaging with flexible barriers

A package is essentially a substitute for a human hand. It frees the hand for other work by 'holding' a handful of possessions—isolating them from other matter and securing against loss. The modern package serves also to identify its contents and often to embellish them. Almost all wares tend to deteriorate under the influence of some extraneous factor; properly selected wrappings will delay this process by excluding that factor.

Primitive wrappings were improvised from available materials such as leaves and skins. Deliberate wrapping in a flexible material only became feasible a century ago, when paper-making machines and new raw materials reduced the cost of paper sufficiently.

The earlier flexible wrappings, apart from metallic foils, were derived from vegetable crops, purified and fluidized by chemical processes. The sheet was deposited from a suspension (paper) or a solution (regenerated cellulose, cellulose esters) and dried. The later wrapping materials have been produced from fossilized vegetable remains (coal and oil) purified and reacted into a solid polymer which is fluidized simply by melting, and extruded into a film by forcing the liquid through a slit. Wrappings are usually produced and supplied in the form of reels, but cut sheets can be obtained where necessary. Melt-extruded films are often produced in tubular form: when cut and sealed simultaneously they then provide bags.

The use of flexible materials has been steadily increasing over the past half century, for packaging foodstuffs and other consumer items, and can justifiably be claimed to have made significant contributions to our way of life—for example to the high standards of food hygiene we currently enjoy, and to the self-service and supermarket methods of retailing which are now so popular. This has arisen because films are able to provide—at a relatively low cost—the essential properties required for the distribution, protection and attractive presentation of consumer products, in both individual and bulk units.

The pre-war industry belonged almost exclusively to regenerated cellulose, the first transparent flexible material, the development of which in the late 1920s enabled the new dimension of display visibility to be added to protective packaging.

Since World War II, the spectrum of flexible barriers has been broadened substantially by the development of a steadily increasing range of materials made from synthetic polymers; those based on polyethylene and polypropylene have been of greatest importance.

Flexible packaging materials provide an alternative solution to the distribution of many types of goods for which crush protection is not important. They function primarily in retaining the goods, separating them from their environment whilst identifying and displaying them to advantage. They are used as containers for liquids, pastes, granules and solids, e.g. strip packs, sachets, bags and sacks; wrappers or liners for packs of other materials, e.g. twist wraps, shrink wraps, stretch wraps, parcels, box liners; and labels and closures, e.g. diaphragm lids on tins, bottles, plastic containers. Flexible packaging materials are used essentially for high-volume machine-packed goods where the forming and closing of the containers is part of the operation of the filling line. The paper sack, the squeeze bottle and the plastic collapsible tube are also examples of the use of flexible materials in packaging. Flexible packaging materials are produced as webs from one or more of a great number of possible starting materials (Tables 9.1 and 9.2). The method of manufacture of the base materials, additives, and the way the components are assembled, all affect the final properties.

There is thus a wide choice of commercial wrapping materials, each of which has its particular value for wrapping certain wares in particular ways. No one flexible wrapping is suitable for all wares or all types of package.

Table 9.1 Flexible packaging materials—possible components.

Plastic films	Non-plastic webs	Coatings and adhesives
Polyethylenes	Papers	Cellulose esters
Polypropylenes	Paper-like webs of mixed	Cellulose ethers
Polyvinylchlorides	cellulose and plastics	Rubber hydrochloride
Polyvinylidenechlorides	Papers made of plastics	Chlorinated rubbers
Polyvinyl acetates	Bonded fibre fabrics	Chlorinated polyolefins
Polyvinyl alcohols	Cloths and scrims	Natural and synthetic rubbers
Polyesters	Spun bonded fabrics	Natural and synthetic waxes
Polycarbonates	Regenerated cellulose films	Natural and synthetic
Polyurethanes	Aluminium and steel foils	bitumens and asphalts
Polystyrenes		Natural and synthetic resins
Polyallomers		Adhesives of all types
Phenoxies		Prime, key, bond or sub coats
Ethylene-vinyl acetate		Latex bound mineral coatings
copolymers		Deposited metal layers
Ethylene-ethyl acrylate		
copolymers		
Fluoro and chloro-fluoro		
hydrocarbon polymers		
Ionomeric copolymers		
Vinyl copolymers		
Block and graft polymers		

Table 9.2 Flexible packaging materials—assembly of components. Adapted from Southam, E.V., *Packaging Technology* 15 (104) Jan. 1969, by courtesy of the author.

Possible forms in which component layers enter the combining operation	Webs, films (continuous or in sheet form)
	Solvent solutions
	Water solutions
	Emulsions
	Dispersions
	Plastisols
	Hot melts
	Resin pellets or powders
	Vapours
	Fibres
	Encapsulated products
Possible methods of assembly of component layers	
(1) Coating	Roller
	Trailing blade
	Air knife
	Rod
	Extrusion
	Co-extrusion
	Curtain
	Dipping
	Brush
	Electrostatic movement
	Magnetic jump
	High vacuum deposition
	Powder fusion
	Spray
	In-situ polymerization
	Polymerization from gas phase
(2) Methods of joining	Hot melt setting
	Adhesive setting
	Heat welding or bonding
	Co-extrusion
	Mechanical interpenetration

Even the size of the pack will help to determine which wrapping is best. In general, no absolute rules can be laid down for selection but, in the pages which follow, there will be an indication of the sort of wrappings which can be used, and the way in which they can be applied, for packing different wares. This indication is in broad principle only, and no responsibility is accepted for the results of trials; but it should enable the user to narrow his choice to a few wrappings. He must make the final selection on questions of price, practical performance and personal preferences.

The styles of wrapping which are possible will first be discussed, then the properties of particular wares which demand certain functions of protectiveness in the wrappings, and finally the wrappings with their individual advantages and limitations. All three of these must be considered together.

A final section considers some incidental processes which can be applied to improve the pack, shrinking, gas-packing, etc.

Styles of wrapping

Hand-wrapping

The simplest styles of wrapping involve applying the sheet to an object and folding or gathering the loose ends as neatly as possible. The loose ends may be twisted (Figure 9.1), pleated (Figure 9.2) or folded (Figure 9.3).

   A simple fold, which uses the minimum amount of film for a slab-shaped object, brings the corners together on one face of the slab (Figure 9.4, bias wrap). More elaborate folds are used for parcels (Figure 9.5a), with or without a 'grocer's fold' (Figure 9.5b) at the overlap, to assist in pulling the wrapping taut.

   With suitable dead-folding or tacky wrapping materials, the folds or twists stay in position and no further closure is necessary. This occurs chiefly with

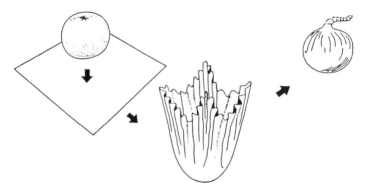

Figure 9.1   Hand-wrapping (twisted ends).

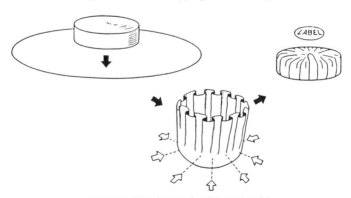

Figure 9.2   Hand-wrapping (pleated ends).

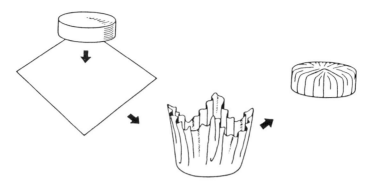

Figure 9.3   Hand-wrapping (folded ends).

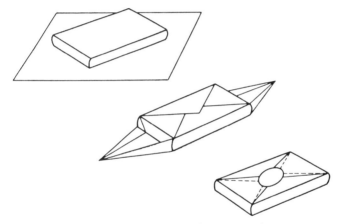

Figure 9.4   Hand-wrapping (the bias wrap).

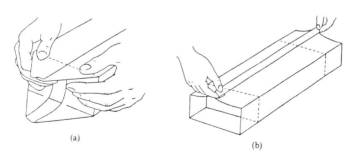

Figure 9.5   (a) Parcel wrap; (b) parcel wrap with 'grocer's fold'.

foils, laminates, thin clinging films and heavily-waxed papers. Other wrappings have to be held in place by an adhesive patch, or by fusing the overlaps together. The adhesive patch can range from a simple piece of pressure-sensitive adhesive tape to a more elaborate printed label with delayed tack.

Heat-sealing may be done by touching the overlaps briefly with a hot iron, or by sliding the pack over a table which has a heated bar let into the surface. The temperature of the bar must be regulated to suit the wrapping used. Distensible films can be stretched tightly over the wrapped object when wrapping by hand. This improves the appearance of the pack, although it may reduce the protection offered by causing thin spots. Temporary distensibility can be produced by heating the wrapping (Figure 9.6).

Tight packs can also be obtained if a film is used which will shrink when the pack is subsequently passed through a warm air tunnel. Shrink wraps are particularly appropriate for fragile or irregularly shaped objects.

Assisted hand-wrapping

When a number of similar packs have to be wrapped, the slow process of hand-wrapping can be assisted by simple devices which eliminate repetitive or slow movements. In particular, the end folds can be completed (after a longitudinal parcel-fold has been made by hand, Figure 9.5) by pushing the pack through a series of folding-ploughs which form the folds mechanically. If heat-sealing plates are placed beyond the ploughs, the packs can be sealed after being pushed through the complete aid as they are made. This is particularly helpful when bulk-overwrapping blocks of small packs. Such aids put the rate of wrapping up from about 4 per minute to about 16, but they depend on a constant supply of similarly shaped wares being ready to wrap.

Bagging

Where the objects to be wrapped together in quantities are not easily stacked, or are of varying shape, it is convenient to use bags. These are

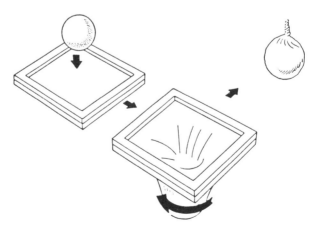

Figure 9.6 Wrapping in distensible film. Temporary distensibility can be produced by heating the wrapping.

virtually prefrabricated wraps, with only one end remaining to be closed. Various forms of bag can be produced, each being particularly convenient for its own range of shapes. The simple pouch (Figure 9.7) suits flat wares (stationery, handkerchiefs, stockings, shirts, etc.) but bulky objects are more easily inserted into bags which can open to a rectangular cross-section. Hand-made 'block-bottom' bags can be made with a truly square cross-section (Figure 9.8) but, if a rectangle is acceptable, machine-made bags are available (Figure 9.9). Several variations are known: with thermoplastic wrapping materials a similar effect can be obtained by making a pouch from gusseted tube (Figure 9.10). Block-bottom bags can also be made by modification of a finished pouch by sealing down the 'ears' which distinguish a distorted flat pouch. Shaped bags may sometimes wrap a specific object econmically so that, for example, a triangular bag (Figure 9.11) can be used for sandwiches.

Filling the bag is usually assisted by mechanical means, such as a funnel attached to weighing scales, or a guide directing the wares into a bag which has already been opened by a jet of air directed at its mouth. For repetitive bagging, the supply of preformed bags can be held in a magazine, with the top one opened by an air jet, or they can be fed from a continuous roll or zig-zag. Wares to be bagged can be preweighed, counted or otherwise premeasured.

Bags cannot usually be closed securely by twisting the ends. For security in transport, the top of the bag can be folded over and secured by an adhesive tape or a label, or a printed cardboard 'header' can be stapled over the mouth of the bag. More simply, the mouth of the bag can be gathered together and held in place by a twist of adhesive tape or of wire. Moulded thermoplastic clips perform the same function and are conveniently re-usable, but they are usually more expensive than simple tapes. Bags made from a thermoplastic material, or from a material having a thermoplastic coating, can be heat-sealed by heated jaws or between two endless belts of heated metal. The

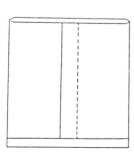

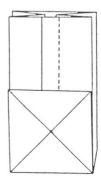

Figure 9.7 Bagging—the simple pouch.   **Figure 9.8** Bagging—hand-made 'block-bottom' bag with square cross-section.

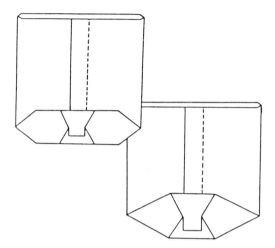

Figure 9.9   Block-bottom machine-made bags of rectangular section.

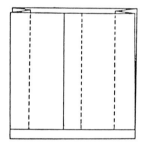

Figure 9.10   Pouch made a from gusseted tube.

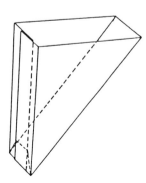

Figure 9.11   Triangular bag.

latter process lends itself to continuous operation on a conveyor belt, which relieves the operator of part of the procedure.

Mechanical wrapping

Mechanical wrapping produces far more packages than hand-wrapping and consumes correspondingly more wrapping material. The previous section is therefore out of proportion to its commercial importance: but it has been given in some detail because the principles of mechanical wrapping are the same.

For wrapping mass-produced articles in a constant flow, automatic wrapping machines replace the manual operator. The speed of packaging is greatly increased, and in the case of small objects which are convenient to feed and wrap, speeds of up to 600 pieces per minute may be achieved by cutting a piece of film, forming it into a tube around the object and twisting the ends of the tube (Figure 9.12).

Rectangular objects lend themselves to mechanized versions of

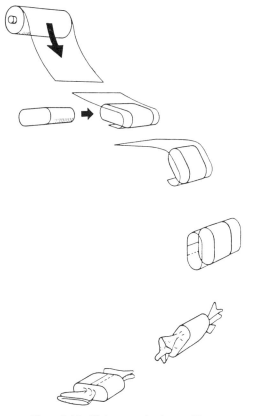

Figure 9.12 Twist wrapping by machine.

the parcel wrap (Figure 9.5). There are several versions which can be selected according to the size and shape of the wares to be wrapped. In each case the principle is broadly similar: a length of wrapping material is drawn or fed from a reel, and cut off. The object to be wrapped is pushed into this length and the ends folded around the object to form a 'tube', with an overlap of 5 to 25 mm. The open ends of the tube are tucked in appropriately, and the overlap and tucks are sealed in place.

Neatly rectangular packs, such as packets of cereals or cigarettes, can be given a fold similar to that of Figure 9.5. One end is tucked in by grippers, and the package is then pushed through pairs of ploughs which fold the other ears in order (Figure 9.13).

When the objects to be wrapped are not constant in size, or are soft, the ends are more accurately folded in by grippers as in Figure 9.14. In order to secure each fold as it is made the ends are then folded down in continuing order, a style more suitable for large loaves of bread.

Modifications of these folds are used for regular but comparatively small objects. If the last 'ear' is folded outwards and underneath, as in Figure 9.15, all heat-seals can be made in one face. The appearance of this style is tidier, and it is also used for dead-folding wraps such as foil laminates, as well as for sugar confectionery (caramels, etc.).

The sealing of the overlap on the bottom face of the parcel in the style of Figures 9.13 and 9.14 can damage the barrier properties of coated wrappings, and with small packets such as cigarettes the overlap is made on one of the narrow edges. The mechanics of this is shown in Figure 9.16, the 'overlap' in this version being made during the later stages.

In all these versions of parcel wrapping it is necessary to feed the wrapping materials from a reel and to cut off lengths for folding over the wares. With very thin flexible wrappings this can be difficult, and careful adjustment of the cut-off knives is essential to prevent misfeeding. Pull-feed of the materials by grippers is also preferable to the simpler method of push-feeding by intermittent rollers. When the end-seals have been formed by

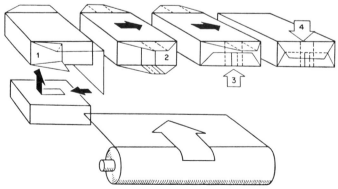

Figure 9.13 Mechanized 'parcel' wrapping.

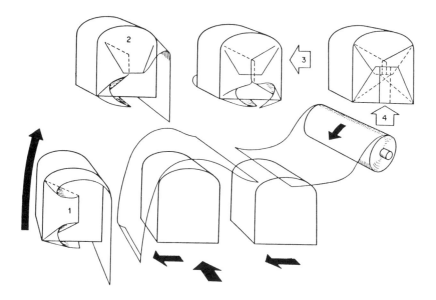

Figure 9.14   Bread wrapping.

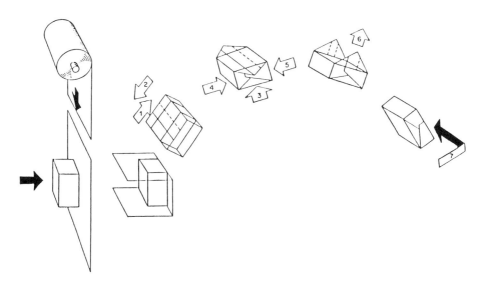

Figure 9.15   'Caramel' wrap.

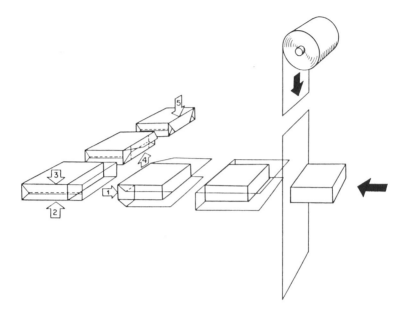

Figure 9.16   Small packet wrap with overlap on the edge.

ploughs, they can be secured by adhesive labels or by heat-sealing if suitable wrappings are used. Pressure-sensitive labels are not easy to dispense at the highest speeds, and it is usual to apply roll-fed printed labels which are coated with a thermoplastic adhesive. The adhesive is formulated to be super-coolable, so that it softens on heating but does not 'set' for a few seconds.

If the assembly of objects to be wrapped is cylindrical, e.g. a pack of round biscuits, it is neater to replace the four folds at the end by a number of pleats, which can be done by rolling the pack between two rows of pleating teeth (Figure 9.17).

The 'parcel' style of pack is not convenient for thin, irregular or numerous objects, and machine wrapping of these is more conveniently done by strip-packing. In one version of this (Figure 9.18), the web of wrapping material is formed into a tube with a heat-sealed horizontal seam. The wares to be packed are fed into the tube, which is sealed across at regular intervals. This method can be used for powdered or granular materials and, if the direction of the transverse seal is alternated at right angles, a tetrahedral pack suitable for liquids is obtained (Figure 9.19). A modification (Figure 9.20) will handle wrappings which have a fusible layer on one side only. It is also suitable for larger packs.

Wrapping materials which are not rigid, or which require special care in heat-sealing, are most easily used in the form of two webs which are sealed together round the edges of the wares. This gives a flat pack (Figure 9.21) when both webs are thin and flexible.

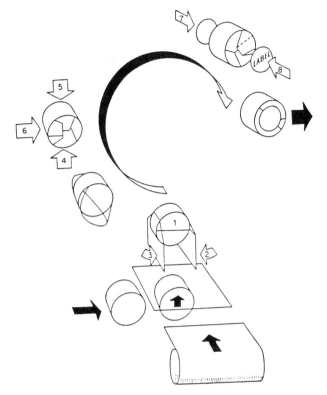

Figure 9.17 Wrapping with pleated ends.

If one of the webs is rigid, a 'skin' pack is obtained (Figure 9.22) and the appearance can be improved by shrinking the thin web so that the wrinkles disappear and the web comes into taut contact with the wares. If both webs are relatively rigid one or both must be preformed to provide receptacles for the wares (Figure 9.23a and b).

Variations of Figure 9.23b will give shaped packs which can be used as containers for liquids, wine, oil, etc., as substitutes for small bottles. A preformed web with thin 'lid' sealed over is a derivative of Figure 9.23a used for pastes such as jam or yoghurt.

The wares to be packed

General

The style of wrapping selected from among those illustrated in the previous section is largely determined by the size, shape and quantity of objects to be packed. The nature of the wares themselves has less

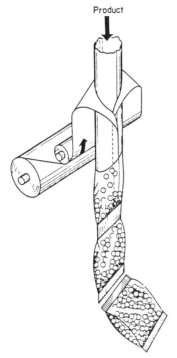

Figure 9.18   Pouch pack.

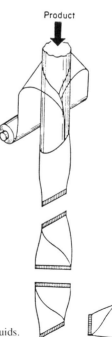

Figure 9.19   Tetrahedral pack for liquids.

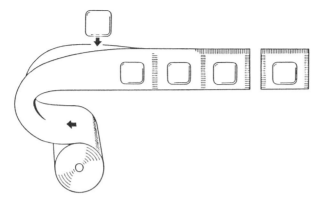

Figure 9.20   Strip pack with fusible layer on one side only (suitable for larger packs).

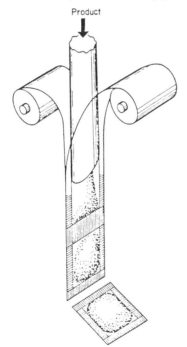

Figure 9.21   Flat pack—both webs thin and flexible.

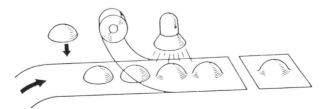

Figure 9.22   Skin pack—one web rigid.

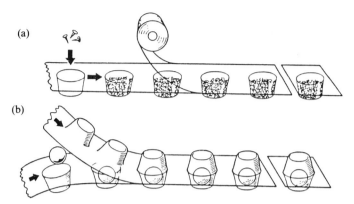

Figure 9.23 (a), (b) When both webs are relatively rigid, one or both must be preformed.

influence on the style of wrapping, although fragile wares may be unsuitable for machines which have coarse feeding mechanisms.

The type of wrapping material to be used is much more critically determined by the nature of the wares to be packed. If the wares are themselves attractive, a glossy transparent wrap may enhance their appearance; if not, an opaque wrap may be preferred. If they are sensitive to light, an opaque or coloured wrap may give the best protection, even at the sacrifice of appearance. There have always to be compromises between display and protection, between preservation and ease of use, and above all between durability and cost.

Some wares are not very susceptible to damage; they need wrapping only for convenience, and almost any decorative wrap can be used. Many wares, however, are wrapped because they might otherwise deteriorate, and a wrapping must be selected to fend off deterioration as long as the wares remain unused. Wares can deteriorate in various ways and, although the mechanisms of deterioration are fairly well understood, the quantitative theory has not yet been developed far enough to select the wraps without final trial under practical conditions.

Dry wares

Many foods sold in a dry form are resistant to microbiological deterioration as long as they remain dry. Biscuits, breakfast foods, boiled sweets, cereals, potato crisps, soup powders and dehydrated foods are all examples of wares which 'keep' so long as they are not allowed to absorb excessive amounts of water from the surrounding air. They would soon do this if they were not wrapped in a barrier material which resists the passage of water vapour. Dry foods which are ready to eat are particularly sensitive to spoilage, because their texture as well as their flavour depends on their dryness.

These dry foods must be wrapped in good water-vapour barriers which are themselves efficiently sealed. Coated paper and cellulose film and, more recently, coated polypropylene film are commonly used for biscuits. The coating is a highly resistant barrier to water vapour, such as a vinylidene chloride copolymer. For wares which are very sensitive to moisture, such as some accelerated freeze dried (AFD) meat or fish products, or where very long shelf-life is needed, laminates of two or more barrier materials (often including aluminium foil) are used.

If sufficient data are available about the moisture content of the wares in equilibrium with atmospheres of different relative humidities, and the initial and critical moisture contents, calculation will show what order of water-vapour resistance will be required for their expected shelf-life. Suitable wrappings can then be tested in the field.

Air-dried foods, such as dried fruit, pulses, pasta, starch, are not so sensitive to small changes in water content, but they can usefully be packed in a wrapping of moderate resistance to water vapour, which will protect them from damage by excessive moisture when they happen to encounter extremes of atmospheric humidity.

Moist wares

These goods lose moisture when exposed to the average atmosphere, and the barrier wrap is needed to prevent a loss of moisture (instead of a gain as was the case with dry wares). Calculation once again can indicate the type of wrapping needed for any required shelf-life. This can be done with cigarettes, tobacco, and certain kinds of flour confectionery.

The problem is more complicated, however, when the wares contain nutrients and water sufficient to support microbial growth. If the wares are moist enough, any mould spores or bacteria present will begin to grow. After two or three days at normal temperatures this would spoil their appearance. Where foods are concerned, microbial growth will also change the flavour, and can often render foods harmful if consumed.

The problem can sometimes be solved for long storage by sterilizing the contents of hermetically sealed packages by heat or by irradiation. Short-term storage can be achieved if the packages are stored in refrigerators at temperatures just above the freezing-point. Chemical preservatives can also prolong the shelf-life, but their use is generally disliked unless they are innocuous.

If none of these methods of preservation is acceptable or practicable, a compromise in packaging must be adopted. A wrapping is selected which will enable the surface of the wares to dry out sufficiently to prevent mould growth, while keeping enough water within the bulk to preserve the texture. Cakes having a low sugar content can be packed in this way. Only a limited life is possible from either cause of deterioration, and the optimal solution to the problem is one which postpones both faults until they develop simultaneously.

More complicated instances occur, for example with bread, which also deteriorates by 'staling'. This is a process which occurs independently of the method of wrapping, and any inexpensive wrap which will give protection for the duration of the period of 'freshness' is good enough for the purpose. The use of permitted preservatives and anti-staling chemicals would lengthen the realizable shelf-life, but the penalty is inevitably a softened crust.

'Breathing' wares

Some wares are sensitive to oxygen, and the action is often increased in conjunction with water vapour.

Iron and steel objects, such as motor spares, tools and surgical instruments, will rust in air if the relative humidity exceeds 40%. The rate of attack is very slow at 40% humidity, but increases rapidly as the relative humidity rises. Such articles could theoretically be protected by packing in a wrapping material which is impervious to oxygen, but so little oxygen is needed to cause noticeable damage that this is impracticable. It is simpler to wrap such hardware in a good water-vapour barrier, and to enclose a small quantity of a desiccant, such as silica gel, inside the package. The desiccant will absorb a considerable amount of water before allowing the humidity inside the pack to rise to the danger point. Volatile corrosion-inhibitors can be enclosed inside the pack instead of desiccant, and a wrapping must then be chosen which will not allow the volatile vapour to escape too quickly.

Oxygen is also harmful to fatty or oily wares such as butter, fried snacks or nuts, and oxidative rancidity develops more quickly under the influence of light. An opaque pack, printed or foil-laminated, will give good protection for these wares. Where the shelf-life is short and the wares (such as potato crisps) more sensitive to other factors, rancidity can usually be ignored. However, an oil-resistant wrapping must be selected.

Oxygen is consumed from the air inside a package by green vegetables, fruits and similar crops which continue to respire and produce carbon dioxide after they are harvested and packed. If the package were to be hermetically sealed, the oxygen would all be consumed, creating conditions favourable for the growth of certain bacteria, and the texture and flavour of the wares would be spoiled. To prevent this, a regular supply of oxygen is necessary, and arrangements must be made for the removal of carbon dioxide. Unfortunately, most vegetables respire at a fairly rapid rate and only porous or perforated wrappings let oxygen and carbon dioxide through quickly enough to preserve the contents. A compromise is again necessary, because excessive ventilation would also encourage the loss of water, with consequent wilting of the vegetables. The usual solution is to bag or pack the wares in an impervious wrapping having three or four holes about 5 mm (¼ in) in diameter, or to leave the pack only partially sealed. It should be stressed, however, that cooled storage during the marketing period is very beneficial for prepacked fruits and vegetables.

Meats

Meat presents a difficult problem in packaging, because it can be spoiled by so many factors. Excess of moisture will encourage microbial growth, while excessive dryness spoils the texture and appearance. Excess of oxygen can change the colour of the meat, but so can inadequate oxygen supply. Even too bright illumination will bleach the colour.

Cured meats are less perishable than fresh meat, but at some sacrifice of the original flavour. They are still sensitive to mould growth unless they are dry or contain added preservative. Semi-cured meats such as bacon are even more sensitive, and even in cool storage they need protection against oxygen (for example, by vacuum packing) and bright light to have a reasonable shelf-life. Fresh meat is most critical and will quickly spoil for one reason or another unless the biochemical processes involved are slowed down by storage at reduced temperatures. Storage at 1 to 3°C will give a shelf-life of a few days if the wrapping is permeable to oxygen, and special grades are made for this purpose. Below this temperature there is a 'forbidden' region where the texture of the meat is destroyed by ice-crystal growth. Prolonged storage calls for quick freezing, with storage at temperature below −15°C and, under these conditions, a wrapping of fairly high resistance to water vapour is needed to prevent freezer-burn by surface desiccation.

Liquids

The main requirements in packs for liquids are leakproofness and resistance to attack or penetration by the liquids packed. Some rigidity, coupled with a flat base so that the pack can stand alone, is also necessary unless a rigid container is supplied for simple packs.

Leakproofness demands a fusible material, or coating, which is thick enough to give an autogenous weld. A layer of low-density polyethylene at least 50 μm thick will provide this, while a second layer of paper, regenerated cellulose or nylon laminated to it, will provide rigidity. The second layer, and the adhesive employed in laminating, covers any pinholes which may be present in the polythene film.

The style of pack can be a tetrahedron (Figure 9.19), a bag (Figure 9.7), or a pillow pack (Figure 9.18) modified by gussets to give a flat bottom. This enables the pack to be stacked for display. The flexible laminate which is used can be printed before the pack is made, so that there is no need to apply a separate label. The packs are generally intended for relatively small quantities of beverages such as milks, wines or fruit juices. Staple commodities such as milk can be packed in simple pillow packs (Figure 9.18), with a rigid container provided for permanent use. These packs eliminate the necessity to return the bottles.

Incidental processes

In the previous sections the styles of packaging and the packaging materials have been outlined. There are various incidental processes which can be applied to improve or enhance the pack.

Preforming

Thermoplastic films which are rigid enough to hold their shape can be preformed for some kinds of package (Figure 9.23). The sheet is heated by radiant heaters until soft, and then drawn down by pressure or vacuum over simple shapes placed on a porous base. Cellulose acetate, polyvinyl chloride, polystyrene, high-density polyethylene and polypropylene can be preformed in this way, and automatic machinery is available for forming, filling and lidding. The 'lid', which is sealed over the top of the package, is usually a thinner sheet often of the same material as the base. Thermo-formed trays are also widely used for the mechanical protection they accord to loose or fragile wares such as chocolates, etc.

Shrink wrapping

As mentioned in the section on hand-wrapping, tight wraps can be made by hand more easily than by machine. Some kinds of wrapping can be applied loosely by machine, and afterwards tightened by shrinkage. The greatest shrinkage is achieved with the thermoplastic films which have been stretched in manufacture, these being heated to a temperature approaching their softening points to cause shrinkage. A brief passage through a tunnel at a thermostatically controlled temperature is the usual way of finishing these packs.

Vacuum packing

Some 'breathing' wares last longer if the packages are evacuated before sealing, so that only small traces of oxygen are left to react with the wares, and this is the case with certain processed meats like bacon. The wrapping must be chosen for high resistance to oxygen and for efficiency of sealing. In fact a laminate is generally used, a ply of aluminium foil and/or a strong film which resists oxygen and a thicker layer of polyethylene for heat-sealing.

The package (Figures 9.18, 9.20 and 9.21) is almost completely sealed. It is then evacuated through a small tube or inside a closed chamber before being finally sealed. Vacuum-packed wares are not perfectly protected, and the shelf-life attainable with perishable food products is strictly limited and usually requires refrigeration at chill temperatures during retail.

Gas packing

The reaction process with oxygen can be slowed down further in some instances (fruit and coffee) if the air in the pack is displaced with an inert gas such as carbon dioxide. The package is first evacuated and then flushed with the gas. If the wares are too fragile to withstand crushing by the evacuated pack (flowers), the atmosphere can simply be displaced by blowing gas in. The same sort of laminate is useful for this purpose as for vacuum packing.

10
Folding boxboard cartons

The folding boxboard carton is familiar as a retail pack in the distribution of food stuffs, confectionery, toiletry and cosmetics, tobacco products, light engineering goods, and toys and games. Prior to 1879, paperboard boxes were made by cutting, folding and gluing with much hand work but, in that year, Robert Gair in America took out a patent for the mechanized process of cutting and creasing board which is essentially the same as that used today. Initially, converted printing machines were used; now, cutting and creasing presses are purpose-designed with outputs to match the present-day demands for precision and productivity.

Definitions

The process of carton making is so versatile that the styles offered are legion. Any definition of folding carton should set it apart from rigid paperboard boxes. Folding cartons are:

1. Made from paperboard of thickness between 300 μm and 1000 μm
2. Delivered in a flat collapsed state for erection at the packaging point.

The thickness limitations set cartons apart from regular slotted fibreboard cases, which are sometimes also called 'cartons' because they are delivered flat. The carton-making process can be used to produce packs in transparent rigid PVC sheet and in E-flute corrugated board. These are again not strictly folding cartons as defined above.

Materials

The boards used for cartons are mainly made on cylinder mould or Inverform machines. Only the thinnest boards are made on single and twin-wire machines.

The two main methods give many possibilities, as all these boards have a ply structure. All plies may be repulped waste as in chipboard. Replacement of the outer ply by better quality pulps gives in succession cream-lined chipboards, and second and first quality white-lined chipboards. Duplex boards without any waste pulp at all are also available. Board made entirely from chemical pulp (solid white boards) are produced for frozen food packaging and other applications where the board must be waxed. The cost of these boards obviously rises with the quality of pulp used.

The external appearance and the printing quality of carton board is greatly enhanced by the use of coatings. Clay and other minerals are coated on to the lined outer surface of the board. The coating is applied either during the board-making operation or subsequently. A limitation of boards which are coated on the board-making machine is that the coating must generally be of a somewhat thinner nature and have a less glossy surface than coatings applied off the machine as a separate operation. The latter include cast coated boards which have a very high gloss and brightness. In many instances, the extra cost of these boards limits their use to the packaging of comparatively expensive goods, such as cosmetics.

Where special properties such as moisture resistance or grease resistance are required, coated and laminated boards are employed. These may be wax-laminated where the wax is the moisture barrier, glassine-lined for grease resistance, or plastic-coated for special properties, including heat-sealing. Paperboard cartons, even if made from plastic-coated boards, unless of special construction, cannot be expected to give much moisture protection to the contents, and are therefore used with inner bags.

Foil-lined boards are also used for various types of carton, not only where protection is of importance but also to give particular display effects.

Chipboard manufactured from repulped waste materials (newsprint, corrugated and solid fibreboard cases, etc.) accounts in its various forms for about one-third of the current carton board consumption in the United Kingdom today. *Plain (unlined) chipboard* is only utilized for the very simple type of carton for holding stocks of materials which do not require any great display. It is almost unprintable in terms of quality, and its strength is somewhat lower than that of any other variety. Some chipboard cartons are produced, however, and are lined with printed paper wrappers.

White-lined chipboard, with its improved appearance, printability and folding qualities, is used in greater quantities than any other grade. It is used for many cartons where the internal appearance is unimportant, and where no contact with foodstuff occurs. Taint is possible when foods containing flour or fats contact chipboard directly. Obviously, if a separate bag is used inside a carton, the contact between its inner surface and the food is prevented and, under these conditions, white-lined chipboard may be usable.

With cartons which are opened repeatedly, the better looking inner face of a *white-lined manilla* or *duplex board* may be required, although white-lined chipboard is often used for these types as well. With certain exceptions, most foods must avoid direct contact with chipboard, so that either pure pulp or duplex boards are used: or alternatively, glassine, sulphite or other types of paper may be laminated to the chipboard side of a white-lined chipboard where foods are concerned.

Pure pulp boards are particularly popular for materials such as flour confectionery and chocolate, where staining of the carton by fats, etc. is possible. In this context also, boards lined with greaseproof paper, vegetable parchment or glassine are often used. Special treatments also exist that can impart particular properties such as grease or water resistance or water repellence to the board.

Wax-laminated white-lined chipboard is widely used for detergent or soap powder cartons if moisture loss or entry is to be avoided. Such waxes are also used as the laminating adhesives for glassine and other greaseproof liners referred to above.

Wax coating on one or both sides of the board can also be carried out at any stage after printing, and waxed cartons are particularly used in the frozen-food industry, although many are employed in an unprinted state, with a printed paper or film overwrap.

As a separate process other materials such as PE, PET or foil may be extruded onto or laminated to the board as required.

Developments in polymeric materials are adding continually to the range of material available combining with paper and paperboard in all forms. Note the increased use of materials such as polyethylene, polypropylene, polyvinylidene chloride and nylon, either in direct combination with board or as ancillary materials. By and large, rigidity, which is one of the prime factors required by all unit retail packages, can most economically be supplied by paper and paperboard, whereas barrier properties, such as moisture resistance, oxygen resistance, and grease resistance, are readily supplied by comparatively thin films of plastics. The combination of the two (rigidity and resistance) in coated paperboard can result in highly protective packages for many goods.

The use of barrier materials in carton board is principally restricted by the inability of the normal types of carton closure to prevent the ingress of moisture directly without passage through the barrier. Because of this, a considerable amount of work has been directed towards the development of one-piece functional cartons, aimed initially at the frozen-food industry where the protective requirement was originally demanded.

Further designs have led to the use of cartons for liquids of various kinds.

The final choice of board is made on consideration of the requirements of the contents, the market image desired and the balance between board strength, expressed as stiffness, and price per square metre. These commercial and cost considerations are combined with those that relate to the conversion qualities of available boards.

Manufacture

Preliminary stage

As a general rule, the manufacture of folding cartons is a 'bespoke' business. Cartons are not manufactured and sold from stock, but only for a specific purpose. All operations from the design to the delivery, including the size of sheet on which the cartons are printed and from which they are cut, are specific solely to the production of one type of carton which has been ordered by the user. Consequently, there is always the necessity to allow sufficient time between the requirement for cartons and the date by which the order must be placed.

The user prepares a design brief, setting out details of the product and any special requirements on protection and method of use, the available machinery for packaging and the part the carton is expected to play in distribution and sale of the contents. This brief guides design agencies, or the design departments of carton makers, in the submission of samples and artwork. Once the sketch and sample have been accepted, and agreement has been reached on the type of carton, the preparatory work can begin. Negotiations will have ensured that an adequate specification for both the construction and the graphic design has been reached. In other words, a specific price has been accepted by the carton user. Such a proposal automatically implies that the method of printing the cartons has been considered, and selected in conformity with requirements of the job, as well as the availability of the equipment.

Four print processes dominate the carton industry: lithography, gravure, letterpress and flexography. Each process requires its own special properties in the board and processing, and the method selected can make considerable differences to the result. Depending on the printing process, work starts on the preparation of the finished artwork. This covers complete working drawings and prints of any pictures that are needed, and the text matter set in type and so on. Full-scale drawings giving the exact dimensions, with provisions for the creasing and cutting, register marks and colour separation details, must all be produced for subsequent handling by the appropriate department of the carton manufacturer. At the same time, the board must be obtained in the correct sheet or reel size and grade for the particular purpose, printing plates or cylinders must be produced, and a cutting and creasing forme prepared. All this preparatory work usually precedes the actual commencement of the job of producing the carton.

Extensive use is now made of computers in the design of carton shapes and graphics. This enables a design to be created, stored, recalled, manipulated and used. Optimization of the layout of multi-ply cartons on sheets is possible and design information can be transmitted electronically between production locations. Design information stored on computer can control the production of the colour separations in the manufacture of printing plates. It is further used in the numeric control of the laser and router equipment for carton cutting and creasing die and makeready production. This router equipment can be adapted to produce cut and creased carton samples for graphic design or even machine trials.

Production stage

The conventional methods of carton making involve first printing the carton board and then cutting and creasing it to allow the subsequent folding to shape, the stripping of any waste material which is not required in the final construction, and finally the finishing operation of putting in the joins where necessary, either by gluing or by stitching.

Printing and cutting and creasing are, in the main, sheet-fed processes but, for certain long-run work, rotary web-fed presses are used. There are in addition web-fed reciprocating-platen cutting and creasing presses which can be combined with printing either before or after cutting.

Printing

All the major printing processes have been used to print cartons. The method preferred is by no means the same in the different countries of the world, but in the United Kingdom probably the majority of medium-sized runs are now printed by offset lithography. As can be seen in Figure 10.1, the lithographic process involves the offsetting of the ink film from the plate onto a rubber blanket, and the transfer of this ink film from the rubber blanket to the surface to be printed.

In the gravure process shown in Figure 10.2, the ink is carried in recesses in the printing surface and transferred to the board when it is pressed in contact with the surface. The process can be either web-fed or sheet-fed, and is used for the longer runs, due to the cost of making the etched recessed printing surface on gravure cylinders. In recent years, however, newer methods of producing gravure cylinders have been developed, and the cost of gravure-printed cartons is now by no means prohibitive in what would in the 1960s have been regarded as short runs.

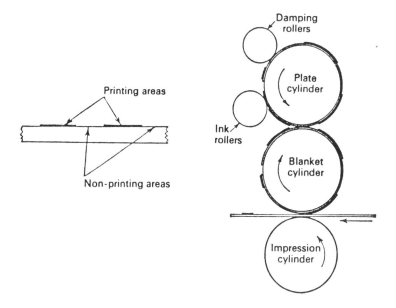

Figure 10.1   Printing by lithography.

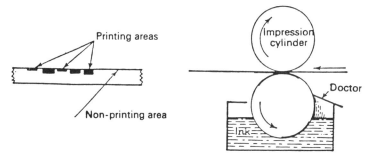

Figure 10.2   Printing from a gravure cylinder.

The letterpress process on the other hand, is generally a direct printing process; the raised surfaces which carry the ink are directly pressed against the surface to be printed (Figure 10.3). During recent years, a process which is referred to variously as dry offset, letterset or indirect letterpress has been used for certain types of carton. In this process, the ink from a relief surface is offset on to a rubber blanket, and the transfer of the ink to the board is achieved by pressing it against the rubber blanket. No water is involved, however, hence the use of the term 'dry'.

Flexographic printing, which is essentially a relief printing process, is largely confined to those cartons where the demands of print quality are less stringent than normal.

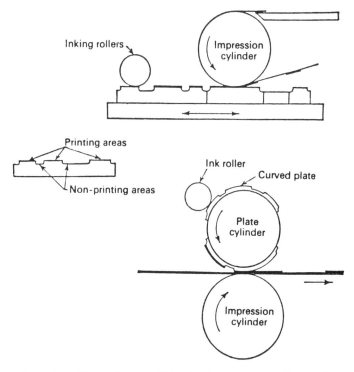

Figure 10.3   Printing from a relief surface by letterpress or flexography.

The carton board must be chosen to make a pack suitable for the particular application, but the printing processes also impose restrictions on the choice of material. For example, in lithography, the high tack of offset inks makes demands on the surface strength (pick resistance) of the board used. The surface reaction on the board (pH) must be controlled to avoid ink-drying difficulties, and the (oil) absorbency of the board surface must be suitable for the inks used, so that the penetration of the ink vehicle into the board takes place at the right speed. Too rapid absorption, or too slow penetration, of the ink vehicle may lead to difficulties, such as powdering of the ink surface or set off on the back of the board. Letterpress printing involves similar but generally somewhat less stringent requirements. In addition, the lithographic process uses water to prevent ink reaching the non-printing areas on the printing plate; this will result in the further requirement that the surface to be printed must be able to withstand the slight wetting which will occur as it passes through the press. This is particularly important where clay-coated boards are used. Although gross deterioration of the board surface is now a very rare occurrence, continual slight leaching of soluble coating material can cause considerable difficulties during the printing process, particularly with multi-colour machines.

The above considerations do not apply to gravure printing, because this uses solvent-based inks which possess negligible tack. The drying of the ink also is largely by evaporation, rather than the partial oxidation of the ink which occurs with most letterpress and litho inks. However, surface irregularities in the board can cause trouble, and result in individual cells in the intaglio surface of the gravure printing cylinder or plate failing to transfer their ink. This causes a deterioration in the print quality (called 'speckle'), and generally means that the property of surface smoothness measured at printing nip pressures is most important.

Whichever printing process is used, the resulting print must have the required resistance to rubbing during the handling that it will subsequently receive in the filling and closing operations on a packaging line. It must also be resistant to the rubbing encountered during transport and distribution. Much can also be done by careful design of the print to keep those areas of the carton surface subjected to rub on machines, etc., as free as possible from ink.

Inks which are non-toxic and give prints of low residual odour must be used on food cartons. Developments in ink formulation and the use of infrared, ultraviolet and electron beam energy sources for drying have reduced the risk of odour from print, and given potential for enhanced gloss and rub resistance. Ink drying times are reduced, compared with air drying, and production rates can be raised.

The cutting and creasing operation

The purpose of the cutting and creasing operation is to enable the carton blank to be detached from the sheet or web with creases correctly located, so that the carton may be readily glued or stitched as required, and subsequently erected to the correct shape.

In the early days of carton making, the press was almost exactly the same as the printing press which produced the printing design. However, instead of a printing plate (which in those days was usually a letterpress printing plate), a cutting and creasing die, consisting of creasing rules and knives to impress the design of cuts and creases upon the board, was attached to the bed of the machine. Nowadays, reciprocating platen presses are used by most carton makers in the United Kingdom, with rotary presses reserved for long runs of carton designs for cereals, cigarettes and detergents.

The formes for platen and flatbed cylinder presses are flat and usually made from special plywood. In a one-piece forme, the sharp-edged cutting rules and round-edged creasing rules are inserted in slots cut in the plywood using fretsaws or more recently laser beams. The alternative, a block die, is made from pieces of plywood, metal or plastics cut to sizes of the carton panels, with the rules clamped between the blocks.

A forme has rows of cutters arranged to correspond to the printed images on the sheet. Accuracy in the panel sizes of the individual cartons and in the

spatial arrangement of cutters in the forme to ±250 μm is considered
necessary for machine-packed cartons. This can most readily be achieved by
using dies made by NC laser cutters.

Figure 10.4 shows a diagram of a forme mounted on one platen of the
press, and a cutting rule contacting the steel of the other platen to sever the
board. The crease is made by the round-edged rule pressing the board into a
groove in the makeready or counter. These creases have two purposes: they
fix the lines at which the board will fold easily, and prevent the board
cracking at the bends.

It is necessary for the board to be partly de-laminated by the crease in
order that a clean roll is formed on the inside when the back of the board is
displaced for folding (see Figure 10.5). The outside of the fold is, naturally,
of very much greater importance, since here any defects would be easily
seen. Both the depth and the width of the crease have to be suited to the
board to obtain this desirable result.

The performance of a carton on a packing line can be influenced by the
quality of the creasing. This may show itself as a high resistance to opening of
the glued carton, which will slow down hand erection or seriously disrupt
automatic filling. Poor crease quality can also produce flaps which spring
open and panels which bow. The key factor is the resistance to folding of the
creases, and there are test methods by which this can be measured and limits
set.

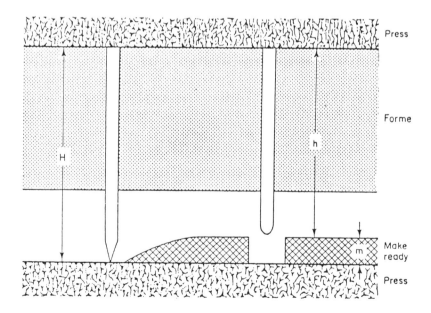

Figure 10.4   The cutting and creasing forme.

Figure 10.5   Stages in the folding of a good crease.

Stripping

The result of the cutting and creasing operation on a sheet of board is to produce a number of cartons held together by an intact front edge and small 'bridges' at various points on the sheet, so that it is still handleable as a sheet and does not fall completely into its component cartons. The next operation is to remove the material which holds the sheet together, a process termed 'stripping'. A certain amount of automatic stripping is done as the sheet passes through the cutting and creasing press; this, in general, removes unwanted central portions of the cut sheet which cannot easily be removed by subsequent operations. Complete stripping is possible on some modern presses, but generally this operation is still performed by hand with the aid of some machine tools such as mechanical hammers. Essentially, the operator uses a rubber-headed hammer and knocks the waste material away from the cartons which are 'torn' out of the sheet in blocks. After brushing the edges to remove fluff and loose material, the blocks are passed on to the next operation.

Finishing

The carton blank may not be treated further by the carton maker if it is destined for certain packing operations. Cartons which are erected and secured by interlocking flaps, or cartons which are erected and their seams glued on the packing machine, are examples, Many cartons are, however, glued by the carton maker along the side seam to form a tube. Prior to gluing, the carton may have passed through various processes, e.g. windowing or waxing.

Alternative means of erecting cartons can be provided by the carton maker. He can print on to areas of the blank, heat-sealing or pressure-sensitive adhesive patches. Cartons using these methods of securing flaps are usually covered by patents, either of their design or of the equipment used in erection.

When a collapsed tube is formed on the side seam gluer, two diagonally opposed creases are folded 180°. The majority of gluers have a pre-folding operation which momentarily folds the other two main creases to between 90° and 180°, and then unfolds them. By this means, the subsequent erection of the carton on the packing line is made easier. The operation is referred to as *pre-breaking the creases*.

It is important to remember that prolonged storage of cartons, particularly under load when stacked in cases, can reduce the effect of the pre-folding operation, and almost return the carton to its original condition. Complaints sometimes occur which can be traced to the fact that cartons have been kept for too long under pressure in stacks, so that the subsequent erecting operation on the forming and filling machine has been unsatisfactory. Modern carton gluers can also make other styles of cartons, such as spot-glued trays and collapsible styles, as well as straight-side-seam glued tubes. Once these operations have been completed, the cartons are packed ready for dispatch to the user. For cartons to work well on the packing line they must not be distorted in store, nor exposed to extremes of temperature and humidity. Therefore, both the bulk containers and the storage conditions need proper selection. In the relevant clauses of section 7 of BS 1133, it states that cartons are best bundled, and paper banding is preferred as it causes less damage.

For transport, bundled cartons arranged on edge in outer containers with separators between the rows and layers will prevent distortion and scuffing damage. A lower cost alternative is overwrapping of the bundles, using paper tape to secure the wrapper.

Advice on storage conditions generally for packaging materials is given in BS 1133, section 2. In particular, for cartons, excessively damp or dry storage should be avoided or the performance of the cartons at all stages will be impaired. The section indicates maximum storage periods of 2–3 months for certain cartons, the critical time depending on the type of board and the sensitivity of the cartoning machine.

Erection, filling and closure of cartons

The purchasing specification agreed for a carton should include a dimensioned drawing of the selected style. Where the style is simple, it may be sufficient to refer to the number in the European Carton Makers Association (ECMA) Code and give the major dimensions in the approved sequence.

Cartons are made in a great variety of styles and are generally named from the method of closure, their shape, their end use or other special features. Nevertheless, while there are literally hundreds of different types of folding carton, the entire range can be considered as having evolved from the two primary concepts of a tube and a tray. We may explain these two terms, tube and tray in the following way: a tube consists of a sheet of paperboard folded over and glued at the edges to form a (rectangular) tube, the ends of which can be sealed or locked in a variety of ways. A tray consists of a sheet of board with all four sides folded at right angles to the main sheet, and locked or adhered together at the corners. One panel of this tray can be extended to form a cover, if desired, which will act as a lid.

Packaging by hand

When the product is made in relatively small numbers or is difficult to load by mechanical means, then manual packaging is a consideration. Certain of the carton styles illustrated in the ECMA Code (and the selection in BS 1133, section 7, subsection 7.3: 1986) lend themselves to this.

Of those with glued side seams, the tuck end style (A111, A112) can be used. The alternative base constructions of quick lock (A420, A430) and the crash lock (A510 *et seq.*) are particularly suitable.

Trays with web corners that erect without adhesive are suitable for hand work (B120), as are four corner spot glued trays that have been collapsed by diagonal perforated folds for storage and delivery.

Mechanization is also possible with some of these styles. Complex special cartons with multiple folds such as may be designed for Easter eggs or toys may also require hand packaging.

Mechanized packaging with carton blanks

Carton styles based on the tray are supplied to the packaging line without gluing of the side seams. Erection and closure of the tray style flat blanks is done mechanically with either hand or mechanical loading of the product.

Two types of machine action are involved. One uses a plunger to force the blank through a shaped aperture such that it is erected into a tray form. These trays are secured at the corners by extensions to the walls which either have tabs which interlock in slits in the adjoining wall or are adhesively bonded to them. The adhesion may be by an applied aqueous or hot-melt adhesive or by heat sealing where plastic-coated board is used for the carton. A number of these machines form part of patented carton systems such as Kliklok and Sprinter.

A second type of machine takes a carton blank and wraps it around either a former or the product and secures it with adhesive. In a number of designs, the resultant carton is similar to the glued end side seam, glued type A220. These cartons may have an internal barrier bag formed on the same machine. Another common design is the hinge lid style, E 530, which may be familiar as the cigarette carton, but is also used for confectionery.

Mechanized packaging with side seam glued cartons

The erection of tuck end or glued end cartons from the collapsed side seam glued state, base closure, filling either manually or by machine and then final closure can be done on intermittent and continuous motion machines at production rates from 60 per min up to 800 per min. Collapsed cartons are loaded into a magazine from where they are extracted mechanically or pneumatically and flaps folded progressively as required by the construction.

Accuracy in carton dimensions including the alignment of the side seam is required for success at all stages of the machine. Crease quality is important for proper flap folding and the minimization of panel and flap bowing. Friction and rub resistance can affect performance on some machines. Carton shape in the flap design may be prescribed by the machine supplier.

Cartons for solid products

Cartoning systems can load the product through the end of the carton or fill the product vertically, and both of these may be performed as continuous processes (Figures 10.6 and 10.7). Cartons may also be supplied for systems which will erect, fill and close either a pre-lined carton or one which makes its own liner on the machine. Both these types of lined carton can be used for vac/gas and vacuum packaging at speeds of up to 60–70 cartons per minute.

Cartons for liquid products

Requirements with respect to impermeability and hygiene are particularly stringent when packing liquid food products. It is therefore often necessary to use other methods than those already indicated in order to obtain satisfactory results. Sealing must naturally be very tight, and there are problems to be solved in connection with filling and how to seal or avoid open-cut edges. The sandwich construction of the two common paper-based laminates used in liquid packaging are shown in Figure 10.8. If no high-gas barrier is required, that material consists of paper with a polyethylene coating on both sides. The paper layer may consist of unbleached, bleached or semi-bleached sulphate pulp, or laminates of these. The paper layer, being responsible for much of the machinability and mechanical properties of the package, requires a high and stable quality. As the total paper consumption

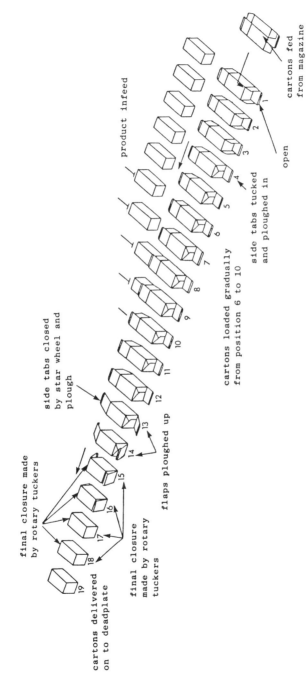

Figure 10.6 Constant-motion cartoning machines: horizontal. Courtesy of Baker Perkins Ltd. and Baker Perkins (Export) Ltd.

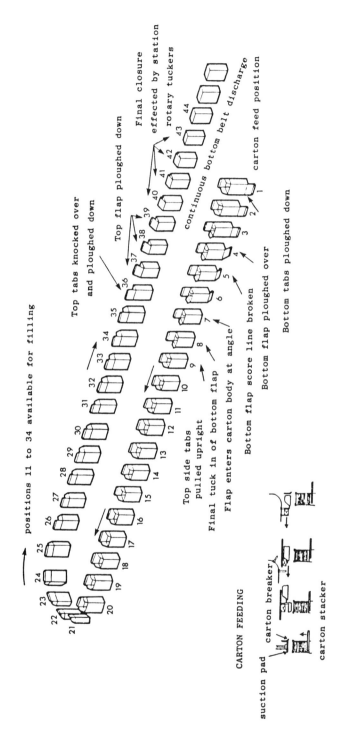

Figure 10.7 Constant-motion cartoning machines: vertical. Courtesy of Baker Perkins Ltd. and Baker Perkins (Export) Ltd.

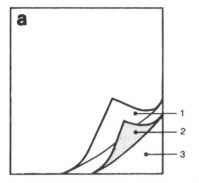 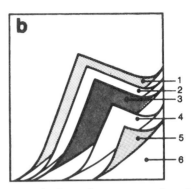

Figure 10.8 The sandwich construction of two common laminates for carton containers is shown. (a) Typical laminate for short-life products like fresh milk consists of (1) exterior PE, (2) paper and (3) interior PE. (b) Typical laminate for long-life products consists of (1) exterior PE, (2) paper, (3) Surlyn, (4) Al-Foil, (5) Surlyn and (6) interior PE.

annually amounts to millions of tonnes, and the production process is highly specialized, only a few major producers can meet these requirements.

A pronounced gas diffusion can occur through raw cut edges of a carton package. Thus it is important, especially for long-life products, that the interior of the package has no raw edges and that the product is as close to being hermetically sealed as possible. Different solutions to how raw edges are protected are shown in Figure 10.9.

The most important liquids at present are milk and milk products and fruit juices. The machines described in the following section may normally be used for both products. There are two main groups of machines, those which work from a reel, form the package and fill it in a continuous operation, and those which start from a pre-manufactured blank.

Figure 10.10 shows the 'TetraPak' principle of forming from a reel, the different stages being combined in one machine. This machine permits carton filling under fully aseptic conditions. Developments in this system have led to the 'TetraBrik' package which is oblong, and not a tetrahedron in shape.

The production of TetraBrik-type packages, whether standard or aseptic, from roll-fed machines follows basically the same principles as for Tetra Standard, but the transverse seams are sealed parallel. The characteristic brick shape is formed after cutting off individual packages from the tube, by folding in the flaps and heat-sealing them (Figure 10.11).

Figure 10.12 shows the PurePak which starts from carton blanks, heat-sealed along one side; output is about 60 per minute.

The filling and sealing procedure of a gable-top type package from a prefabricated blank is schematically shown in Figure 10.13. It starts with feeding of the blank from a magazine. The lay-flat tube is then unfolded and enters a mandrel where the bottom is heated with hot air. The bottom is then folded in accordance with scorelines, and pressure is applied for finishing the

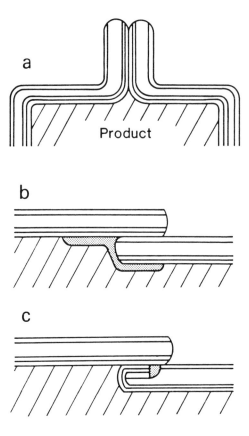

Figure 10.9 Different solutions for protection of raw edges are shown. (a) The common 'fin seal'; (b) an extra strip of plastic overlapping the internal side of the longitudinal seal as in TetraBrik; (c) in the so-called 'skiving' technique, the inner end of the carton is reduced to one-half of the original thickness, folded in and sealed to the outer end.

bottom sealing. Now an open rectangular box, it is removed from the mandrel on to a conveyer, filled with liquid and the top sealed. The top seal is performed with hot air and pressure.

Machines for the packaging of liquids are frequently not purchased outright but rented, because of the special working conditions and the importance of perfect servicing. In such instances, a down payment plus a certain sum per filled carton, and/or a rental charge is made. Important features of such packages for liquids—apart from economy in manufacture and handling and absolute leakproofness even under difficult transport conditions—are the necessity for aseptic filling, simple transport containers to permit easy stacking, and easy opening and closing of the units.

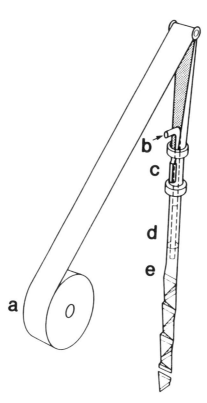

Figure 10.10 The principles of making tetrahedral packages from a reel. (a) Packaging material on a reel is fed to the machine and formed into a tube by a longitudinal seam (b). The liquid enters the tube from (b); filling is below surface level (d). Finally, the package is transversely sealed (e) and is given its shape.

Common carton styles and their uses

Bearing in mind that virtually all cartons are produced on one or other of these principles, we can consider some typical styles. Undoubtedly, the style accounting for the largest carton tonnage is the familiar glue-end carton (Figure 10.14a). This is produced from a single blank, and is delivered by the manufacturer in the form of a collapsed rectangular tube with four flaps at each end for sealing. The style is particularly suitable for high-speed automatic packaging, and is used for granular materials, including common foodstuffs, soap and detergent powders.

The tuck-end carton (Figure 10.14b, c, d and f) is widely used where a reclosure is required. This is often the case with many household goods, clothing items, toys, etc., where customer inspection may be desirable. With this style, one of the four flaps of the glue-end style is omitted, and on the

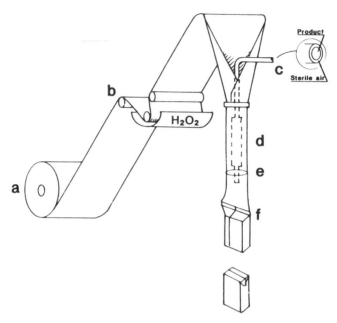

Figure 10.11 The TetraBrik aseptic concept is shown. (a) Packaging material on a reel is fed to the machine passing through a hydrogen peroxide bath (b) and formed into a tube by a longitudinal seam. The tube heater zone is indicated by (d). The liquid is admitted through (c) and filling and sealing is performed below surface level (e). Finally the package is given its final shape (f).

opposite flap is provided an extension which can be tucked into the body of the carton once the tube has been produced. This reclosure facility also makes it popular where repeated dispensing of things such as tablets, drawing pins and other small articles may be required.

Heavier articles often require a safer reclosure device than is provided by the tuck-end style, and for these many forms of the lock-end carton (Figure 10.14e) have been designed in which a tongue or tongues on one of the main flaps engages with corresponding slits in the opposite flap. All the above styles of carton, as will be readily realized, are based on the tube.

The shell-and-slide carton (Figure 10.15), which consists of an outer tube (shell or hull) and a tray-like slide with tuck-in ends, is familiar to all of us in the cigarette pack. The recent tendency to replace this shell and slide with a so-called 'flip-top' carton in the cigarette industry is an example of the modern tendency to reduce all types of carton to a one-piece blank. Shell-and-slide cartons are particularly suited to the frequent dispensing of solid objects which substantially fill the carton at the beginning. The flip-top carton is obviously also suited to this purpose.

The largest single remaining class of cartons is based on the various types

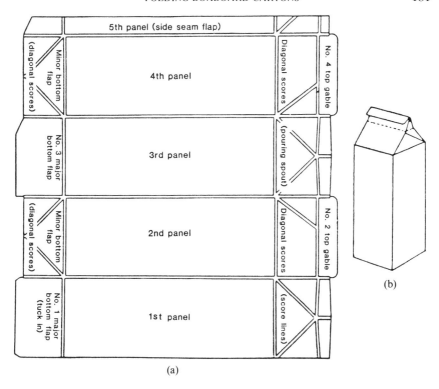

(a)

Figure 10.12 (a) An unfolded gable-top blank is shown. Scorelines necessary for folding are indicated. (b) PurePak carton.

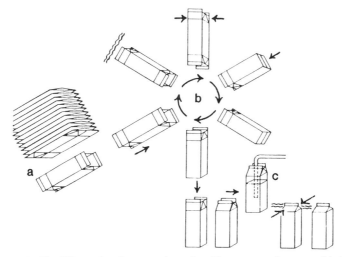

Figure 10.13 The filling and sealing procedure of a gable-top carton from a prefabricated blank is shown. (a) The blanks are fed to the machine from a magazine, unfolded and introduced into a mandrel (b) where the bottom is heated, folded according to scores, and heat-sealed. Now an open box, it is removed from the mandrel, filled (c) and top-sealed by means of heat and pressure.

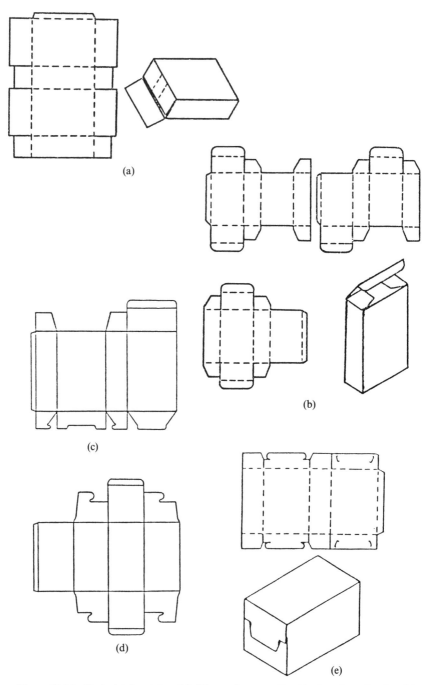

Figure 10.14 Typical tube styles. (a) Glue-end carton, flat blank and partly closed box. (b) Three shapes of blank for a tuck-end carton. (c) Lock-bottom construction (flat blank). (d) Combination tuck-end and lock-end closure (flat blank). (e) Lock-end carton and blank.

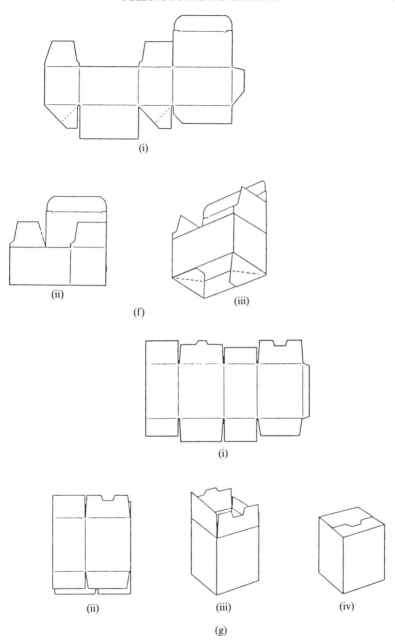

Figure 10.14 (*continued*) (f) Automatic-bottom carton: (i) blank; (ii) carton glued for shipping; (iii) bottom view of erected carton. (g) Sealed-end carton: (i) blank; (ii) side-seam shell; (iii) carton erected for loading; (iv) filled and sealed carton.

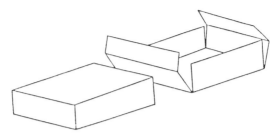

Figure 10.15   Shell and slide.

of tray (Figure 10.16). These consist of an unbroken bottom with the sides folded at right angles and secured either by spot-gluing (Figure 10.16a, b, d), stitching, locking (Figure 10.16c, e) or folding over (Figure 10.16f) in some manner to complete the tray. Their principal advantage is that they have a larger area for initial loading, and they also provide the same larger area for visibility of the contents where this is required. They are thus particularly useful where it is difficult to insert the contents through the ends. They find wide uses for fruit, biscuits and cakes, and are often overwrapped with a transparent film. The addition of a hinged lid to the basic tray (Figure 10.16e) forms a fully enclosed one-piece carton which is again used for similar types of products.

A number of the tied carton systems, in which the user purchases cartons to be filled on equipment leased from the carton maker, give packs of the tray type.

The lid-tray type consists essentially of two trays, one of which is made slightly bigger than the other, to act as a lid. This is sometimes referred to as a *semi-rigid box*. This particular type of carton is often used for packing shirts (Figure 10.17) and other clothing which folds to a rather small thickness with a large area and needs display of some description. These trays are provided with considerable rigidity where necessary by utilizing fold-over ends and by gluing them. This is particularly useful where heavy goods are concerned, or where the goods are easily crushable, and where rigidity in the walls of the cartons is essential. The retailing of boots and shoes is a good example of the use of this type of carton.

Window cartons have become part of present-day marketing and are produced by having some area in a main panel cut away and covered (usually with a transparent film) to provide greater visibility of the contents. The window style can be incorporated into virtually any style of carton. The limitations are generally on the actual type of window that can be used.

All of the carton styles mentioned above are commonly produced with folds and creases which are based on straight lines. During the recent past, the ability to handle curved creases has grown, and a considerable amount of experimentation on shapes based on the use of curved creasing rules is in

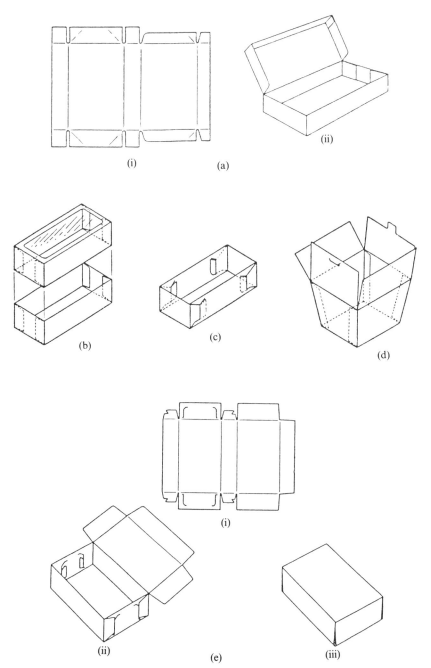

Figure 10.16 Typical tray styles. (a) Six-corner carton: (i) blank; (ii) carton erected for loading. (b) Two-piece tray with transparent window. (c) Lock-corner tray. (d) Glued-corner tapered tray with locking top flaps. (e) Locked-corner hinge-cover carton: (i) blank; (ii) carton erected for loading; (iii) filled and sealed carton.

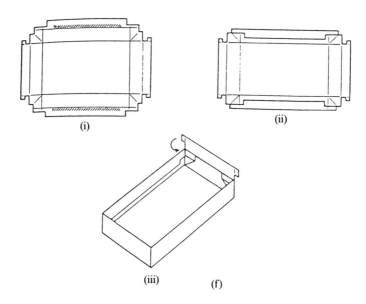

(i)   (ii)

(iii)   (f)

Figure 10.16 (*continued*)   (f) Web-corner tray: (i) blank; (ii) side walls glued; (iii) final panel folding.

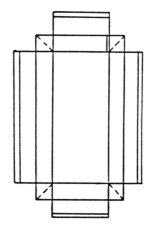

Figure 10.17   Blank for one half (lid) of a shirt box. The other half is similar in construction but just slightly smaller when made up.

progress. This can give considerable scope for display purposes, but usually entails complications in terms of the packing of the retail units into their outer cases.

Where enhanced barrier properties are required, a number of patented styles are available, incorporating a loose liner which forms a bag or pouch inside the carton on the filling machine. These pouches can be liquid-tight. With appropriate machinery, liquid-tight cartons can be obtained in single-wall construction.

Special constructions

This classification may be employed for cartons that do not fit tray or tube descriptions, or that represent sufficient departures from normal tray or tube practice. The bottle carrier carton in Figure 10.18 is one example of the many wrap-around carrier cartons for multi-packs of bottles, cans, or plastic tubs. These cartons are either locked or glued after being wrapped around the primary packages.

Advantages of carton systems

Each of the cartoning systems has particular applications, and usually a feature which is patented. In the early systems, this was often a locking tab and slit (Kliklok and Sprinter for example) which could be incorporated in various ways to give a lock-corner tray, with or without an integral lid also closed with a lock. Variants on this lidded tray style used adhesive sealed flaps to make the closure (e.g. Diotite).

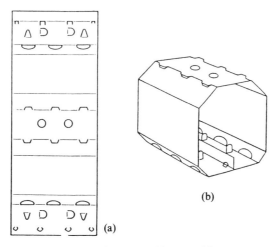

Figure 10.18   Bottle-carrier carton: (a) blank; (b) erected carton.

The erection of these cartons was on relatively slow-speed intermittent-motion machines, often coupled to simple hand or machine filling lines. When higher packing speeds (over one per second) are needed, a continuous-motion fully automatic machine is usually required, and vertical and horizontal cartoning systems using end-loading cartons were developed. These use side-seam glued blanks. Although the resulting cartons can be sift-proof, they are neither liquid- nor gas-tight.

The first approach to a liquid-tight carton erected carton blanks, and then waxed them by dipping. Later these could be pre-waxed and heat-sealed into shape (e.g. Perga and Pure Pak). More recently, in the TetraPak and BrikPak systems, the cartons were made and filled continuously from a reel of pre-printed plastic-coated board. Developments here have led to the possibility of aseptic filling.

The incorporation of a liner which can be sealed to form a bag-in-box can give a carton able to contain a liquid and provide gas and water vapour protection to the contents. Examples of developments of this style are the Cekatainer, Hermetet and Pemplex systems (Figures 10.19 and 10.20). Further developments have incorporated various tear-off, easy open and dispensing devices.

A 'cartoning system' is in every sense a package deal. The carton user negotiates with one source for the cartons, the machinery on which to erect, fill and close them, and the servicing of that machinery. Many of these cartoning systems are obtained through carton suppliers rather than machinery manufacturers.

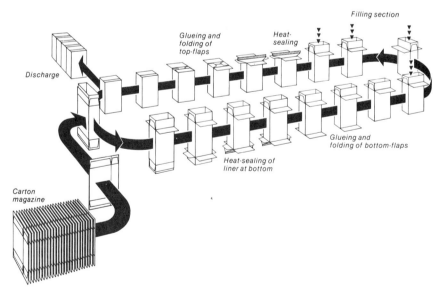

Figure 10.19   Carton system for pre-lined cartons. Courtesy of Pembroke Cartons Ltd.

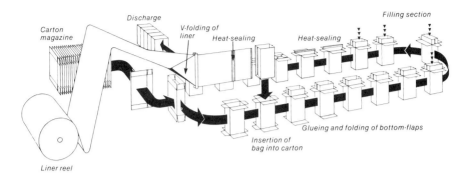

Figure 10.20 Carton system for lining cartons with bags made in line. Courtesy of Pembroke Cartons Ltd.

The pack styles available as carton systems extend from simple creased blanks erected into lock-corner trays, through side-seam glue-end loading styles, to cartons with an inner liner that can contain liquids or function as vacuum packs. A range of single-wall cartons often made from the web are also available for liquid packing.

The machinery available for handling cartons can vary from simple hand-fed erecting machines (which are then filled by hand on a conveyer and closed by hand by a simple tucking operation) to a sophisticated two- or three-headed erecting machine, one or more automatic filling stations and closing lines, coupled with film overwrapping equipment and means for packing the cartons directly into cases for dispatch. More recently, developments in shrink wrapping have led to collating a number of cartons into units for dispatch by these methods.

Whatever the system employed, three main operations are performed. First, *forming or erecting the container*. Material may be fed to the carton erection point as a continous web, as a flat carton blank, or as a folded carton flat with a manufacturer's joint secured. Second, *loading or filling the container*. In the continuous web-fed and the top-loading systems, the container has only one face open at the moment of filling; with an end-loading system it is possible with certain products to load the carton and close both ends afterwards. The third operation is *closing and sealing*.

In addition to these main operations, cartoning systems may be required to carry out secondary operations such as handling paper liners, embossing codes and inserting leaflets. All these can be performed along with the three main operations on manual, semi-automatic or fully automatic lines.

When forming, filling and closing must be carried out on one machine in a single operation, fully automatic machinery is usually used. The division between semi-automatic and fully automatic machines relates principally to how the filling or loading of the carton is carried out. If loading is done directly into the cartons, even though an operator inserts it into the infeed conveyor by hand, the system is classed as automatic. If the rest of the operation is automatic but the load is inserted directly into the carton by hand, the system is described as semi-automatic. Most systems requiring high speeds use continuous-motion machines; lower speed systems usually use intermittent-motion machines. The latter can also be of advantage where the nature of the product demands a stationary carton at the moment of filling.

From the point of view of the package user, there are several advantages in utilizing a cartoning system. Firstly, there is a simpler administration operation involved. One company is responsible for the supply and servicing of both machinery and cartons. Since the particular supplier of cartons will normally be producing them for the type of machine used in the system on a relatively large scale, he should be familiar with the detailed requirements of the carton blanks or flats, and the limitations of the carton handling machinery concerned. The package user need not, however, be completely tied to one carton supplier as most of the systems available are licensed for more than one production unit. In some instances there may be two or three sources of supply within one group of companies, and with other systems, competitive companies who are both licensees for a particular system can provide alternative sources of supply.

The tied cartoning system is principally applicable where versatility and flexibility rather than full automation are needed, and this is the requirement with many food lines. A cartoning system using leased equipment has the further advantage that the capital outlay on machinery is avoided in situations where the packing requirements of a company may change from year to year.

Factors to be considered in a cartoning development project

Product factors

Obviously, as with all packs, the nature of the product and the way it is presented for sale will be prime considerations in the selection of a system from the alternatives available.

1. What sort of protection does the product itself require from moisture, from oxygen, from outside odours, etc.? Is it greasy, or wet or otherwise able to affect the board from which the carton is made?
2. What sort of variations are likely to be expected in the product itself? How easy will it be to control the size of the product, and within what limits?
3. How can the product best be filled into the carton? What experience is there of packing similar products?
4. What further processes follow filling? Do the cartons have to be overwrapped for extra moisture protection or deep frozen for example?

Machinery considerations

5. What is the required production rate of the machine, both immediately after installation and at the expected peak of production?
6. What is the number of package sizes which may be involved?
7. What is the frequency of size changes?
8. Will the system have to cater for alterations of the product type or number within the size variation anticipated?
9. What are the labour requirements, how do they vary with production rate, and can they be met locally?
10. What sort of space is required for putting in the machinery, bearing in mind possible future increases in production?

The choice of board to meet the product protection requirements can have an influence on the efficient functioning of the cartoning line, feeding, erecting and closing the packs.

Marketing considerations

11. Can the graphic design and print requirements of marketing the product be met with the system?
12. How will the carton affect distribution and consumer acceptance of the product?

General considerations

13. Does the proposed new equipment have to link up with any other equipment already in existence, such as overwrapping machines, or case packers or filling heads?
14. What sort of advice can be obtained from the carton supplier on such other ancillary equipment with which he may have had experience with his cartons on other lines?

When we consider the possibilities of making all these many styles in a variety of board qualities and rigidities, and of their decoration by printing and lamination, the versatility of the carton as a retail pack can well be understood,

Carton specification and quality control

The specification agreed between carton user and supplier can be as simple or complex as the application demands. The carton maker will always seek to deliver a commercially acceptable product, but the critical requirements will obviously differ between a small order for single-colour printed cake cartons, hand-erected at a local bakery and a multi-million carton order for running on high-speed packaging machinery.

Table 10.1   Types of board for specific uses.

Type of board	Principal uses
Unlined chipboard	Rigid and folding boxes, for soaps, detergents, hardware, electrical goods, boots, shoes; also showcards, stationery, tubes and bookbinding
Cream-lined chipboard	Folding-box used in food, pharmaceutical and clothing industries
White-lined chipboard (WLC) (2 grades, No 1 and No 2)	One of the most popular grades of board for economical folding cartons; used for detergents, clothing (e.g. shirts and hosiery), toilet tissues, some foodstuffs (e.g. breakfast cereals), hardware, toys and games
Machine-coated white-lined chipboard	Markets as for WLC but where superior finish is required for greater printability and surface appeal
Triplex board	Some cheese cartons and a few other foods, pharmaceuticals; decreasing usage
Duplex folding carton board	Cigarettes, frozen foods, dried foods, pharmaceuticals, cosmetics, toiletries, baked foods such as biscuits, cakes, etc.
Duplex folding carton board, coated	As uncoated board where a high finish for greater printability and superior quality is needed.
Solid white food board	Frozen and speciality foods, cosmetics, pharmaceuticals
Laminated boards	Specialized applications: frozen foods, cakes, oils, fats, high quality prestige packs for whisky, cosmetics, etc.

Besides production checks to see that the carton meets the customer's requirements, the carton maker has to make quality acceptance tests on his raw materials to check their suitability for the converting processes and ability to give the desired quality.

Printing quality may be specified as colour matches to specimen trios of prints, the target and the acceptable variants either side and, for example, a minimum level of rub resistance.

Creasing quality starts with the forme and the checking of the dimensions of every die on it. It must be appreciated that the final carton size will depend on other factors, including the sensitivity of board dimensions to change with variations in the moisture content of the board, which may be affected by the relative humidity of the surrounding atmosphere. When carton production starts, cartons are examined for clean cutting, crack-free folding, and correct assembly of the carton. If necessary, crease stiffness can be measured. These checks continue throughout the run, to see that wear of the makeready does not cause a deterioration in crease quality.

Where the carton incorporates an opening device relying on perforations, as do many soap powder packs, a maximum value for the burst strength of these perforations may be included in the specification, to prevent complaints of difficulty of opening.

During side-seam gluing, samples will be examined, and checks make for extraneous glue spots inside the tube, and for good adhesion and skewing of the side seam. Gluer adjustments may be made so that cartons meet special critical requirements, such as even stacking when loaded into a packing-machine hopper.

Many of the tests that may be made on board, inks, print, and cartons are covered by industry and national standards. Early discussion of the carton application between supplier and user is desirable, so that the necessary quality can be established and appropriate tests selected. A sample purchasing specification is shown in Table 10.2.

Table 10.2 A purchasing specification for ordering cartons.[a]

JOLLY FOODS LIMITED	No. 25 F
TITLE: PAPERBOARD CARTON PACKAGING MATERIAL	Date issued: March 1st, 1980
APPLICABLE TO: NOMINAL 500 g JOLLY CEREAL FLAKES	

A. *Supplier*
   See approved list

B. *Carton style*
   One piece, side seam glued paperboard carton with four flaps at each end
   Two outer flaps at each end to be equal lengh and square cut
   Two inner flaps at each end, cut away design, not meeting

C. *Materials of construction*
   1. Paperboard   Duplex folding boxboard suitable for food packing (375 gm^{-2} Blacks Ltd)
   2. Adhesive   Normal moisture reistant type used in carton manufacture

D. *Carton construction*
   Carton blank (printed surface uppermost, reading correct).
   Manufacturer's join: the glue lap to be to the right of the right side panel and glued to the inside of the back panel.

E. *Dimensions*
   Dimensions are measurements from centre of one crease to the centre of the next crease, measured on the inside surface of the carton blank
     Length   152.5 mm
     Width   76 mm
     Height   216.5 mm front panel[b] 214.5 mm sides and back panel[b]
   Manufacturer's join glue lap to be 12.5 mm wide

F. *Prefolding*
   To be prefolded along each vertical crease during manufacture to facilitate easy erection

G. *Method of flat fold*
   Folded glued carton when viewed from the back to have back panel to the right of the manufacturer's join with print reading correct and join central

H. *Printing*
   Design: as copy provided
     Supplier's design proofs to be supplied for approval and each design to be given a Jolly Design Number
     Design location to be shown on supplier's design proof
     Print bleeds on to the flaps to be within 2 mm of flap creases and cut edges.

Table 10.2 *cont'd*

Inks
  Type: to be low odour, rub resistant and light stable
  Colour: see range of approved colour samples
  Appearance: to be sharp and clear with the register free from readily noticeable
  inaccuracies

Varnish
  To be overprinted with a rub-resistant low odour varnish on all print areas except print
  on inner flaps adjacent to horizontal creases

Identification
  Manufacturer's Code and Die Station Number to be printed on bottom flap of left side
  panel; manufacturer's name flash may only appear on this flap
  Jolly Design Number and Print Edition Number to be printed on bottom flap of right side
  panel

J. *Packaging*
  Cartons be be banded in units of 100
  Cartons to be edge packed into corrugated cases with three units of 100 per case
  The units to be banded with bands of glazed kraft paper at least 100 mm wide and fixed by
  adhesive tape; the positions of glue laps on adjacent bundles to be reversed; cases to be
  palletized five layers high as shown in diagrams
  Pallet load to withstand multi-pallet stacking (maximum four pallets high); carton load to
  be tight-strapped by Whites 64 to pallet in four positions a shown in photographs
  Edge protectors to be used between strapping and carton load
  A single trip wooden pallet close boarded on the top deck should be used with dimensions
  that match the pallet load; irrespective of pallet bottom design it must have at least three
  stringers each at least 125 mm wide
  Stacking pattern; see diagram (not shown here)

K. *Identification*
  A printed carton to be fixed on each side of pallet load and each pallet to be clearly labelled
  on each entry face with the following:
    SUPPLIER:
    JOLLY ORDER NO:
    JOLLY DESIGN NO:
    QUANTITY:
    DATE OF MANUFACTURE:
    DATE OF DESPATCH:

aWith appropriate modification this style of spec. can be employed for almost any type of
packaging material
bThe front panel creases to be staggered equally at top and base. See Jolly Drawing No. 25/f
March, 1980 for detailed dimensions of carton blank.
(*Note:* not shown in this text).

11
Rigid and cylindrical boxes and composite containers

Rigid boxes

Rigid-paperboard boxes are also called 'setup' boxes. Unlike folding cartons, they are delivered to the packager, set up and ready to use.

Introduction

The first known use for a paper box was for packing tea. The word box generally means a receptacle with stiff sides as distinguished from a basket, and the name arose because they were first made from the wood of the Box tree.

Boxes for gifts became popular many years ago when Roman priests encouraged people to send presents during the seasons of rejoicing. The paperboard box of today originated in the 16th century with the invention of paperboard. In Europe, one of the earliest types of paper boxes was commonly known as a band box. These were decorated by hand and used to carry bands and ruffles worn by the Cavaliers and Ladies of the court. It was not until 1844 that setup boxes were manufactured in the United States.

A Colonel Andrew Dennison found that making boxes by hand was tedious work and developed the Dennison Machine to cut the blanks. The Colonel's invention was revolutionary, but until the Civil War, most consumer products were still packaged in paper bags or wrapped in paper. There were only about 40 boxmakers in the United States and most boxes were made by hand. For these 40 craftsmen, boxmaking was merely an adjunct to other lines of business, which varied from printing to the manufacture of the consumer goods they would eventually package in the boxes.

In 1875, John T. Robison, who had worked with Colonel Dennison and others, developed the first modern scoring machine, corner cutter, and shears. These three machines still form the machinery basis for most box shops, but it was not until the end of World War II that significant progress was made to improve the production of machinery for the industry.

Markets

The rigid (setup) box has stood the test of time and competes well within its selected markets. It protects, it builds images, it displays the contents and it sells. It is difficult to define the principal market for rigid boxes, but they are used in the United Kingdom principally by hosiery and footwear trades, for small hand tools, and for the expensive perfume and cosmetic markets.

The major markets for rigid boxes in the United States are listed in Table 11.1. Notice that the figures tend to fluctuate from year to year. Rigid boxes compete well in some of these markets but with changing technologies, industry forecasters predict a decline in rigid box use for department stores, textiles, personal accessories, and hardware and household items. To offset this decline, they expect an increase in use by manufacturers of computer software, confections, educational material and electronic supplies.

It is impossible to lay down hard-and-fast rules in rigid box making. Conditions, layout, etc., vary so much from one firm to another, that methods tried and found successful by one company may be hopeless in another set of conditions.

Materials

Board. The choice of board used depends very largely on the size of the box, and the type and weight of the contents, the intricacies of manufacture, and the degree of accuracy required.

Table 11.1   Use survey of industry sales, percent of total market.

Industry	1975	1977	1979	1981	1983
Textiles, apparel and hosiery	7.5	4.8	4.5	5.5	3.4
Department stores and speciality shops	9.9	10.6	8.3	7.1	7.2
Cosmetics and soaps	1.8	5.1	4.3	3.9	1.4
Confections	13.3	8.1	15.7	18.0	17.2
Stationery and office supplies	9.7	13.2	15.0	9.3	19.1
Jewellery and silverware	5.2	14.0	7.6	9.7	6.3
Photographic products and supplies	2.2	2.7	3.5	4.3	3.6
Shoes and leather	4.8	1.1	0.4	0.2	0.4
Drugs, chemicals, and pharmaceuticals	6.9	7.3	13.1	7.8	6.7
Toys and games	2.3	2.6	1.4	1.8	2.2
Hardware and household supplies	7.0	6.2	3.4	4.0	6.7
Food and beverages	2.9	0.3	5.4	3.9	1.5
Sporting goods	1.9	1.0	1.7	1.2	0.4
Other major customers					
Electronics	3.4	2.6	4.3	3.5	3.4
Educational	3.9	6.5	3.8	4.4	2.7
Other	12.2	7.1	1.1	7.0	7.9
Miscellaneous customers	5.1	6.8	6.5	8.4	9.9
Total	100	100	100	100	100

Source: National Paper Box and Packaging Association.

Originally two main types were used. Firstly, Dutch strawboard was the traditional material used by the majority of box makers engaged in the mass production of machine-made boxes. It is a stiff material, and makes a good rigid box. However, because of various drawbacks (which include its propensity to absorb moisture, to vary in size, and also to give off a 'musty' smell when damp), box makers looked for alternative materials.

The original chipboard offered by British board mills, although not as rigid as the Dutch strawboard, offered the advantage of not shrinking to quite the same extent as strawboard. It therefore was more reliable for ensuring that the correct fit was obtained between boxes and lids, and for minimizing the acute warping of shallow lids. The main disadvantage was the lack of rigidity, and often a thicker board was required to obtain a similar rigidity to the Dutch strawboard. This led the British board mills to produce a new grade of board known as *rigid board*, as a substitute for Dutch strawboard. It was intended to combine the advantages of both Dutch strawboard and the home-produced chipboards. This board is now widely used in the United Kingdom and is available in thicknesses ('calipers') ranging from 500 to 2900 microns (μm)

No strict rules can be drawn for the rigid box industry as to the type of board to be used. Generally speaking, the box maker will decide on the best board for the particular job in question, and this may vary from a plain unlined rigid board, through the complete range to high-class food-quality duplex board

Covering papers. The covering papers used are numerous, though generally in the rigid box industry there are a number of standard types used. These include enamels, flints, tints, leatherettes, and many others. The covers can, of course, be printed, embossed, or gold blocked and, as with board, the type of paper used depends on what the customer requires. The main consideration for the box maker is to know the behavioural characteristics of the various types of papers, and how to overcome them if they have adverse effects on production.

Having therefore selected the type of board, and decided the thickness and weight suitable for the job, the box-making operations can begin.

Manufacturing stages

Cutting. First of all the board must be cut into the correct blank size. This is usually done for the straightforward type of rigid box on a rotary box cutter and scorer.

There are several types of machine available with hand or automatic feed, straight-line or right-angle. Straight-line machines cut and score the board in one direction at a time, passing through the machine. Right-angle machines carry out the cut-and-score operation in both directions in one pass through the machine. The scoring operation is, however, usually carried out in the same way on all machines.

A scoring-wheel (Figure 11.1) which has a cutting edge, is fixed at a pre-determined distance from a roller, and the board is passed between the two. The distance between the roller and the cutting wheel is such that the cutting edge penetrates only half-way through the board. Cutting can be carried out either by setting a scoring wheel sufficiently low to cut right through the board or on some machines it is done by a shear cutting action, brought about by two wheels partially overlapping each other. This produces a piece of board, the overall size of which will correctly make up into the box, and on which there are two sets of scores, one set at right angles to the other (Figure 11.2). These scores divide the box into its length, width and depth dimensions. This operation, and that of corner cutting which follows it, may also be carried out on a cutting and creasing press using a metal die, in a similar manner to that employed for folding cartons. Using this method results in an improved quality of blank, as well as alleviating the need for the separate process of corner punching. The blanks processed by this method make up into a better rigid box, as the improved accuracy allows for better production on the quad staying and automatic covering machines. Consequently there is a strong tendency for rigid box makers to move away from the standard rotary cutting and scoring machines to using hand-fed and semi-automatic cutting and creasing platens. However, on smallish quantities it is often not economic to pay for the cost of the metal die and the additional cost of makeready of the cutting and creasing press. In addition, especially on thin boards, the difficulty of pressing a length of metal rule into a piece of cardboard (which may vary in thickness) just sufficiently to give the effect of a score, can be problematic.

Corner cutting. The next operation is to cut out the corners of the blank (Figure 11.2) to allow the sides and ends to be folded at right angles to each other, and to the bottom of the box. It is essential that extreme accuracy be maintained in this cutting operation, otherwise the box will not make up

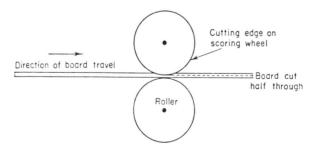

Figure 11.1 A scoring wheel.

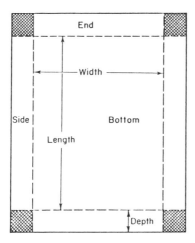

Figure 11.2   Board showing two sets of scoring lines at right angles to one another.

square. It is done on a machine which can be designed either to cut one or two corners at the same time. Nowadays there are machines available which will cut out the corners at the same time as they fasten them together with stay paper.

Corner staying. The object here is to stick gummed paper around each corner, thus making the box rigid. There are various types of machines to do this job. The simplest machine, which will only stay one corner at a time, is hand-operated—the operator having to fold up the sides and ends of each box by hand. The stay paper used can either be gummed paper tape, made sticky by the application of water, or a thermoplastic coated paper which is heat-activated. The choice of paper depends entirely on the type of machine. Thermoplastic stay papers have advantages, particularly when staying a deep box (say 100 mm or more) and when the operator is learning the job. With gummed paper, the rate of consumption of stay paper can be so rapid that insufficient time is available for the tape to develop the necessary tack to enable it to stick properly; whilst at the other end of the scale, when operating the machine, a trainee may be slow or fumble a box to such an extent that the paper dries out before it is applied. However, it is not always possible to obtain sufficiently thick 'thermo' stay paper to meet all customer requirements. The quad stayer is virtually four single stayers combined, with the addition of an automatic feed. Here also there is a choice of machine using either water or heat as the means of creating tack. One point which must be borne in mind when using a quad stayer which offsets to a certain extent some of its advantages, is that a separate forme or block is required for every size produced, although adjustable formes can be obtained. This block is used for forming the blank and is normally made of wood, metal or a plastic material. (There is also a device which may be built in to new quad stayers or added to existing machines in some instances, which will corner-cut all four corners of the blank before folding and staying them.) Finally, the boxes are nested one within another automatically, straight off the quad. The box is now ready to be covered with paper.

Alternatively, in instances where the box size is small and where strength is not required, the work can be delivered to the covering machine with the sides bent and not corner-stayed. The strength of the box is obtained by the covering paper holding the sections together. The blanks can either be bent by hand or by an automatic bending machine, which can be linked directly to the automatic covering machine.

Paper slotting. The covering paper must be processed so that when wrapped around the box on the automatic covering machine, it produces a neat and tidy turn-in. This process is called *mitring*, and a label punch which operates on a similar principle to a corner punch is used.

Firstly the cover is guillotined to a size which allows for a turn-in. The cover is, therefore, normally about 22 mm larger than the dimensions of the blank. Secondly the cover is mitred (Figure 11.3).

This enables the cover to be turned in automatically on the machine and at the same time produces a suitable finish. This process is necessary for producing covered boxes on an automatic covering machine.

Box covering: simple lid/tray boxes. The board has been cut and corner-stayed, and the paper mitred. The next operation is to amalgamate the two,

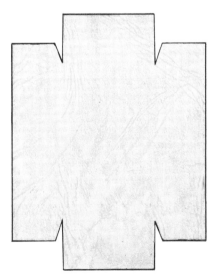

Figure 11.3 Shape of box cover paper, showing mitring.

in other words to cover the box with the paper. Most machines for this job work on the same principles, with similar basic movements.

The paper cover is first glued on its under side. This can be done either by an operator manually passing the paper over a glue roller and placing it on a table, glue side up, or as is more frequent on an automatic gluer. In this a stack of covers is placed in the gluing machine, which picks them up one at a time and feeds them over a glue roller, from which they pass on to a long moving belt, which is perforated to allow suction to hold the paper in position and flat. It also helps to prevent the covers from curling due to the application of glue, which would make it more difficult for the operator and machine to handle them. The length of the belt, coupled with the timing of the automatic gluer, is designed to create a sufficient time-lag between the application of the glue and the use of the paper for the required degree of tack to develop.

When the cover reaches the operator, a corner-stayed box is placed on it. The accuracy of positioning the box is important, for on this largely depends the quality of the final product. The box with the cover stuck to the bottom is then placed by the operator on the forming block fitted to the machine. A separate block is required for each size run on the machine, and may again be made of wood, metal or plastic.

The forming block with the box on it (Figure 11.4a) passes down into the machine, where the two long sides of the paper are folded up against the side of the box (Figure 11.4b). The four end-flaps of the long paper sides are then bent round at right angles against the ends of the box (Figure 11.4c). Next the two short sides of the cover are rolled up against the ends of the box (Figure 11.4d). Then the 11 mm of projection paper all round (the paper is normally 22 mm larger in both directions than the board) is turned inwards on all the four sides, at right angles to the side of the box. This is possible because the top and middle blocks on the machine separate at this stage, and the paper is inserted between the two blocks. After this, the top block comes down to join the middle block again thus turning all four edges of the cover inside the box. At the same time, equal pressure is applied externally to all four sides to ensure perfect adhesion. Finally, the block rises up out of the machine, bringing with it the covered box, which is ejected on to the packing table (Figure 11.5) where it may be thumbholed or 'lidded up', or have any other ancillary operation performed on it that may be required.

Thumbholing can be carried out with hand punch, a treadle-operated punch, or a fully automatic machine where the box is fed straight into the thumbholer off the covering machine without being handled by an operator.

There are even more advanced machines than this. On one, the boxes are automatically fed after they have been placed on the cover by hand, but the basic movements are the same.

Another system allows for the linking up of a quad staying machine with an automatic covering machine, with an intermediate machine taking the

Box Covering

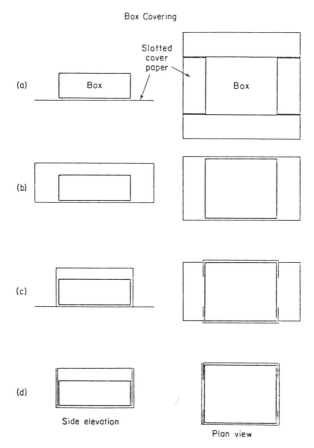

Figure 11.4 Box covering. (a) Box stuck to slotted cover paper. (b) Long sides of cover paper folded up and stuck to long sides of the box. (c) Four end-flaps bent round at right angles. (d) Two short cover sides rolled up.

corner-stayed boxes from the quad and spotting them automatically on the covering paper prior to the wrapping unit. This requires no direct operator, and the machine can run at well over 2000 boxes per hour.

Other than purely hand-made boxes, of which there are still many produced, there are many instances in a rigid box factory where hand work is necessary for the finishing off of automatically made boxes. This includes the production of hinged lid boxes, where a calico or paper hinge is applied to the lid, or where racks or other platforms must be fitted to boxes.

There has been an increase in the demand for rigid boxes which contain vacuum-formed plastic fittings; these must also be applied by hand.

Flanged and envelope-wrapped work. Apart from performing the normal tight-wrapped work already described, some machines are capable of

Figure 11.5   The box-covering process. The packing table, showing the box on slotted cover paper (as in Figure 11.4) and the machine with a forming block.

producing variations in the form of flanged and envelope (or loose) wrapped boxes.

For a flanged box, a flange card (with or without a pad to give a domed effect) is cut and stuck to the scored box, either before or after corner staying. (The domed effect can also be obtained by moulding the board on special machinery which is steam or electrically heated.) The covers are slotted slightly differently. The operation of gluing the cover and placing the box on to it then follows the same pattern as already outlined, the only difference being that the rollers used for turning up the sides and ends of the covering paper have a longitudinal recess in them to accommodate the projection of the flange card; and instead of a wooden bottom block, a brass or similar material plate is used. This supports and protects the cardboard flange, and gives a sharp edge to the paper where it is wrapped around the projection.

For envelope-wrapped work, the covers are not slotted at all, and what is more, they are glued only along all four edges. The box, therefore, cannot be stuck to the cover initially, because there is no glue in contact with the bottom or sides of the box. The paper is centred on the machine against two lays (side and end) and the box is put on to the block. The block then descends, taking the paper with it into the machine. This movement, which in the tight-wrapped box turns in the ears of the cover, folds the envelope-wrapped cover into two 'pleats' at each end. Finally the cover is stuck to the board all the way along the turn-in. The glue may be applied to the cover by a simple hand stencil or, alternatively, automatic gluers are available which are provided with an attachment which wipes the glue off the application roller mechanically everywhere, except where it is to be applied to the edges of the cover.

Another variation is to machine-cover a flanged box so that the cover ends about halfway up the outside of the box, in other words forms a capped flange cover, and then band a piece of contrasting paper around the sides, either by hand or on a banding machine.

Hand-made boxes. The hand-making of fancy boxes is a craft, and requires a great deal of experience and training before it can be satisfactorily performed, particularly the more intricate types. Round boxes are still sometimes made by hand, although mechanization is utilized for the production of large quantities.

Wire-stitched boxes. There is another category of rigid box which must be mentioned, and that is the wire-stitched box. This is the crudest form of box, and its method of manufacture is very simple, the board being guillotined, bent instead of scored, slotted instead of corner cut, and then stitched up to make it rigid. It is purely a utilitarian type of box, useful as a means of transport, and to a certain extent for protecting loose articles.

Problems of rigid boxmaking

Variations in the dimensions of board due to shrinkage or expansion. As far as possible all materials should be reasonably dry, and in equilibrium with the works atmosphere before use, to prevent shrinkage or expansion. The most satisfactory method is to expose the material to the workshop atmosphere for as long as possible before use, but away from any sources of direct heat. This process can be accelerated by hanging the board in specially constructed racks, often suspended from the ceiling.

Curl of paper due to application of aqueous adhesives. Curl is often due to the disproportionate expansion of one surface of the paper compared with the other, brought about by the application of moisture, with or without heat, when glue is applied. Varnished papers often curl worst of all, due to the coating of varnish being less affected than the plain paper. As yet there is no satisfactory method of controlling curl on all occasions. The automatic gluer with the suction belt helps considerably, but is not always enough. There are several other factors, all of which are relevant on occasions in dealing with curling troubles, for example:

1. The adhesive must not be too hot (approx. 60°C).
2. The water content of the adhesive must be correct.
3. The grain of the paper also affects the curl. On a long deep box it is better to have short-grain covers, i.e. covers where the shorter dimension is cut in the machine direction, so that the paper, which will tend to curl more with the grain than across it, will have the tendency minimized on the longest measurement.
4. The time-lag between applying the adhesive and handling the cover must be

sufficient to allow adequate penetration of moisture. It is common knowledge that paper will often curl rapidly when one side of it is moistened, and then flatten out again after a further period.

Bowing of boxes. Bowing is often due to damp board being used or, alternatively, due to the cover paper shrinking to a greater extent than the board. Board as thick as possible should be used, to offer maximum resistance to the tension produced. Another cause can be the gluing of different-substance papers to either side of a piece of board. Thus when lining the inside of a board with a fancy paper, and then covering the outside with another material, the substances of the two papers used should be as near to one another as possible, so that the pull exerted on one side is equalled by the resistance offered by the opposite side. Lidded boxes, particularly shallow ones, may have lengths of board folded up 'zig-zag' fashion placed on edge inside the box before lidding up. This will help to prevent any inward bowing as the box dries out.

Tearing of papers. Customers who insist on having boxes covered with weak material because it is cheaper are often deluding themselves. The boxmaker may point this out when the order is placed, but knows that he will experience higher spoilage than normal, and makes an appropriate allowance for wastage.

The rigid box as a package

Disadvantages

1. Relatively high cost compared with folding cartons. This gap tends to widen as labour costs increase, due to the higher labour content of the rigid box compared with the folding carton.
2. High transport costs because of high volume of air held within the box.
3. The same factor also applies to storage, rigid boxes require much space.
4. Rigid boxes cannot be printed pictorially to the same extent as cartons. For example, it is essential to avoid tight-register printing between the sides and the ends of a box since, as the box must be sited manually on the paper, variations are inevitable.
5. A relatively slow rate of production coupled with storage problems make it difficult for a manufacturer to stockpile big quantities in readiness for a large scale advertising campaign for instance.

Advantages

1. The possibilities with regard to the use of covering papers are infinite. By judicious choice of papers, very pleasing and satisfying results can be obtained.
2. The same applies with regard to the variations possible in the design of a rigid box, compared with a folding carton.
3. The protection against crushing rigid boxes afford their contents is nearly always greater than that provided by folding cartons.

4. In conjunction with this last point, the greater rigidity allows the boxes to be stacked to a greater height in the warehouse.
5. For the smaller manufacturer, or the producer of the exclusive type of article, the rigid box is valuable because it can be obtained economically in much smaller quantities than printed cartons, since the initial costs are very low.
6. The most important of all considerations is that the rigid box can be made to sell and present its contents in a fashion which cannot be equalled. It can be displayed on the shop counter, with the lid open, increasing the sales appeal of the article inside. It can be so designed and covered as to be suggestive of a quality and standard of luxury which is difficult to achieve by other types of container.

Rigid plastic boxes

The simpler types of plastic boxes are very like their corresponding paperboard counterparts in both design and construction. The most widely used are probably those made of general purpose polystyrene by injection moulding. Others are made by direct extrusion into tubes and from extruded sheet by thermoforming or by fabrication.

Injection moulded boxes are produced in many sizes, shapes, styles and can have hinged lids or a telescope lid closure. Square, rectangular, round and oval shapes are standard and almost any shape may be had at a price. Polystyrene is not the only plastic available; PVC and cellulose acetate are all used where transparency is required, and high impact PS can be used where visibility is not needed. Polypropylene of course provides the advantage of having a one piece 'living' hinge built in to the construction. They may be decorated by hot stamping and by silkscreen as well as by labelling.

Extruded boxes can also be made in many shapes starting from extruded plastic tubing. The simplest and most usual cross-section is circular. The extruded tube is cut to the correct length, a permanent bottom is attached and a removable friction lid/closure is applied.

Thermoformed boxes are made from a sheet of plastic by most of the standard thermoforming processes (polypropylene sheet is used if an integral hinged lid is required). Trays with a paperboard cover are also possible as are paperboard bases with a transparent plastic lid.

Fabricated boxes are made from sheet plastic (cellulose acetate, PS, PVC or polyester). The sheet is cut, scored, folded and glued in much the same way as folding boxboard cartons or rigid paperboard boxes are handled.

Table 11.2 gives some uses for rigid plastic boxes.

Table 11.2   Uses for rigid plastic boxes.

Hardware items: drill bits, screws, nuts and bolts, etc.
Fishing accessories, hooks, flies, etc.
Writing instruments
Games
Toys
Gifts including orchids and flowers for corsages
Cosmetics and toiletries
Costume jewellery, watches, etc.

Cylindrical boxes and composite containers

These classes of packages have paperboard walls or bodies and closures of board, plastics or metal. These walls of paper and cartonboard with as necessary layers of plastic film or metallic foil are produced by a winding process. Cylindrical boxes, have closures formed by beading of the walls and insertion of card ends of paperboard. Composite containers have metal ends seamed on as in can manufacture. Both types of packages may use plastic dispensing closures.

Cylindrical boxes with beaded sides had their origins in the 'Littlejohn Drum' named after its inventor John Bradbury Robinson of Chesterfield.

The manufacture of *composite containers* or composite (fibre) cans as they are called in the United States started towards the end of the 1920s to provide a cheaper package than the metal can. This has now changed and they have found a specific niche in the packaging spectrum.

As the name implies a composite container is made from more than one material. In practice the composite container can be circular or rectangular in cross-section; have a paperboard or plastic body; be fitted with tinplate, aluminium or plastics ends and use inside liners and outside labels. They are supplied for filling with one end fitted and the other loose to be fitted by the packer; the end fitted by the container maker is more often than not the closure/opening end, leaving the packer to apply the more simple bottom end.

The most common types used today are spirally wound paperboard bodies of circular cross-section, with some kind of barrier lining and two metal ends.

Manufacturing processes

Spirally wound walls or bodies (Figures 11.6a and 11.7). Only cylindrical shapes can be produced by this method, yet in a multitude of sizes. To form the tubular body, two or more plies of board are superimposed and glued together around a stationary cylindrical mandrel in a spiral manner. Each

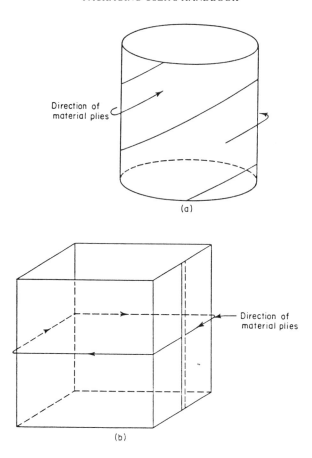

Figure 11.6   (a) Spiral winding. (b) Convolute or straight winding.

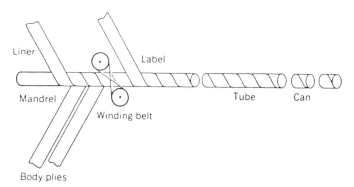

Figure 11.7   Making a spiral-wound body.

ply is applied at an angle, which varies according to the width of the board used and the diameter of container required, and the tube produced is conveyed along the mandrel by means of a driven rubber belt which is wrapped spirally around the mandrel. When the requisite length of tube is achieved, it is cut off automatically by a revolving saw and ejected from the mandrel.

Naturally the weakest part of the container is where the plies meet and, to reduce the possibility of damage when external pressures are exerted, it is essential to ensure that all joints are well butted and the various plies overlapped in the correct position.

A cylindrical tube can only be formed by having all the plies in the wall construction overlapping. This interlacing of adjacent layers of board provides an adequate bond, and the two-ply constructions of this type are suitable for many purposes.

Convolutely wound bodies (Figures 11.6b and 11.8). In contrast to the spiral technique a wider variety of shapes, including round, square, triangular, rectangular and oval, can be produced by straight winding. In this process, the board from the reel is passed over gluing rollers and fed into a gripper in the winding mandrel. This mandrel then rotates, pulling the required number of wraps of material around itself to form a tube. After each rotational cycle of the machine, the board web is guillotined to separate the tube from the parent reel. An ejector pushes the partly formed tubes along the mandrel for the subsequent operations of rolling, labelling and reforming. The reel of board used is usually of a width equal to the height of several individual bodies, and for non-cylindrical work is slit down to correct individual box height on the feed table of the machine. These separate webs are wound together and produce multiple bodies for each cycle of the machine. For round work, the full web width can be used throughout the winding operation, so producing a long tube which can be gang-labelled instead of individually labelled, and finally cut into individual bodies at the reforming stage.

The strength of a composite container is an important factor to consider, and can be improved by increasing the number of plies or by using thicker board. Where wall thickness limitations apply, to provide more strength it is

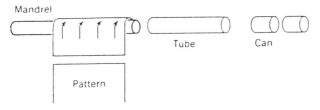

Figure 11.8   Making a convolute body.

better to use more plies of thinner material to achieve the required thickness. The standard board calipers used range from 250 μm to 500 μm, but occasionally use is made of material as thick as 750 μm.

Labelling. A plain board container is not particularly attractive, and almost invariably the body is labelled to give it the appropriate customer appeal. There are two ways of labelling composite containers: during the winding process, or afterwards as a separate operation.

The first method is operatively cheaper, but the initial gravure cylinder cost for printing the labels is high, and therefore it is only economic for large quantities. However, flexographic printing, as opposed to the more expensive rotogravure process, has overcome this problem to some extent, and shorter runs have become viable propositions where relatively straightforward graphics are involved. Intricate half-tone designs with fine registration tolerances could not be achieved. Dimensions and angles are extremely important on spirally applied labels. When originating designs, the layout angle must coincide with the angle at which the label is wound on to the tube, so that the print finishes in the horizontal plane.

Spirally applied labels can be printed either with a random or with a registered design. Random printed spiral labels present little difficulty, whereas registered designs have to be carefully prepared and laid out, and consequently tend to be more expensive.

A more significant advantage of the random printed spiral label is that, as the design is repetitive around the body, the print can be read from any angle in a horizontal plane. Random printed labels may not appear at first sight to be very attractive, but attractiveness is not the only selling point; labels are made to be read, and this is an outstanding characteristic of this type of label.

The second method, known as flat or 'gang' labelling necessitates an additional operation where the gang of several individual label units is fed into a machine, coated with adhesive, and applied automatically to a length of tubing. This type of labelling is more expensive because of this extra operation, although origination costs are considerably less, and relatively small quantities can be economically produced. Most paper labels for this application are printed letterpress or litho, but gravure-printed foil labels can create really impressive embossed designs.

Irrespective of the method employed, the governing factor of cost is the detail of the actual design, the number of colours involved, and the quantity. A basic principle is that the greater the quantity, the lower is the unit cost. This is particularly so with labelled composites, where there can be a considerable price change effected by quantity variation.

Body and functional barrier materials

It has now been established that there are two basic types of composites, spiral and convolute, but each of these can be further sub-divided. Dependent upon the purposes for which a composite is intended, it can be classified as *functional* or *non-functional*. The latter category includes all containers which are required to offer mechanical protection only.

When considering the potential of a composite pack, the question of function is of first importance, and it is the material construction which determines this.

Every packaged article at some time is subjected to the risk of mechanical damage, or contact with environments that can cause deterioration. The prime function of any container is to protect the contents, during distribution, from the manufacturer's premises to the consumer's. Since it may take several months to complete the distribution cycle, it is important that the container should withstand this prolonged handling. The physical characteristics of the product, the design of closure, the shelf-life required and the price are all relevant factors which influence the correct construction.

Various materials are employed in the manufacture of composites, and the next section describes those in common use.

Body materials

Paperboard. The primary strength of the composite can is derived from its body construction, which is usually paperboard. Body strength in composite cans is an attribute that has improved over the years and can be varied to meet many application demands.

In the early years of composite can development, it was common for can manufacturers to start with a readily available body stock such as kraft linerboard or tube-grade chip. These boards are adequate for most applications, but new boards with special qualities have also been developed for more demanding end-use requirements.

Chipboard. Most composites are fabricated from this type of paperboard which is used where high strength is not paramount. Chipboard has no special functional properties; it is not a barrier material and will not protect against atmospheric influences, but it does provide mechanical protection. It is suitable for containers which hold non-foodstuffs, such as small engineering parts, scouring powders and insecticides. It is possible to use white-lined chipboard for some foodstuffs where flavour contamination could occur in contact with unlined board. The white lining also gives a clean appearance at low cost, but does not give any protection against moisture. For decorative purposes chipboard can be lined in a variety of colours.

Kraft paper. Probably the most dramatic change has been the use of kraft paper as a protective packaging material. Although this material is light, it still imparts sufficient strength to composites with thin walls. Kraft paper readily absorbs moisture from the surrounding atmosphere, but this can be prevented by treating it with compounds and various laminations which increase its protective properties. It is heat-resistant up to 150°C.

Adhesives. The adhesives and coatings used in the manufacture of composite containers have also been improved to provide better heat and water resistance plus increased operating efficiency. Most product applications require precision gluing equipment to control the amount and position of the adhesive on each web. The most commonly used adhesives are listed in Table 11.3.

Lining materials

During recent years new lining materials have been developed as a result of technical research, and it is now possible for a composite to provide a greater degree of product protection than originally thought possible using the standard functional linings described below.

Pure vegetable parchment. Pure vegetable parchment will withstand immersion in boiling water and is especially resistant to oils, greases, mild acids and alkalis. It is free from odour or taste, and has the great advantage of high strength when wet, advantages which have created a good market for its use in composites for liquids and fresh produce.

Wax laminates. Paper/wax/tissue laminates have a low water-vapour transmission rate. They are suitable for products requiring protection from moisture loss or gain, such as biscuits, sweets, and certain pharmaceutical tablets and powders. They provide limited protection from the flavour and odour aspect and very little resistance to grease. Their use has fallen considerably in recent years.

Table 11.3   Adhesives used in composite manufacture.

Poly(vinyl alcohol–acetate) blends	Good initial tack, good runnability, moderate to good water resistance
Dextrin	Fast tack, poor water resistance
Animal glue	Good tack, vulnerable to insect attack
Polyethylene	Requires heat, good dry bond, moderate water resistance
Hot melts	Require heat, difficult to handle, good water resistance, good water bond

Aluminium foil. Aluminium foil is the best and most economic liner for composite containers in terms of cost and performance. Two standard gauges are usually used for composites but it is available in other gauges. The foil is often laminated to paper, which reinforces it, and in some instances other 'carrier' materials impart heat sealability. Coatings and print can also be applied to foil and, where necessary, it can be suitably lacquered to prevent corrosion. Its main advantages are non-toxicity, opacity, impermeability to water vapour, freedom from flavour and odour contamination, and resistance to oil and grease. In most instances, some form of laminated aluminium foil is used in composite manufacture, and plain unsupported foil is not used for the body, only in connection with the top closure as will be seen later.

Glassine. Glassine is a supercalendered smooth dense, highly beaten paper which was developed for purely functional purposes for a wide variety of greasy and oily products. It is made primarily from chemical wood pulps and, when lacquered or laminated, is highly resistant to transmission of water vapour. The fibrous nature of ordinary paper is almost removed by the processing, leaving a non-porous sheet which provides, in addition to grease resistance, a degree of odour resistance and protection from bacterial infiltration. Glassine, made in various colours, is consequently used as a protective liner for cereals, tea, coffee, bread, grated cheese, butter, soap, tobacco products, chemicals and oily metal parts.

Glassine/foil/glassine. This is a triple lamination of pure aluminium foil sandwiched between two layers of glassine. Both materials used separately have good barrier characteristics so naturally in laminated form they provide a material of exceptional quality. Highly volatile solvent-based products such as adhesives and sealants, can be packed in composites with this liner.

Silicone release-coated paper. This makes an internal lining material which is particularly useful for hot-filled sticky bituminous and resinous products and hot-melt adhesives. Any product of an adherent nature can be packed in these papers and, when required, freely extracted without any attachment to the body wall. If necessary, the composite can be torn down the spiral butt joint in order to strip the body away from the product. In certain instances the contents shrink on cooling, and can therefore be tipped out as a solid block.

Polyethylene LDPE coated paper. Polyethylene extrusion coated paper is used in various gauges, but mainly 150 gauge polyethylene (low-density) on bleached kraft. It is suitable for products with high moisture content. Containers with a polyethylene/paper liner ply can be made by utilizing a patented process of extrusion which seals the join or overlap whilst the tube is being wound. Such containers are adequate for packing products like dairy cream, frozen liquid eggs, and various water-based emulsions.

Polyethylene coated aluminium foil. Polyethylene coated aluminium foil can be used in conjunction with the extrusion sealing process. Hot-melt adhesive lap sealing can also produce composite containers which are liquid-tight and which have extremely low moisture-vapour transmission rates. For very hygroscopic products, this is an ideal barrier lamination, and it is particularly practical for such products as dried milk and powdered drinking chocolate.

Vapour corrosion inhibitor (VCI) paper. This paper can be incorporated as an internal ply in composite containers used for steel and iron products, such as roller bearings, small precision instruments and gear shafts, where there is a risk of rusting.

Other barriers

Other materials which provide some form of barrier include paraffin wax, with which the containers can be impregnated, and bitumen adhesives applied between the layers of board. Paraffin wax provides a high degree of water-vapour resistance and increases rigidity, while bitumen adhesive provides both liquid water and water-vapour resistance.

In some containers, instead of applying a barrier lining as an extra ply of material, it can be sprayed onto the inner wall. During the final stages of making the body, as the end component is seamed on, a fine pressurized jet of quick-drying 'flushing compound' is sprayed inside. As the container rotates, centrifugal force ensures that the compound is evenly distributed on the wall. By this method, a coating is also applied to the component seam, giving extra protection against seepage where the metal is attached to the board.

Virtually any mechanical or functional barrier material which can be laminated to or coated on to paper or foil can be used in the manufacture of composite containers. This potential has recently been developed to extend the range of uses of the package. LDPE, HDPE, PP, non-oriented, PVdC coated and oriented, ionomer and PVdC coatings and polyester films have all been used for specific purposes, particularly for liquid products.

Types of closure

By definition, a composite container is any fabricated pack with a board body and possibly one or more tinplate or plastic components. Let us now examine the functions of the base and dispensing closures.

We have a tube open at both ends and as with metal cans one end is closed by the container maker and the other by the packer after filling.

Composites almost always require some type of reclosure and/or a dispensing device. In general, closures which are applied to metal cans are commonly used for composites too, but there are certain additional methods which are peculiar to composites.

Every closure used has some definable feature which, if expressed correctly by the customer, can help to convey his exact requirements in terms which the supplier understands.

The three basic types of closures are similar to those on metal cans, but apart from these, there are others which are applicable only to composites.

Generally composites are supplied with the bottom component already in place and with the top closure loose—to be applied after the container has been filled. It is important to differentiate between closures which require no seaming by the customer and those which do. First of all, let us consider those which require no additional seaming operation, where the customer is only required to place on the lid.

Beading of cylindrical boxes. After labelling and the cutting of tubes into individual container bodies, either one or both ends may be beaded. Pressure on the wall end is progressively applied by a rotating heated shaped die that causes the edge to curl inwards and over to form a bead. In making cylindrical boxes, a circular base card may be inserted prior to beading.

Post-beading deformation of the body into, for example, an oval or heart shape is possible. Insertion of an appropriately shaped based card secures the new cross-section shape.

Base cards may be friction fit or secured by adhesive application. Base materials may be paperboard, coated or laminated as the package use dictates.

Seaming of composite container closures. A standard end (Figure 11.9) is, as the component is called, clenched on to the body wall and not interlocked with the body as in double seaming. The flanges of the component are initially flat or slightly curled, with sharp edges. During the seaming operation, these edges are turned inwards and compressed against the board wall, whereby the end is attached to the body (Figure 11.10). The resultant joint or 'seam' is quite firm and will withstand considerable internal and external pressure.

Figure 11.9   Cross-section of an end.

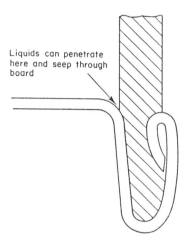

Liquids can penetrate
here and seep through
board

Figure 11.10   A standard end seam.

Under normal conditions an ordinary end-seam is quite adequate for most purposes, but where composites have to withstand severe external pressure, then it is better to use a double-seam end as there is less chance of the end becoming separated from the body. Another feature of the double seam is that it prevents seepage of liquid contents, since the raw (cut) edge of the board is protected within the enclosed seam.

To produce a *double seamed end* it is necessary for the base of the body to be 'flanged' (Figure 11.11a) so that the engaging component can grip the body and fold it over upon itself during the seaming operation. Now, instead of the metal merely pressing against the body, the two are interlocked together in a tight curl (Figure 11.11b).

A distinct advantage of the double seam is that it is much more rigid and difficult to separate from the body.

Dispensing closures. In this category, there are three different types of closure.

The *slip (on) lid* is a rather loose type of closure, the rim of the lid fitting over the outside of the body. Since the closure is not air-tight, where there is a possibility of deterioration due to ingress of water vapour, it is advisable to seal the container after filling by securing the lid with tape; but this detracts from the appearance. An alternative method of firmly securing the lid is to over-label a plain composite after filling with the lid in position. As the rim or 'skirt' of a slip lid is usually about 13 mm deep, the label encircles this part of the lid and fixes it securely. This operation is only achieved successfully with a raw-edged lid. It is appreciably more difficult if a curled-edge slip lid is

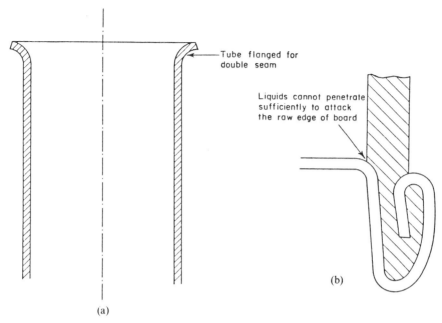

Tube flanged for
double seam

Liquids cannot penetrate
sufficiently to attack
the raw edge of board

(b)

(a)

Figure 11.11   Double seaming.

used. The advantage of a curled-edge lid is that it enables easier removal
because of improved grip on the lip of the curl.

A refinement of the ordinary slip lid is the 'captive' slip lid which is
peculiar to composites. It is an ordinary curled-edge slip lid which is
reshaped by the customer after filling, whereby the component is clenched
into the body wall during the closing operation. Since the lid is beaded on to
the body, rendering a tight fit, this obviates the need for taping, and after
first removal the lid can be replaced by 'cicking' back over the body head
(Figures 11.12a, b and c).

A *plug (in) lid* (composites only), as the name implies, is inserted into the
aperture at one end of the body, which completely closes the container. The
flush engagement of the sides of the lid with the internal container wall
produces a tight-fitting closure which can, however, be opened with
moderate finger leverage. To accomplish this, plug closures have curled
flanges, which also prevent them being pushed wholly into the container. If
an exact cubic capacity is required, eliminating ullage, it is essential that the
countersunk depth of the lid is taken into consideration, otherwise the
overall effective depth will be insufficient (Figure 11.13).

Both plug (in) and slip (on) lids have certain characteristics which make it
very difficult to decide which type to adopt. They are both fitted by assembly
on to the composite after filling, and are not sufficiently air-tight to offer any
functional protection. There is very little advantage in using one type as

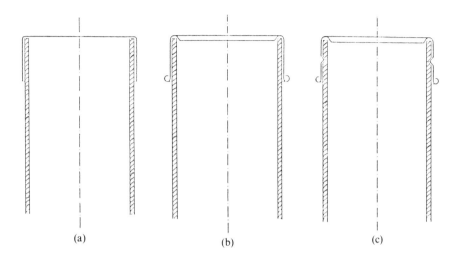

Figure 11.12   Slip (on) lids.

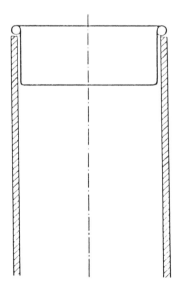

Figure 11.13   Plug (in) lid.

opposed to the other, and choice can really be decided upon only after considering the conditions the composite has to withstand. Under normal conditions both function equally well, but probably the plug lid is slightly more secure, as it is held in position by the body wall. As a general guide, slip lids are usually cheaper than plug lids, although this may not always be true, depending upon the methods of manufacture.

Figure 11.14   Ring and cap assembly.

The *lever lid* (as for tins) is always used in conjunction with a ring which is seamed on to the top end of the container (Figure 11.14). The composite can be supplied with the bottom-end component and the top-end ring seamed on and, after filling, the customer merely has to insert the lid into the ring. Where this type of closure is used on food containers, particularly those which are dispensing packs, it is advisable to incorporate a safety rim on the ring. Bulk catering packs, for example, where the contents may be dispensed by hand, invariably have this type of 'safety' ring to prevent injury when extracting one's hand from within the ring aperture (Figure 11.15).

One optional feature of the lever lid and ring closure is a diaphragm which may be incorporated to act as an extra seal. Here, the container is supplied with the lever lid, ring and diaphragm assembly seamed on, and the bottom end open. The customer fills through the bottom, and then seams the bottom end on. In order to remove the contents, it is necessary not only to lever off the lid, but then to puncture the diaphragm. The diaphragm, which may be paper, parchment, or aluminium foil, operates as a functional barrier, the tamper-evident feature being a secondary characteristic.

These closures have an inherent disadvantage, however, which can be an inconvenience at times. Everyone will know how difficult it can be to lever open a paint can unless a suitable implement is available. However, food-can lever closures do not require the same degree of pressure to release the lids. A recent development has now considerably reduced this problem, and the new modified-profile 'lever lid' closure (Figure 11.16) can be easily opened by slight twist leverage with a coin under the bead of the lid. Equally, it can be re-closed simply by exerting hand pressure on the lid to snap it over the bead contour of the ring to give an air-tight seal.

Screw thread engagement of the lid. This type of closure is no longer extensively used on composite containers, as its importance has gradually diminished with the introduction of plastic dispenser devices. There is one design of screw thread closure which is used, however, and this is the

edge folded to
give 'safety ring'

Figure 11.15   Safety ring and cap assembly.

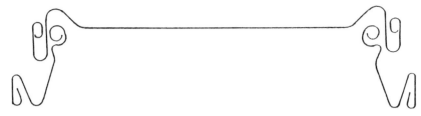

Figure 11.16 New profile ring and cap assembly.

threaded sprinkler neck and screw cap with assembled wad which is used on talcum powder composites. This type of closure is similar to the ring used in the lever lid and ring assembly, but with a much smaller aperture, through which the customer fills the container. As the composite already has both ends seamed on prior to filling, all the packer has to do is to insert the neck and cap unit into the ring, and the pack is complete.

No more need be said on screw thread components in this context as the matter is discussed in the section on metal cans.

One significant characteristic of the closures mentioned so far is that the packer is not involved in any seaming operation once the container is filled, with the exception of the lever lid assembly, which after all is an optional feature depending upon the packer's circumstances.

The *semi-perforated top* is an end component which has been partially perforated in a number of positions to facilitate puncturing by the user. No metal is removed initially; it is only scored to prevent leakage when filled prior to use. This top is not commonly used and is gradually being replaced by the fully perforated top.

The *fully-perforated top* is an end component which is completely pierced in a number of positions, i.e. metal is totally removed. The holes are then sealed by a removable board disc or piece of adhesive tape. Sometimes an additional dispensing feature is inserted into the fully-perforated top, such as the plastic sprinkler top used for talcum powder.

The *string opening container* is a development peculiar to convolute containers only, where the opening device is inserted automatically during the winding operation. (A circumferential thread of string cannot be introduced into a spirally manufactured body because of the angle at which the board is applied on to the mandrel.) Once the container has been filled and finally sealed with a standard end by the customer, it can be opened by pulling the circumferential thread which perforates the label, whereby the two sections of the container can be separated. Since this style cannot possibly be opened without affecting the external appearance, it possesses an inherent tamper-evident feature.

Membrane closures. A similar closure to a lever ring component with a foil or paper membrane and plastic reclosable snap-on lid has been developed in recent years. When the membrane has been pierced and removed, the pack can be closed by replacing the lid over the protruding ring seam. It is only styling and easier opening facility that differentiates this from a conventional diaphragmed ring and lid closure.

Some membranes are heat-sealed to a polyethylene coated liner, thereby eliminating the need for a metal rim under which a seal is attached to the body. When the membrane has been removed, a plastic plug-in lid closure reseals the full aperture. This form of closure has a limited application, as it can be used only when a heat-sealable inner lining is incorporated in the body construction.

Generally these closures are fitted to the container by the supplier, and the customer is required to assemble the base component after filling, thus giving a tamper-evident unit.

Usage

There is a wide variety of components available, but we are limited to some extent by the functional performance of the individual types. Actually the product determines the most suitable closure. Composites with two seamed-on components are more practical for small engineering spares and industrial components. Both types of perforated closure are suitable for scouring powders and insecticides. Plug-lid closures are used for sports goods, toys and novelties. Slip-lid closures are used for some confectionery, and ring-and-cap closures for custard and milk powders.

In addition to the many tinplate closures, there is also a wide range of plastic dispensing and pouring devices which can either be inserted into or glued on to the metal components.

Other constructions

Sleeve-type composites. Included in this section are insecticide puffer packs, balloon inflators, ammunition cases, pharmaceutical calculators, and telescope sleeves. This type of container, as the name implies, consists of two, three or more separate cylindrical tubes which are assembled together to form one complete sleeved unit. The inner portions have an overall external diameter marginally less than the internal bore of the outer tubes, so that a compact unit is formed when all the integral parts are united.

A sleeve-type container is both practical and necessary, yet serves two completely different purposes, protective and operative.

Generally, composites have an average body wall of three plies of material, which is quite adequate to withstand normal handling and fulfil most requirements, but as always, there are exceptions to the rule. Certain containers must endure considerable stress, to provide suitable protection to

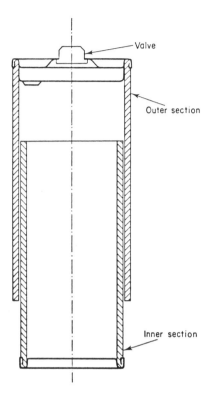

Figure 11.17   Inflator pack: sleeve-type composite.

their contents, and in some instances sleeved containers are advantageous. Where durability of the pack is critical, a sleeved container (which is reinforced by the interlacing sections forming a double or treble-thickness wall) will suffice.

Exceptionally strong containers can be fabricated in this basic style. By assembly of component parts, it is possible to make containers with walls over 35 mm thick. Such constructions, fitted with special board collars, metal discs and injection-moulded retaining rings are used by the Services for ammunition cases.

From an operative viewpoint, it is essential to have sleeve containers for puffer, inflator and many other kinds of novelty packs. An inflator pack is merely one open-ended container inverted into another which is fitted with a valve assembly; it operates in the same way as an ordinary pump (Figure 11.17).

Three part composites (shouldered boxes) also exist. The top and bottom sections of the same diameter are joined by an inner tube extending above the top of the bottom tube.

Composite or metal container?—the choice

A composite is a versatile container, strong and rigid, yet lightweight and easily disposable. It is relatively inexpensive, attractive and can be used on orthodox can-filling equipment. This type of pack has many advantages to offer, but it cannot be categorically distinguished from an all-tinplate container. In some instances the two are complementary, yet for other products one would be incompatible with the other.

Generally speaking, the performance of metal cans against composites cannot be determined, as there are so many incomparable aspects. Each has its attributes and, in the final analysis, the decision will be based upon which of these qualities has the greater influence on what is considered the most important factor in marketing the product.

Basic cost factors

The cost of a composite container is determined, to a large extent, by the requirements of the commodity packed. A thorough investigation is conducted on every new project to determine the correct liner material, adhesive, tinplate lacquer and coating, required to fulfil these requirements. After all, performance is of first importance, size and shape being secondary. The product governs the cost of all the material used; the size influences the production speed; and sales appeal imposes the cost factor of decoration. To arrive at a satisfactory situation, we must relate performance, size, style and shape with cost.

Materials and construction. At present, of the two different types of composite, it is cheaper to produce spiral than convolute or straight-wound composites on a comparable basis. Convolute-winding machinery has much greater bearing on size limitation and cost than spiral-winding equipment, since it can only accommodate a narrow web width. Although all non-cylindrical composites are more costly to produce than cylindrical spiral ones, this premium may be counterbalanced by the savings in transportation costs and storage space.

With the many varied operations and special processes involved in production, it is very difficult to lay down any definite principles upon which we could base the cost structure. Since the material used represents a major part of the total, its cost serves adequately as a general guide.

The board forming the body is applied from a standard-width reel, which itself is slit from a larger reel. As the board for any particular composite is cut to a uniform width, there is no shred or trim allowance necessary, and the total board allocated forms the bodies without much wastage.

The size and shape of a composite has an indirect influence on cost in connection with the tinplate closure. On most cylindrical composites there are one or two closure components. Each of these circular components is formed from square or rectangular sheets of tinplate, and consequently there is a percentage of waste material which becomes increasingly more expensive as the diameter becomes greater. In theory, square or rectangular components are more economic propositions, but this advantage is offset by the present technical limitations of non-cylindrical composites.

The most economic situation to aim for is one where the cost of the pack does not outweigh its usefulness, and in this respect it is essential to arrive at the most suitable balance between material cost and performance. As tinplate is more expensive than board, it is better to use more board and less metal, provided circumstances will allow it. For example, the board cost in a tall narrow composite could be almost the same as or slightly more than the tinplate cost, whereas in a squat composite the board cost could represent as little as a quarter of the tinplate cost. As a general rule, it is cheaper to have a tall composite, where there is more board used than tinplate, rather than have a squat composite where the tinplate cost is disproportionate in relation to the board cost. By careful correlation it is possible to assess an equilibrium cost, whereby the two factors determine the optimum size of a composite relative to material cost.

Measurement of composite containers

Once a customer has decided upon a particular item, it is essential that the necessary information is conveyed to the supplier in precise terms. To express the dimensions in standard units of measurement is not sufficient, because the composite maker's interpretation of this information may be completely different from that which was intended. In general, composites are defined by the base and height dimensions only, with the base details given first (see Figures 11.18a, b and c).

The basic diameter is taken as an internal measurement, excluding the wall and seam thicknesses. This exclusion is merely for the supplier's benefit, as to include various wall thicknesses would make measurement extremely tedious, since sizes would have to be defined to the nearest thousandth of an inch. Most base dimensions are expressed in fractions of an inch as an internal measurement, but sometimes dimensions are described in the trade by the standard digit system. If the diameter of a composite is expressed as a series of numbers, then it can be assumed that this is a digit system measurement, and the diameter is taken as an overall measurement across the seams. Under this system the last two digits denote sixteenths of an inch, and the preceding digits denote the whole inches; e.g. 211 means $2\frac{11}{16}$ in, 502 means $5\frac{1}{8}$ in, 700 means 7 in.

The body height dimension is taken over the top and bottom seams, and described as a deep external measurement. Component protrusions beyond

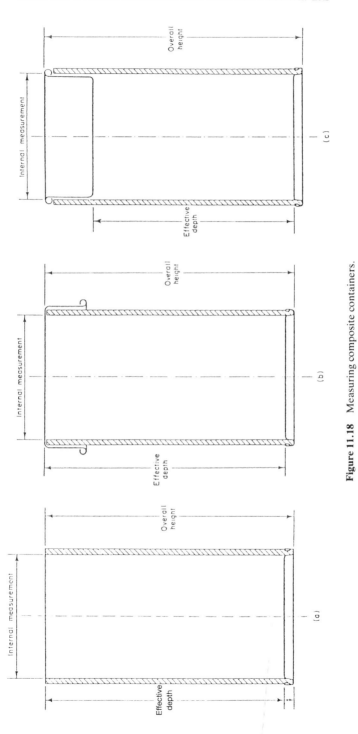

Figure 11.18 Measuring composite containers.

these seams are not taken into account, and this is a very important point to consider when calculating packing-case sizes, storage space and distribution costs. For example, a talcum powder composite may have an external depth of 4½ in, but an overall depth of 5⅜ in from the base joint to the top of the plastics sprinkler cap.

To express the exact depth of a composite container can be confusing, especially where the cubic capacity is critical, so the following principles should be borne in mind. The diameter or base dimensions are always taken as internal measurements, but the depth can be expressed in two ways according to the style of container. A composite container, i.e. a container with a metal end seamed on, has an external depth measured over the component seams, whereas a board tube (core) without any components is described by its overall body length.

Another peculiar characteristic of composite measurement to observe is the countersink depth of certain components. Slip (on) lid closures do not affect the internal depth of a container, but the plug (in) lid does reduce the effective internal depth by as much as 15 mm in some instances. Other components do not make such internal depth alterations, but some standard ends are panelled, and this does affect the packing of certain products because of the uneven surface.

12
Metal cans

Almost every food product has, at some time or another, been packed in a metal can. It is accepted as commonplace that tins of paint are found in the shops and cans of oil at the petrol stations. Worldwide, the metal can is used to convey and distribute the necessities and luxuries of life.

Not many years ago there were some who forecast the demise of the tinplate container, claiming it would be displaced by newer plastic materials or lightweight glass, and that new ways of processing food would no longer require the can in the form we know it.

The metal food can continues to be made in ever-increasing quantities, and the new materials and processes have found their places alongside it for those products for which they are best suited.

Some metal containers have, however, passed out of common use. Cigarette tins are now a rarity in the United Kingdom, and instant coffee is now almost entirely in glass jars for retail distribution.

These paragraphs have used the words *tin* and *can* almost indiscriminately to refer to various metal containers. The words have come to mean much the same, and the word *can* is more widely used than in the past. The more subtle differences in meaning may be ignored, and the words are used almost synonymously in the rest of this chapter. Almost but not quite. A hinged-lid cigar tin is surely never a *can* and an all-aluminium beer can is difficult to accept as a *tin*.

Tinplate is still the most common raw material for tins and cans. This is mild steel sheet with a very thin layer of tin on each of its surfaces. Blackplate is not often used, but it is most easily described as tinplate without the tin, i.e. mild steel sheet, available in the same range of thicknesses as tinplate. Tin-free steel is like tinplate, but the tin has been replaced by other corrosion-resistant metals such as chromium. Aluminium is being used in increasing quantities.

Tinplate and aluminium allow designers scope to produce intriguing effects through the use of surface finishes which can be produced with lacquers and varnishes. An example is the well-known Golden Syrup can where the tinplate, lacquered with a transparent gold lacquer, shines through to create a golden glow.

Some cans for food can be produced with an internal liner, often in white, to enhance the appearance of the food.

Combining the metal can with other materials such as plastic or foils can create an entirely new image. Cans for powdered milk, for example, have an aluminium foil tagger sealed on the closure ring conferring pilfer resistance and easy opening as well as an impression of hygiene. Plastic lids fitted over normal end seams and used for re-closure after opening are both decorative as well as functional.

Historical background

Until about 1970 the soldered three-piece can was standard. Since then much development has taken place. The construction of the three-piece can, however, has not changed fundamentally in over 150 years. A flat rectangular sheet of tinplate was formed into a cylinder and the resulting join was soldered to form a side seam and circular discs of tinplate were then mechanically secured to flanges made at the ends of the cylinder by a rolled seam, known as a double seam. One end was fitted by the can maker and the other by the packer after filling.

Advances have been made in techniques but not in principle. Since about 1985, solder has been eliminated almost everywhere by electrical resistance welding and laser welding is a prospect for the future. Automation has increased the speed of manufacture and better, more secure mechanical joints have been developed. In addition to the improved three-piece can, two-piece cans have been developed and today there are three major types of open top food and beverage cans in use;

1. The three-piece welded can (Figure 12.1)
2. The two-piece drawn and redrawn (DRD) can (Figure 12.2)
3. The two-piece drawn and wall-ironed (DWI) can (Figure 12.3)

Easy opening ends, first for beverages and then with full aperture for foods, have been developed and are now made out of steel as well as aluminium where the first developments took place. The steel easy-open ends are becoming more popular now that the initial problems have been largely overcome.

Any survey of these more recent technical developments shows that the driving forces behind them are:

1. A reduction in cost by using less material (e.g. thinner plate)
2. Lower processing costs by the use of automated and higher speed production lines
3. The provision of greater convenience for the user (easy opening etc.)
4. Better presentation (e.g. new shapes, improved printing and graphics and the use of metal/plastic in combination).

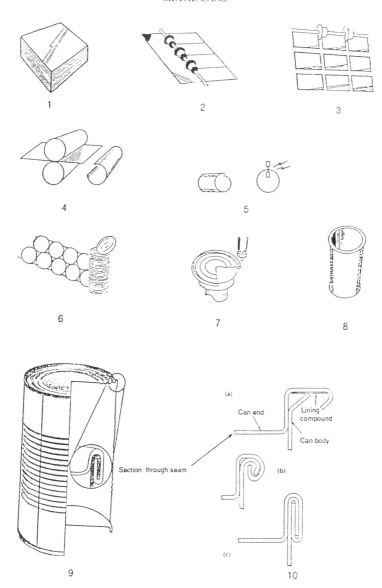

Figure 12.1 The three-piece electrically welded food can. (1) Starts from a stack of tinplate sheets which are coated with a lacquer, dried and stoved in ovens for 20 min. Different lacquers are used for different products, (2 & 3) the lacquered sheets are first cut into strips and then the strips are cut to the correct blank size for the can bodies, (4) the body blanks are now rolled into cylinders and (5) the two edges are welded together electrically. The area adjacent to the join is again coated with lacquer and oven dried, (6) other sheets have meanwhile been cut into circular blanks for the ends of the can, (7) these can end blanks are curled at the rims and a sealing compound flowed into the curl, (8) a lip (or flange) is now formed at both ends of the welded body cylinder and (9) the end is now seamed to the body to give a can ready for filling and closing by the packer who seams another end on to the body after filling, (10) formation of a double seam, (a) end placed on body, (b) seam part-formed, (c) finished seam.

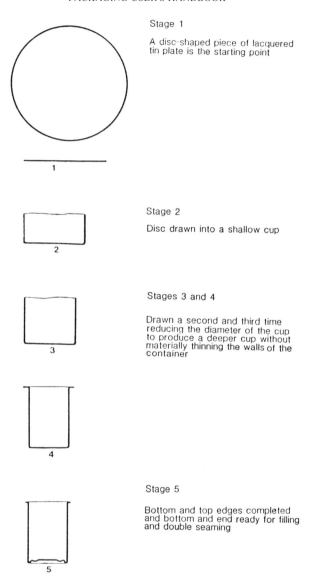

Stage 1

A disc-shaped piece of lacquered
tin plate is the starting point

Stage 2

Disc drawn into a shallow cup

Stages 3 and 4

Drawn a second and third time
reducing the diameter of the cup
to produce a deeper cup without
materially thinning the walls of the
container

Stage 5

Bottom and top edges completed
and bottom and end ready for filling
and double seaming

Figure 12.2 The drawn and redrawn (DRD) can.

Making metal boxes is all about producing good joints in sheet metal and
there are three basic types of body construction: the built-up body, the
hooked or seamed corner body and the seamless or drawn body
construction. Let us discuss these in more detail.

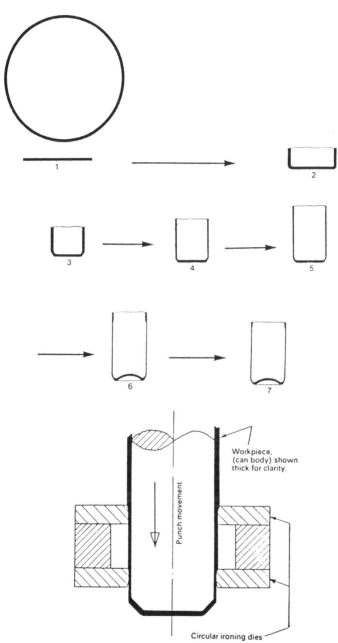

Figure 12.3 The two-piece drawn and wall-ironed (DWI) can. The cup-shaped component is first drawn from a round blank. The cup is forced through the ironing rings by the punch. Only two rings are shown, but three or more may be used in practice. Each ring reduces the wall thickness by between 20% and 35%. Stage 1: circular metal blank drawn into a shallow cup shape; stages 2, 3 and 4: the cup is wall ironed into the can shape; stages 5, 6 and 7: top trimmed and bottom finished, can cleaned and printed and walls finished; now ready for filling and closing.

The built-up construction

Built-up body construction is the form of construction in which a metal container is produced by using more than one piece of material and joining the several pieces together with suitable seams. This distinguishes the method from that which produces a solid-drawn body having no seams, and clearly having been produced from a single piece of material.

The open-top food can is the most familiar use of the built-up form of construction. It is called *open top* because it is supplied to the canner with the top of the can open to receive the product. Top end components are supplied separately to be attached by the canner.

To manufacture this style of can, a rectangular blank of tinplate is mechanically wrapped around a mandrel to form a cylinder. Where the two ends of the blank meet, an interlocked seam is normally made, and this is soldered to make it leak proof, before the bottom end is secured by a double seam. Figure 12.4 shows a section through an interlocked side seam (a) when the side seam 'hooks' are first brought together around the mandrel and (b) after they are hammered flat to secure the seam just prior to soldering.

Figure 12.1 illustrates the stages in the formation of the so-called *double seam* which secures the bottom end to the body. An identical method is used by the canner to secure the top end to the can after filling. Reference to Figure 12.4 shows that an interlocked side seam has four thicknesses of material, and these would need to be bent over into no less than eight thicknesses before the can body is seamed to the bottom (or top) end. This would be quite unacceptable, not only because of the unsightly bump it would leave in the double seam at its junction with the side seam, but because the massive local distortion in the end seam at that point would be a source of leakage.

To avoid this, the body blank is prepared for seaming by having specially shaped notches cut in each of its corners so that, instead of the interlocked seam being formed over the whole length of the can body, the extreme ends are simply overlapped.

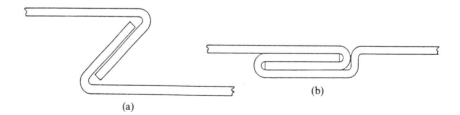

(a)   (b)

Figure 12.4   Interlocked side seam (a) before hammering and (b) the finished seam.

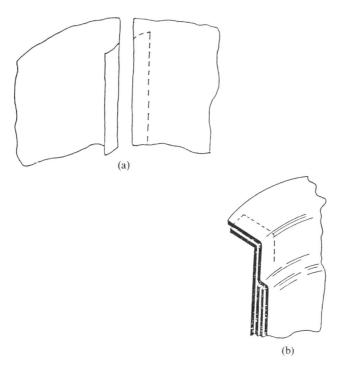

Figure 12.5 Transition from four thicknesses to two at the end of the interlocked side seam. (a) Top of can body notched away to make a lap. (b) Longitudinal section through the interlocked seam showing the lap joint at the top.

Figure 12.5a shows what is meant by notches, and Figure 12.5b shows a section along the length of an interlocked side seam, near one end where the flange has been formed to prepare for double-seaming at an end. The transition from four to two thicknesses of material can be seen.

Where the two thicknesses of material simply overlap, they are secured by soldering, this being done at the same time as solder is applied to the interlocked part of the seam. The area is called the *lap section* of the seam.

Despite this way of reducing the thickness of the side seam at each end, the intersection between the side seam and the end double seam remains a potential source of leakage. To eliminate this possibility, a resilient compound is applied to the whole peripheral area of the can end before seaming. This lining compound fills any small interstices which might otherwise occur when the end double seam is made. Figure 12.6 shows where this compound is applied to a can end. Strictly speaking, the lining compound is hardly necessary at any position except where the side seam intersects, but it is easier in high-speed production to apply the compound to the whole periphery of the end.

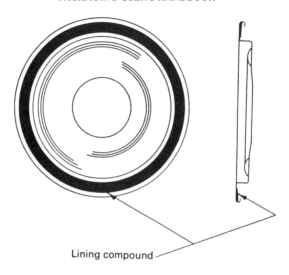

Lining compound

Figure 12.6　Placement of lining compound in can end.

Treatment of can seams

End double seams

The introduction of end lining compound to assist in sealing double seams is referred to above. The constituents of can-end lining compounds can be varied to suit various circumstances and methods of application, but are mostly natural or synthetic rubbers which are either dispersed in water or dissolved in a suitable solvent. The compound employed is squirted through a small nozzle into the annular area of the end which will subsequently become part of the double seam (Figure 12.6). The end is rotated beneath the nozzle so as to ensure an even spread of compound, which is quite fluid when applied. If water-based compounds are used, the ends are passed through an oven which cures the compound by vulcanizing it after the water has been driven off. The process is irreversible, i.e. the compound cannot be rendered fluid again by warming or by the addition of common solvents once it has passed through the curing operation.

Solvent-based compounds can be allowed to cure in the air. Can double seams may also be soldered, but this is a practice which is no longer common. It is to be seen on some containers which serve unusual purposes in engineering, such as oil or brake fluid tanks or filter bodies, some of which are required in sufficient quantities to be able to enjoy the economies of can manufacturing techniques. Some large tins, of 5 litres capacity and above, have their end seams soldered, particularly if they may be called upon to hold penetrating or volatile fluids for long periods in an unpredictable or hostile environment.

Interlocked side seam

The most common means of sealing against leakage is to use solder. Means have been developed over many years to undertake side-seam soldering reliably and at very high speed on automatic equipment. The interlocked seam can be made so that the surfaces of the tinplate in the folds are not quite in contact with each other. Hot solder is therefore drawn into the small space by capillary action and, once there, is solidified by being cooled under air blasts or by contact with the atmosphere.

The high costs of tin and lead which go to make solder have stimulated a search for other ways of sealing side seams. Soldering also calls for the use of energy to pre-heat the cans and melt the solder.

An early alternative was rubber solution. If applied to the body blank before the interlocked side seam is hammered flat, it can be effective in filling the joint. Many alternatives and more elaborate materials have been tried, all being generally referred to as *solutions*.

Other materials, which can also be applied to the edges of the body blank just before (or just after) the hooks are formed, are designed to adhere to the metal much more strongly than does solution, and to set harder than solution. Some of these materials, commonly referred to as *cements*, are thermoplastic and can be applied through nozzles as jets of hot liquid. Once placed on the metal, they are immediately trapped and hammered in the interlocked side seam.

Although there is a variety of solutions and cements available to the can maker, the distinction between the two material categories should be stated. Solutions fill the interstices of the interlocked side seam, adhering only strongly enough to stay in place. Cements not only fill the gaps which might cause leakage, they add something to the mechanical strength of the seam by the strength of their bond to the metal surface.

Other types of seams and joints

End seams

The previous section has referred at some length to the double seam and Figure 12.7 shows another way of securing ends to bodies on built-up tins, instead of the double seam. The double seam is by far the most common way of securing ends to bodies, not only because it is fundamentally easier than other techniques, but because high-speed machinery is available for its production as a result of the exacting technical demands of the food canning industry.

The single seam has its uses and is employed mainly where mechanical strength and liquid-tightness are not vital. Its simpler form makes it attractive for use in tins of unusual cross-section, and it is often to be found on talcum powder tins.

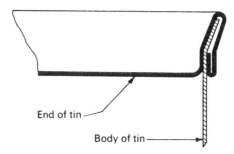

Figure 12.7 The single seam, another way of securing ends to tins.

Figure 12.8 shows two other types of side seam to be compared with the conventional interlocked seam already discussed at some length. Figure 12.8a shows a variety of the interlocked seam, and this is known as a 'Mennen' seam or powder seam. The usual interlocked seam is strong in tension, i.e. it is well able to resist the effects of internal pressure within the can, or any other mechanical forces tending to expand the can from within. It is not as strong in resisting those forces which might tend to crush a tin from the outside. That is where the Mennen seam becomes useful. It is often used on talcum powder tins which have shoulder components tightly fitting on the outside which tend to crush the tin.

The lap seam shown in Figure 12.8b looks deceptively simple, but becomes complicated by the ways in which the two overlapping edges are secured to each other. Soldering is probably the oldest technique, but solder is not a strong material and is not well able to withstand high stresses continuously applied for long periods, such as occur when a can is pressurized or under vacuum. The can maker has, therefore, resorted to other materials; in particular, modern polymeric materials which are able to bond metal to metal and which are often thermoplastic. This makes them capable of hot application to the edges of the body blank by jet or roller, but they solidify by rapid cooling as soon as the overlapped edges are pressed together. This method of side seam formation can be employed on metals other than tinplate, such as tin-free steel or aluminium, both of which defy soldering by traditional techniques.

Figure 12.8 Other types of side seam. (a) Mennen side seam; (b) lap side seam.

Lap joints also lend themselves to welding. Resistance welding is used and, although such joints may appear to be continuously welded throughout their length, X-ray examination shows that they are formed by a very large number of overlapping stitch-like welds which give the outward appearance and practical effects of a single continuous weld. Such joints are mechanically strong in tension and compression.

The question of material economy is of ever-increasing importance, and an examination of the various diagrams will show that the sheet metal required to produce a tin of a given peripheral length will vary with the type of seam chosen. The hooks of an interlocked or Mennen side seam are between 3 mm and 2.5 mm long, and the overlap in an adhesively bonded or welded-lap side seam is 3–4 mm. In order of material economy, the seams fall in the following order:

1. Lap seam—most economical
2. Interlock seam
3. Mennen seam—least economical

Cans without side seams

Much of the space in earlier sections has been taken up in discussing the various ways in which seams may be produced and prevented from leaking. Clearly there would be some advantage if seams could be partly or wholly dispensed with.

Because of the advanced manufacturing techniques which have been developed for high-speed can making, labour costs are far less significant than material costs. Any way of saving material is worth exploring. A can of a particular height and diameter will require a minimum area of sheet metal for its construction. The area can be minimized by reducing the number and complexity of the joints and, once that is achieved, attention falls upon the thickness.

Many tins and cans could have certain of their parts reduced in thickness without undue risk to their main purpose of conveying and preserving their contents. This is particularly true of the bodies of cylindrical cans. Two difficulties have stood in the way of this. Firstly, it is not so easy to produce very thin tinplate with the same assurances of quality as can be given for thicker sheet. Secondly, such thin sheet as we are now contemplating would have the 'feel' of thick paper and would behave in unacceptable ways on high-speed machinery, mainly because of aerodynamic effects which are seldom noticeable on conventional thick material. It is not uncommon to find can-making processes which require tinplate sheets or flat blanks to be pushed through the air at speeds exceeding 350 miles per hour (560 kph).

The drawn and wall-ironed can

These and other reasons led to the development of the drawn and wall-ironed can, often abbreviated to DWI. Although the tooling and machinery to make such cans calls for high precision and advanced technology, the process is fairly simple to understand. It is illustrated in Figure 12.3. A shallow tinplate cup is first produced by the long established technique of drawing from a metal blank. This cup is then forced through several rings or dies, each slightly smaller than the previous one. The punch which does this supports the inside of the cup to stop it collapsing, and consequently the cylindrical side wall is progressively reduced in thickness. After passing through the last dies, the can is removed from the punch and is automatically carried away to be cleaned and printed. Because the material is reduced in thickness, it is increased in area, and the can is therefore made taller as it passes through the die rings. A trimming operation follows the wall-ironing, and this ensures that every can is the same height, with a well-finished open top which will eventually receive the top end component.

Although the process takes a little time to describe, it occurs so rapidly as to appear instantaneous. The high speed and the considerable reduction in wall thickness achieved, demand that copious amounts of lubricant be used in the press operations, and these must be removed by thorough cleaning. Tin on the surface of tinplate acts as a lubricant itself and assists the operation but, by using tools which have been properly prepared, the process will also handle aluminium.

DWI cans have very thin body walls where the material has been ironed, but the bases remain substantially the same thickness as the stock material fed to the process. Whilst much is gained in the way of material economy by using this process, the can maker is faced with the difficulty of printing on the cylindrical surface of the can after manufacture, whereas conventional can making allows printing to be carried out on flat sheets, which are cut and formed into cans after printing.

Solid-drawn containers

Solid-drawn tins are still used to some extent for packing tobacco under vacuum, and in other shapes for packing fish, such as sardines. They have had many other uses, and in the last year or two a variation of the solid-drawn tin, the drawn redrawn can (DRD) (see Figure 12.3) has become of major importance in the packaging of food and drink.

The lid of a round slip-lid tin is solid-drawn from a single piece of material, without the use of seams or joints. The first stage in the manufacture of a DWI can (Figure 12.3) requires that a plain round cup should be drawn from a circular disc or blank of raw material.

In general engineering, the process of drawing cup-shaped components from flat material is common, and consequently the fundamentals of the process are well described in engineering textbooks. For a detailed exposition of the subject or knowledge of the mathematics involved, these should be consulted. For our present purpose it is sufficient to examine those matters which bear upon the solid-drawn component as a tin or part of a tin.

The round vacuum tin

Figure 12.9 illustrates the main features of a round vacuum tin. With any kind of slip-lid tin, a perfect seal between lid and body is almost impossible to achieve, because

1. A rolled body is not smoothly rounded and thus prevents a good seal being made on the diameter.
2. The presence of a vertical side-seam creates a major irregularity preventing a good seal.

Both of these shortcomings are avoided in a solid-drawn tin. Even so, when the top edge of the body is rolled over (or *curled* to use the tin boxmakers' term), minute irregularities can still exist which would stop the tin holding a vacuum. To overcome this, a small amount of resilient lining compound is introduced into the lid of the tin to form a gasket near the periphery, into which the edge of the body can become embedded. After this precaution, the tin becomes a highly satisfactory vacuum pack which can be opened with a large coin.

Twisting a coin in the groove provided merely lifts the lid from the body locally (near the coin), and allows air to enter until the internal pressure equals that of the atmosphere, and the lid can then be lifted off. This point is worth making because it is easy to overlook the magnitude of the force

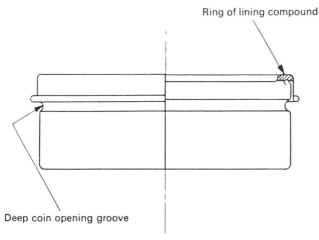

Ring of lining compound

Deep coin opening groove

Figure 12.9   Round vacuum tobacco tin (half section).

acting to keep the lid closely on a vacuum-packed tin. The following calculation illustrates the point. Given that the diameter of the tin is 68 mm and the pressure difference between the inside of the tin and the outside atmosphere is 0.9 bar (1 bar = 10^5 N/m^2 or roughly 1 atmosphere). Note that the achievement of a perfect vacuum within the tin would give a pressure difference exactly equal to atmospheric pressure. In practice, a near perfect vacuum is difficult to achieve and also unnecessary. The area of the lid which is subjected to the pressure difference is

$$\pi \times (34)^2 \text{ mm}^2 \quad \text{or} \quad \frac{\pi \times 34^2}{10^6} \text{ m}^2$$

The total force acting on the lid because of atmospheric pressure and tending to keep it shut is therefore

$$\frac{0.9 \times 10^5 \times \pi \times 34^2}{10^6} \approx 328 \text{ N}$$

This force would be produced by standing a weight of 33.4 kg (74 lb) on top of the lid.

Such a force, acting on the lid of the tin to keep it closed, makes for a most secure closure, in the mechanical sense. The air-tight seal prevents the contents from deteriorating by drying out or by losing flavour or aroma.

Some popular tobaccos sell steadily and, with proper retail stock rotation and frequent replenishments, the vacuum tin seems hardly necessary and might be replaced by a less costly and less air-tight container. Other brands sell more slowly or sales are erratic because of advertising or other promotional devices. Even with popular brands, it is not always possible to guarantee perfect stock rotation, and the tin provides a valuable protection against deterioration brought about by chance or by accidental long storage. There is also much to be said for a standardized container which allows a tidy approach to retail display and demands no special precautions by the shopkeeper.

Manufacturing limitations of deep-drawn tins

When a deep-drawn tin is produced, the metal from which it is formed undergoes great strain because of the way in which it is redistributed. Figure 12.10 illustrates this for a round tin. If a flat metal disc is printed or marked with a series of parallel lines when in the flat state, the lines will take up a new pattern, as shown, when a round deep-drawn cup-shaped component is made from it.

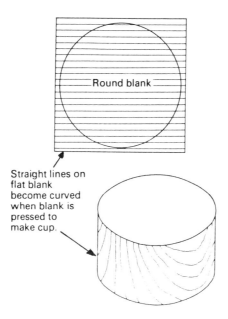

Round blank

Straight lines on
flat blank
become curved
when blank is
pressed to
make cup.

Figure 12.10 Distortion which occurs on drawing a metal cup.

It may be of interest to explain that, despite the major redistribution of material which takes place in the drawing operation, the area of the sheet of metal remains substantially unchanged, i.e. the metal blank from which a drawn component can just be made will have the same total area as the finished component. There are minute differences which occur in practice, but for most purposes the assumption of area equality is a safe one. This distinguishes drawn components from the component or can body made by the DWI process. With DWI there is a very substantial increase in the total surface area of the metal during the process.

Returning to the drawing process, an examination of Figure 12.10 should suggest that, for a given diameter of finished component, the deeper the draw the more severe will be the distortion of the top edge. Also as the ratio of depth to diameter increases, the forces involved in pressing the article to shape become greater and may exceed what the sheet metal will stand, thus leading to fracture during manufacture. Distortion of the surface affects printing and must be allowed for in advance. Possibly more important is the fact that surface coatings, applied to the metal to produce a uniform coloured effect or to supplement the protection from corrosion given by the coating, are less pliable than the metal. Surface coatings tend to craze under severe distortion, and their appearance becomes more matte than in those areas where the strains are less. In extreme circumstances the adhesion between coating and metal can break down altogether and leave plain metal showing through.

There are, therefore, two ways in which the depth to diameter ratio of a drawn component is limited; by severe strain which can cause fracture, and by surface disturbance which can make the appearance visually unacceptable. The strains may be minimized by drawing the cup-shaped component in more than one operation, i.e. by starting off with a shallow cup, and then subjecting it to more drawing operations, each of which will increase its depth and reduce its diameter. This can produce extremely deep, small-diameter components by working through several easy stages, and the technique is quite common in general engineering. It is expensive both because of the manufacturing time it takes to perform the numerous operations, and because of the amount of manufacturing plant it employs.

The degree to which diameter reduction can be achieved depends on the surface friction between the tool and the properties of the material as influenced by lubrication. Figure 12.11 shows one of the relationships.

Limitations of drawing round tins

The manufacturing limitations for deep-drawn tins are really concerned with what can be done in a *single drawing operation*. For *round tins*, a safe limit is a depth to diameter ratio of 1:4 which means a depth of only 1/4 of the diameter. Having made that statement, it must be noted that ratios as extreme as 1:1 have been achieved but this is with material the thickness and physical properties of which have been carefully selected, and where the manufacturing operation has been specially designed to achieve the required result, as with DRD cans.

Limitations of drawing rectangular tins

Rectangular solid-drawn tins present a special case which is illustrated in Figure 12.12. If the rectangular body is divided into its geometric parts, it is

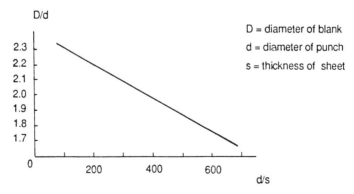

Figure 12.11   Limiting draw ratio. The maximum blank diameter which can be drawn without metal failure. Courtesy of International Tin Plate Conference, 1976.

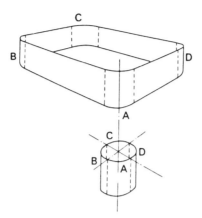

Figure 12.12 Significance of corner radii on a rectangular drawn box. The straight sides of a drawn rectangular box are simply bent into position. Drawing takes place on the radiused corners. The severity of the operation may be judged by assuming the four corners are brought together to form a small round tin.

seen that the straight sides are formed by simply folding them through an angle of 90° to the base. No drawing takes place along the straight sides, there is no redistribution of material as there is in the side wall of a round tin.

The rounded corners are drawn, and each can be considered as one fourth of a small round tin. Because the corners are drawn, they impose the acceptable limitation on depth. The depth of a rectangular drawn box with rounded corners is normally limited to:

Depth = 2×corner radius

If the four corners could be brought together as in Figure 12.12, it is possible to talk in terms of the diameter of the tin so formed, and the depth limitation would become

Depth = diameter, because diameter = 2×corner radius

It is interesting to note that, even for rectangular tins, the 1:1 depth to diameter ratio previously suggested as the absolute limit for deep-drawn work, using a single drawing operation, is again observed.

The whole subject of solid-drawn tins is complex and important, not only because of the usefulness of tins made by this method, but because of the extent to which the manufacturing technique is employed in can making and tin box making generally. The references to solid-drawn tins could just as easily have been to solid-drawn components. The lid of a slip-lid tin is solid-drawn; so is the shoulder of an oval talcum powder tin. Even the end for a conventional food can is solid-drawn, although the draw depth is so shallow as to be almost insignificant. More significant is the drawing of metal involved in making a cone-shaped top for an aerosol can, and a particularly complex example is a screw neck made as an integral part of the top of an oblong pourer-type tin.

Much of what has been said regarding solid-drawn tins is therefore of importance to the manufacture of separate components which form parts of tins and cans.

General-line built-up tins

The term *general line* refers to all of the many types and styles of tinplate container which cannot be described as open-top cans. Factories tend to specialize in one or the other. Although some manufacturing techniques are common to both, the management style required to run a general-line business is different to that which is appropriate to open-top can making.

To attempt to catalogue all of the tins available would be space-consuming and is unnecessary. Reference to trade literature will assist the reader, as will a study of those British Standards which relate to tin box making. Some types of tin are, however, of fundamental importance, and will now be described and discussed.

The slip-lid tin

This type serves in the home as a cake or biscuit tin and has aided the retail sale of many dry commodities. It is not usually employed for liquid products.

The body and bottom end are secured to each other by a conventional double seam, and the body side seam is most often an interlocked seam which may be soldered, doped or cemented. There is no technical reason why the side seam should not be a welded or bonded lap seam. The bottom end seam may contain compound or, like the side seam, be soldered. Consequently, the tin without its lid is able to hold the finest dry products without leaking or sifting. The lid, by definition, slips on the body (Figure 12.13). It is not secured by solder or any other sealing medium, because it is meant to be put on and taken off with ease, and without resorting to any tools or mechanical aids. The lid is *solid drawn*, i.e. it is made from a single piece of sheet metal formed by pressing through a steel ring or die by a punch. The process is examined more thoroughly in the section dealing with solid drawn tins. The lid has no seam or joints, and is thus well able to prevent the passage of liquids, moisture vapour or finely divided products.

What tends to let down the slip-lid tin is the simple friction fit between lid and body. This is unsealed so that liquids, gases and fine solid particles can find their way in and out of the tin. Nevertheless, for some products, the slip-lid tin is ideal. An example is individually wrapped sweets, which need both

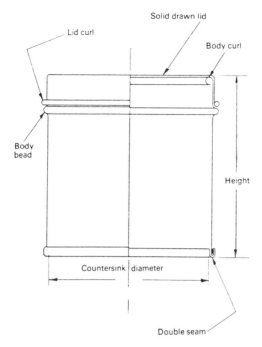

Figure 12.13   Slip-lid tin (half section).

atmospheric and mechanical protection. The individual wrapping provides the protection from the atmosphere, and the tin augments this and provides protection from mechanical damage. Perhaps most important, a tin presents an opportunity to use lithographic printing to great advantage for display and advertising.

Many ways can be found to improve the protective capability of the slip-lid tin. The problem is usually a matter of preventing oxygen and moisture vapour from entering the tin. A very common approach is to use self-adhesive tape to cover the joint between lid and body, and machines are available which enable this to be done rapidly.

Another method is to secure an adhesive-coated impermeable diaphragm to the top edge of the tin body by heat. Aluminium foil is a suitable barrier, and so is paper coated to make it gas and moisture-vapour resistant.

It is not uncommon to find slip-lid tins completely overwrapped with transparent film. This, too, contributes to protecting the contents from moisture during transport, storage and display. It is assumed that with all these measures the removal of the sealing material by the user presents little hazard to the contents, which are consumed in a relatively short time following initial opening.

These remarks about slip-lid tins serve to underline certain principles which can be applied to other types of metal container.

1. In choosing which type of tin to use, the total package must be considered. This means that account should be taken of
   a. any primary wrapping, actually in contact with the product
   b. any lining or cushioning material within the tin
   c. the tin itself
   d. any sealing tape or overwrapping material outside the tin
   e. any other outer packaging, such as a fibreboard case with or without cushioning material, used to transport one or several tins to the point of sale.
2. The tin as a protective package can often be adapted to new uses by employment of materials and methods evolved in other areas of technology.
3. It is important to define the packaging requirement. Leaving aside matters of display for the moment, the question is often a matter of how much protection the cost budget will allow.

The lever-lid tin

When discussing the slip-lid tin it was said that a difficulty arises in making a highly efficient seal between lid and body and that it is common to use measures such as overwrapping to provide improved protection to moisture- or oxygen-sensitive products. The reason for the inadequate seal lies in the method of body construction. This must be rolled or bent to its cylindrical shape, and is impossible to achieve with a guarantee of absolute smoothness to the exterior body. Instead of bending into a smooth cylinder, the material will often develop numerous small flats producing, instead of cylinder, a polygon with a large number of sides. Shortcomings of the process and local differences in raw material will produce larger irregularities. In addition there is the side seam which, however it is made, will present an irregularity which cannot be sealed by a slip lid well enough to contain liquid.

To achieve the quality of seal necessary to contain liquids or exclude moisture and gases, a further component must be added. This component is the *lever ring* (Figure 12.14). This is a seamless component made by drawing the metal from a flat sheet. It is seamed to the top of the can in the same way as the bottom end is fitted. Because it is seamless, the lever ring enables another drawn seamless component, the lever lid, to be fitted to it with good assurance of an effective seal.

Many styles of lever ring and lever lid have appeared on the market, some having particular technical merit and others being different just for the sake of a change. No lever ring is really easy to make. The aperture in the ring and the diameter of the lid must both be made to close engineering limits, if a satisfactory and consistent fit is to be achieved between any ring and any lid out of many thousands. In fact, the tolerances on the diameters of the ring and lid for a 1 litre paint tin are typically as follows:

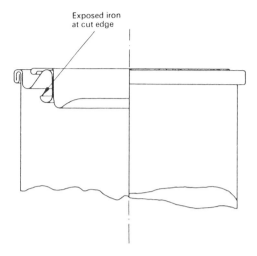

Exposed iron
at cut edge

Figure 12.14   Lever lid and ring (half section) showing cut edge of ring.

Ring diameter: 100.18 mm   Tolerance +0.02 mm
                                         −0.08 mm

Lid diameter: 100.41 mm   Tolerance +0.05 mm
                                         −0.05 mm

No lever ring is easy to make, but it is easier to finish with the cut edge of the tinplate around the aperture on the inside of the can rather than the outside (Figure 12.14). This is unacceptable for water-based products which will attack the iron exposed at the edge and not covered by tin in the same way as the rest of the surface. This explains the justification of one variation in design of the lever rings where the exposed cut edge of the metal finishes on the outside of the can and, although exposed to the atmosphere, will be attacked only very slowly compared with the rate of corrosion it would suffer if within the container.

Some manufacturers consider that the lever-lid paint can is much more difficult to make than the conventional food-can. Certainly it has four components against the food can's three, and the precision necessary to make the closure components has already been noted. The argument is, however, based on the packaging requirements of the lever-lid paint can which must be;

1. Easy to fill (through a large aperture)
2. Easy to close at high speed
3. Easy to open without specially designed tools
4. Easy to re-close and seal after use

When in storage and on display, the can must prevent evaporation of volatile solvents. Effective re-closure after use has to be achieved in the presence of paint or other products which will have contaminated the sealing surfaces. Quite a demanding set of requirements.

The lever-lid paint can is sometimes criticized for its lack of security. Because of the demanding requirements on the can and, particularly because it must be easy to open, it will not withstand unlimited rough handling. If a 5-litre can of emulsion paint is dropped on its side from a height of 1 m, the lid is almost certain to be forced off by the sudden surge of pressure caused by impact with the ground.

Ways of preventing or reducing the risk of lid removal are available. Spring clips can be fitted to help retain the lid, or an overall cover or capsule may be spun under the double seam. Spots of solder applied to three or four points on the periphery of the lid do the same job as spring clips. All of these are more or less effective in adding to security, and more or less effective in irritating the user who wants to use the contents. It may be significant that the vast majority of paint cans are sold without any added security devices, and very little paint seems to be spilled in retail shops or the pavements outside.

Oblong pourer tins

Before leaving the subject of built-up tins, some mention must be made of the oblong pourer tin. It is another important example of a difficult task being attempted because it is justified by non-technical considerations. Figure 12.15 shows the general style of such tins which are commonly available in capacities from 125 ml to 5 litres, the smaller capacities being supplied without handles.

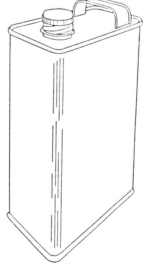

Figure 12.15   Five-litre oblong pourer tin.

Round tins would be stronger and easier to make, and could well be cheaper, and yet this general style of oblong tin enjoys worldwide popularity.

The flat display area on the sides and easy storage afforded by the rectangular shape seem to outweigh other factors. This type of tin is often chosen for products which are easily able to penetrate the most minute leaks, and the consistency of mechanized production enables the tin to meet these demands.

The difficulty in making an oblong built-up tin is concerned with the corners. When making a double-seamed round tin, the manufacturer employs what might be described as a *metal spinning operation*. Although the extent to which material is redisposed in making a double seam is limited, the process is similar to that employed by silversmiths and others in making deep cup-shaped articles from flat sheet. Such techniques benefit from speed and smoothness of operation. The smoothness is lost when an oblong tin is made. At the radius connecting one straight side to another, mechanisms must undergo violent changes of direction in forming the seam. This causes uneven wear of bearings and gears, and results in loss of quality which must be countered or forestalled by maintenance activity.

This example is used simply to indicate the nature of the technical difficulty. It must be accepted that round tool and machine components are easier to make with precision than are oblong items, and this is the root of the difficulty presented in the manufacture of an oblong tin.

An in-between method: the locked-corner tin

The previous section discussed built-up tins which are made of several pieces of material put together in such a way as to form a hollow container. An earlier section explores the features of so-called solid drawn tins.

Between these two methods of construction there lies a type of tin known usually as a locked corner tin or sometimes a sunk and seamed tin. In tin box making, 'sinking' refers to the making of a depression within a component, or to the manufacture of a dish or cup-shaped article from a flat sheet or blank.

The locked-corner tin is a rectangular tin, the body of which is made from a single blank. It is distinguished by the fact that the body has sharp corners, not rounded or radiused like oblong built-up tins, and each corner has an interlocked seam. Figure 12.16 shows both the blank and the finished body. The body is made by pushing the blank through a die which is itself rectangular and of a size to match the required outside dimensions of the body. The four corners of the die are, however, of a very special form and they trap the edges of the blank as it is pushed through by a punch, in such a way as to form an interlocked seam at each corner.

Examination of the sketch of the blank shown in Figure 12.16 should be

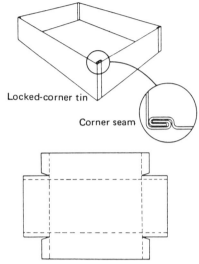

Locked-corner tin

Corner seam

Blank of locked-corner tin with lines of bend
shown dotted

Figure 12.16   Locked-corner tin.

sufficient to suggest that, in terms of material use, this manufacturing technique is most economical for shallow tins, and grows progressively less economical as the body depth is increased. This is simply because deep tins require the blank to have a large, almost square piece cut away from each corner. The material cut away is called a 'notch', and the deeper the tin the bigger it has to be. Material cut away in the notch is discarded and therefore wasted.

Despite the requirements for carefully made tools and the loss of material from body notches, locked-corner tins are used when sharp square corners are required within the tin, as when packing cigarettes and some types of biscuit. They may be fitted with a solid-drawn slip lid or hinged as circumstances demand.

Economics

As with other manufactured articles, the cost of making a can is made up of

1. Direct material cost
2. Direct labour cost
3. Overheads including indirect materials, labour, depreciation, factory rent, rates and cost of energy, etc.

In general-line tin box making, direct labour costs may be between 5%

and 10% with material 30–60% of total selling price. The remainder is overhead and profit. In high-speed can making, direct labour is an even lower percentage, but overheads tend to be higher because of the cost of the complex machinery involved.

Assuming that, as a rough guide, material will approximate to 50–60% of the cost of any tin or can, it is thus the largest identifiable single cost. Much ingenuity has been displayed by engineers in trying to achieve the greatest possible economy in the consumption of tinplate or other box-making material, and considerable sums of money have been invested to minimize waste in making cans by high-speed methods. The same pressure for economy has not had much effect on the selection of tins for various products. These are determined by the market needs, what will sell, or what is acceptable as distinct from what is most economical.

The ideal geometry

The problem seems to be a matter of containing the greatest volume by the least area of sheet material. If that were all there were to it, every tin would be a sphere, because a spherical shape is the most economical way of using sheet material to contain a given volume. In practice, of course, we are confined to using tins which are either cylinders or cubes (or in the more general case, rectangular prisms).

Assuming that the tin has no seams, no joints and no overlapping edges, in the way that a slip lid overlaps the body, then if cylindrical, the tin is like a short section cut from a seamless tube and fitted with two ends, each of which is a simple disc of diameter just equal to the diameter of the tube. For such a container, it can be shown that the maximum volume is contained by the minimum area when (a) for a cylindrical container, height equals diameter; and (b) for a rectangular container, all sides are equal.

As in many situations, such mathematical perfection is upset by practical considerations. Not the least of these is that circular components are cut for the cylinder ends from square pieces of tinplate (Figure 12.17). This is wasteful because more than 22% of the square piece will be thrown away as shred, and less than 78% used to make a lid or bottom end.

It is also necessary to allow for a seam in the body of the tin and, with a slip-lid tin, there is an overlap of the lid on the body to consider, as well as the material in the bottom end seam. Taking account of all of this, the most economical proportions for a slip-lid tin are such that the height exceeds the diameter slightly. As a rough guide, a height to diameter ratio of 1.2:1 might be about right. The exact proportions can, of course, be calculated in each instance; they will vary slightly from small tins to large ones and be determined by the exact amount of material employed in seams and the overlap of the slip lid on 'he body.

Contrast these ideal proportions for economy with tins used for retail sale.

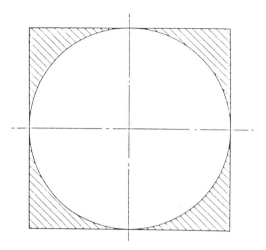

Figure 12.17   Round disc taken from a square.

Some tins do have proportions approaching the ideal for material economy, others are far less economical. Typical of the latter are large round biscuit tins which are popular because of their re-use value in the home. Here, considerations such as display value and utility outweigh material economy in the designer's order of importance.

It is almost impossible to find a cubical container in use today, and the popular oblong pourer-type tins used for oils, solvents and many other liquid products, are a long way from being of economic proportions. The reader might like to make his own list of reasons for this, noting in passing that a cube is also uncommon in architecture.

Getting the most out of available material

A large part of the can makers' work is concerned with cutting round discs from sheets of tinplate or aluminium. The obvious way to do this is shown in Figure 12.18a. The strip, cut from a sheet of tinplate, is fed to a stamping press and the discs are each cut out in quick succesion.

In practice, the tool fitted to the press is designed to produce, say, can ends, and one almost-finished end would be formed at each stroke of the press; but it is convenient to talk in terms of discs, because we are only concerned with the plane geometry of the problem for the moment.

It is not practicable to arrange cutting so that the edges of the discs just break through the edge of the strip. Instead, a small margin must be left along the edges of the strip and between each disc. The material which remains after the discs have been cut out is called *shred*. In Figure 12.18a the

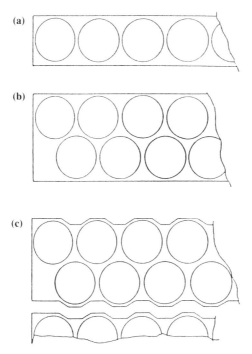

Figure 12.18 Progressively economical ways of cutting discs from strips. (a) Single row from plain strip. (b) Staggered layout on plain strip. (c) Scroll sheared layout.

shred which remains is about 25% of the original sheet. In other words only 75% of the original strip finishes as discs.

Such a wasteful arrangement is to be avoided if at all possible. Figure 12.18b shows a rather better arrangement, where the discs have been staggered, and two rows are cut from a single wider strip. The result here is that about 75.5% of the material goes into discs.

The principle of staggering the pattern of discs is a sound one, but most of the waste arises along the edges of the parallel strip. Figure 12.18c shows how the can maker reduces this. A large sheet of tinplate is cut into strips, with zig-zag edges as shown, and the strip is then fed longitudinally into the stamping press. This system achieves the use of 82% of the strip. This is a much better result, and its only weakness is that because squared sheets of tinplate are usually involved there is still waste along the straight edges. Even that has now been overcome by using coils of tinplate instead of sheets and cutting the zig-zag pattern strip from continuous strip, fed from a coil. In this way, an overall 82% usage is very nearly achieved.

The zig-zag edge is not as easy to cut as a simple straight edge, and requires a carefully made pair of shear blades and a particular type of shearing machine to accept them. The machine is called a Scroll Shear, and the blades it uses are expensive to produce and maintain. Equipment of this kind can only be justified by the long and continuous use which stems from standardization of containers. Food cans, beverage cans, aerosol cans and other such containers, for which there is massive demand, are made only in certain heights and diameters which are the subject of national and international standards. Because of the UK membership of the EEC and our metrication programme, the various standards are undergoing changes. Such changes could be expensive in new tooling or changes to tooling, and cannot be entered into hastily or without proper consideration of their merits. To the manufacturer, a standard is only of value if it remains substantially unchanged for long periods.

Measuring and describing tins

This has undergone substantial change recently for three inter-related reasons. These are the UK entry into the EEC, metrication and increased activity by the International Standards Organisation (ISO). Can users and other readers may still meet the obsolete terminology, however, and it is necessary therefore to make a brief reference to the old method.

The unit of measurement was $\frac{1}{16}$ in and this unit was used in two ways. A simple slip-lid tin with the bottom end seamed on, could be described as $700 \times 3\frac{5}{16}$ in. The diameter was and still is stated first and consequently the 700 refers to the diameter. It means 7 in exactly, i.e. that the diameter is neither $6\frac{15}{16}$ in nor $7\frac{1}{16}$ in. Other sizes stated in this way might be 307 ($3\frac{7}{16}$ in) or 404 ($4\frac{1}{4}$ in). Although the word exactly has been used, what is meant is 'closer to 7 in than $6\frac{15}{16}$ or $7\frac{1}{16}$ in'. But what is closer to 7 in? In this instance the diameter is measured outside the double seam of the bottom end. When the three-digit notation is used, e.g. 700 or 404, etc., it is understood that a diameter accepted as standard is referred to. If the diameter was $7\frac{1}{4}$ in, then it would have been written as $7\frac{1}{4}$ in because the trade had not accepted such a diameter as standard.

Whereas the digital method refers to the diameter outside the double seam, a size shown as inches and fractions ($7\frac{1}{4}$ in) refers to the diameter inside the recess of the end—within the *countersink*, to use the trade term. On any tin, the actual difference between the countersink diameter and the diameter outside the seam is approximately $\frac{1}{8}$ in (see Figure 12.19).

To return to our slip-lid tin, then, it measures 7 in diameter over the double seam and it is $3\frac{5}{16}$ in tall, the convention here being that the height is the overall height of the tin, without the lid fitted.

The digital notation is only employed to demonstrate the height of tins and cans when they are established as standard sizes. So much for the past. Now measurements are in millimetres and, although the descriptions look similar, they mean something different, e.g. a large rectangular tin might be described as $231 \times 217 \times 118$. This means that the tin has base dimensions of

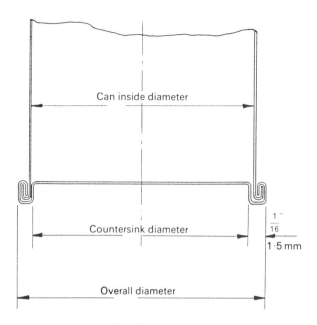

Figure 12.19   Diameters of a can.

231 mm × 217 mm and is 118 mm tall. A round lever-lid paint tin is described as a 176 diameter, 5-litre lever-lid tin.

Generally, the diameter in millimetres is stated, followed by the nominal volume. If the reference is to an oblong pourer-type tin, forming part of a standard range, the base dimensions are given followed by the nominal volume. As with the rectangular tin of the first example above, the larger of the two base dimensions is given first.

Tins of non-standard height or non-standard volume, give both the diameter and the height in millimetres: e.g. 176×120. This is the method which has also been adopted in the United Kingdom for describing cans with both ends seamed on such as food, beer and beverage cans.

The use of metric units is becoming commonplace, but the dimensions do not always refer to the same features as with the old digital method, using a unit of one sixteenth of an inch. For height, there is not much change. The height of a food can is the external height, measured over both top and bottom seams. This is rounded up or down to the nearest millimetre.

With cross-sections, an attempt is now made to use dimensions as close as possible to the inside diameter of round tins, or the inside length and width of rectangular tins. This is easy for the manufacturer, who knows what size of can he is trying to make, and can obtain precise measurements from the mandrels or other tools over which the can bodies are formed. It is not so easy for the can user who may not be equipped to measure the inside diameter of a can, even if it happened to be rigid enough and uniform enough to permit this. His problem might be a matter of identifying the size of can from two or three possible standard sizes. If the can is filled and has both ends seamed on, access to the interior in order to measure its diameter is not even possible.

A method is available which overcomes this difficulty. For round cans with double seamed end, it is sufficient to measure the diameter inside the countersink of the end and round-up the measurement to the nearest whole millimetre. The same procedure will also serve to establish the size of a rectangular container.

When other types of tin or can are encountered, it will usually be sufficient to accept the nominal dimension of a cross-section as the actual inside dimension to the nearest millimetre.

Closures

Compromise in design must always be accepted, no matter whether the object is a motor-car, a central-heating system, or a suit of clothes. More often than not, the extent to which our wishes are achieved depends upon cost. The useful life of a motor-car, the reliability of a central-heating system, or the workmanship in a suit, are limited by cost. The compromise, in all instances, is dependent on how much we can afford.

With closures, cost is one of the limiting factors, but there are others, and the perfect closure possibly does not exist. If it did, it might easily pass unnoticed, because no-one would complain about it. The lever lid and ring, used on paint cans, is a good example of compromise. The lid must enter the ring easily, but it must also remain there, even when the can receives a degree of rough handling. None of the product must leak from the closure, but the lid must come off without difficulty, with only the twist of a blunt screw driver. The lid must also go back on again and make a seal, even when the ring is covered with paint, and these functions must be achieved at the least possible cost.

Had the lever lid and ring not been developed over a long period of time, it is difficult to believe that it could be introduced today. No packaging engineer or designer would be in any hurry to accept such a demanding set of requirements.

Because of this need for compromise it is worth while to analyse some of the basic properties of closures. Such analysis leads to an understanding which helps in selecting or specifying what is needed for a product or for a new packaging requirement. There are two broad categories of closures:

1. The 'once-only' closure with no re-closing facility
2. The re-closable closure

The once-only closure

Some typical examples are:

1. The open-top food can opened by a can-opener
2. A beverage can with an easy-opening end or pierced by a special opener
3. The engine-oil can with a tear-off strip covering both a pouring hole and a vent hole.

Such closures are cheap to produce, provide good resistance to leakage, and are suitable for their particular purposes. They include the 'frangible diaphragm' beloved of patent specification writers. Some require an opening tool, typically the can opener.

A common factor is that this type of closure is useful only for those products which will be dispensed and completely used from the can within a short time of it being opened.

The re-closable closure

The lever-lid and ring is clearly in this category, and so is the screw cap and neck. Many other closures are now available involving plastic spouts, pourers and other dispensing devices, and the development of these continues at such a pace that to list them will hardly be of service to the reader. What does seem to be important are the principles involved. All re-closable closures have three properties in common:

1. Friction
2. Resilience
3. Amplification of effort

Friction. It is friction which keeps a cork in a glass bottle, friction between the cork and the inside neck of the glass bottle. Friction also retains a lever lid in its ring against the internal pressure which may build up because of heat from the outside, or against a sudden surge of internal pressure resulting from dropping the tin.

Note that the friction between two tinplate surfaces can be markedly changed by application of lacquer to one or both surfaces.

With lever-lid tins, it is sometimes necessary to have the inside of the tin protected from corrosion by a surface lacquer. The lacquer also has to be applied to the lever ring and the inside surface of the lever lid. Care must be taken to ensure that the application of such lacquer has not reduced the friction between the surfaces to an unacceptable level, which will limit the security of the closure.

Only friction holds a screw cap tight on its associated neck. The way in which the friction is used best becomes clearer as we examine the other two properties involved.

Resilience. When a cork, a stopper or a lever lid is forced into the aperture designed to receive it, some resistance is met, initially, which is greater than that which opposes further entry of the plug. This initial resistance is produced by the compression of the plug, the expansion of the aperture, or both. With a cork in a glass bottle neck, the change of size is almost all due to the cork. The change in size of a tinplate lever lid and ring is due to what the engineer describes as the interference between the lid and the ring. The lid is slightly reduced in diameter and the ring is slightly enlarged and, if the lid is removed, the two components will tend to revert to their original size. That is what *resilience* means—the ability to change diameter in contact, and subsequently revert to the original dimensions when the two parts of the closure are freed from each other.

Resilience is found in the wad of a screw cap, which thus acts as a spring between the inside of a cap and the top of a screw neck as shown in Figure 12.20. It is the wad that makes the necessary seal, of course, but it has the additional function to provide a resilient member between the neck and the cap. Without it, the cap would be either tight or loose. When tight, the slightest rotation in the unscrewing direction would make it loose, and it would be insecure because of this. Any slight shock or continuous vibration in transit, or even sudden changes in temperature, could cause the cap to loosen on its neck.

The presence of a resilient wad aids security. The cap must unscrew through a reasonable angle from the fully tightened condition to reach a point where the previously compressed wad expands to something like its original uncompressed condition. Because of this range of movement, a cap with a wad is better able to withstand transit hazards than one without. Perhaps more significant, in practice, is that the most resilient or most elastic wad will be the one which provides the best security, other factors being equal.

Amplification of effort. This property is easier to appreciate in some closures

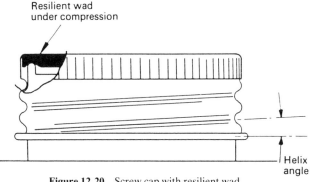

Figure 12.20   Screw cap with resilient wad.

than in others. With screw caps, the helix angle of the thread provides the amplification (Figure 12.20). The *helix angle* is the angle the thread appears to make with the horizontal plane when the cap is held with the axis vertical—the way it normally appears on the top of the tin. The smaller this angle, the greater the amplification of effort, i.e. the greater the force produced to compress the wad for a given effort applied to tighten the cap. Practical considerations in the manufacture of screw threads in thin sheet metal prevent this angle becoming very small. Within the range of angles available, however, the force compressing the wad is substantially greater than the force applied by hand or by machine to the periphery of the cap when it is tightened.

A note for engineering readers is necessary at this point. The way in which the rotation of a screw cap on a neck or a nut on a bolt produces a higher axial force than is applied to the periphery of the cap or nut is strictly due to what the engineer calls *mechanical advantage*. The reason we do not use this established term is that with other forms of closure, such as the lever lid and ring, there is amplification of effort which is not strictly mechanical advantage.

It was noted when discussing resilience that a lever lid pressed into its associated ring will cause the ring to expand, and the lid to be compressed. When in the closed condition, therefore, the forces are held in balance, the ring exerting radial forces tending to crush the cap, and the cap exerting equal outward forces to expand the ring.

These forces exist because of the resilience of the two metal components. It happens that the total radial forces acting in this way are greater than the force required to push the cap into the ring. It is evident that the greater the radial forces the better is the closure able to prevent leakage. Amplification of effort exists, therefore, because the total of the radial forces produced is greater than the force required to press the cap vertically into the ring.

Decoration of cans and tins

Typical cans are shown in Figure 12.21.

There are four methods in use for the external decoration of cans. These are:

1. Offset lithography
2. Dry offset printing
3. Silk screen printing
4. Paper labelling

In addition to and in support of offset lithography and silk printing, roller coating is extensively used. This is the method which is also employed to apply protective coatings to that side of the metal which will form the inside of the can.

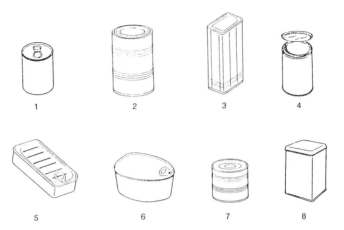

Figure 12.21   Typical cans and their special features. (1) Beer beverage DWI, (2) three-piece open top, (3) oblong, key opening, (4) paint can, (5) sardine, (6) ham: (7) two-piece DRD, (8) biscuit tin with lid.

Offset lithography

Lithography has its roots in artists' methods of reproducing their work. For hand methods of reproduction, the picture to be reproduced is drawn on the surface of fine-grained stone slab using wax crayons. When the stone is moistened, water is absorbed by the unwaxed areas and repelled by the lines and marks drawn by the artist. Subsequent application of greasy ink to the whole stone by means of rollers, pads, or other means deposits ink only in the waxed areas and completely repels it in the wet areas. If a sheet of paper, board, or other suitable material is then pressed hard against the stone surface, the ink image will be transferred to the paper or board. Repeating the procedure with several stones provides the means to achieve multi-colour reproduction. The technique is still in use by artists to achieve small-quantity reproduction of their work.

Stones were originally used in tin printing. Because tinplate is non-absorbent and because its stiffness prevents its being brought into continous close contact with the stone, some way had to be found to transfer the coloured image from the stone to the metal. Paperboard of a particular grade and thickness was found to be successful. The board was pressed against the stone, from which it collected ink in the shape of the required image, and the inked surface was then pressed against the surface of the tinplate so that the image was once again transferred, this time from the board to the tinplate.

When an intermediate web or blanket such as the paperboard sheet is used in this way, it is said to 'offset' the image. Thus the process obtains its name, *offset lithography*. Many years ago, the paperboard blanket was replaced by a fine-grain rubber sheet, and the stone slab by a metal plate which could be bent and stretched around the surface of a cylinder. These and other innovations permitted the full mechanization of the process, so that in the latest modern equipment, sheets more than 1 metre square are printed in multi-colour machines at the rate of 3000–5000 sheets per hour.

Drying the ink. Because tinplate, aluminium and other metal sheets are non-absorbent, the inks must be dried on the surface rather than allowed to soak into it as they do on paper. Consequently, freshly printed sheets must be held out of contact with each other and passed through an oven before they can be stacked, or handled through the making-up operations. The ovens are normally gas-fired, and the products of combustion (mixed with air) are forced over the printed sheets to carry away the solvents and harden the resins which carry the coloured pigments.

Other special inks have been introduced which make the space-consuming ovens unnecessary. These inks are of a different formulation, and contain much less volatile solvents. The resin in which the pigments are supported is capable of being polymerized by exposure to ultraviolet light. If the intensity of the light is strong enough, the exposure time is only a fraction of a second; the ink is then dry, and ready to receive a subsequent colour, superimposed where necessary. Alternatively, a clear varnish can be applied immediately following the exposure to radiation. The differences between the two types of ink are illustrated in the formulae shown in Table 12.1.

Principles of colour reproduction. Any given design will start as original

Table 12.1

A formula for a typical sheetfed offset ink for paperboard might be:

Organic pigment	20%
Phenolic–hydrocarbon resin varnish	40%
Drying oil alkyd	10%
Hydrocarbon solvent (500–600°F or 260–316°C)	25.5%
Wax	2.5%
Cobalt drier	1%
Manganese drier	1%

A formula for a simple radiation-curing metal decorating ink could be:

Organic pigment	15%
Acrylate oligomer	40%
Acrylate monomer	30%
Photo-initiator and sensitizer	8%
Wax	3%
Tack reducer	4%

artwork in a studio. In order to reproduce it this artwork must be separated into its component colours. This is done using a camera and the appropriate number of colour filters; usually red, green and blue; to produce a series of transparencies.

Colour reproduction is then achieved by using half-tone printing. Each of the component colours (usually yellow, cyan and magenta plus black) of the original artwork are reproduced by printing, using a separate printing plate for each colour, as precisely registered dot-images one after the other on the particular substrate.

Thus the full colour illustration on the printed sheet is created in the brain of the observer since the printed dots remain separated on the substrate and do not actually mix.

Over the last 30 years many advances have been made in the technology of colour separation and printing and the use of electronic scanning and computer analysis now permit these processes to be made with great precision and stored digitally on a computer disc in a relatively short period of time.

Limitations and pitfalls. Offset lithography is a versatile process, and it can achieve many useful and striking effects, if properly used. Like other processes it has its limitations, but these can usually be avoided with thought and planning.

The process does not like fine detail. If a choice of typeface is possible, it is best to select one which has an absence of serifs and contains clear open letters. Such a choice will ensure pleasing results over long printing runs.

Many designs can be achieved by printing three colours and black on a white background. There is a commercial pressure to minimize the number of colours employed, because the cost of printing is almost entirely proportional to the number of colours used in the printing run (the number of passes). However, it sometimes happens that, regardless of cost, an effect is required which necessitates six or even more printing passes. From the earlier explanation it will be evident that to print six colours may require six journeys through a hot oven. This means that the initial background coating, which is often white, will have to pass through the oven eight times in all, because one pass is necessary to dry the white itself and, when all else is done, an overall varnish will be applied which again requires a pass through the oven. It is not surprising that the white background may become yellowed or discoloured because of all the cooking it receives. Generally it is best to make a genuine effort to minimize the number of colours employed, cost considerations apart.

Other things to avoid are connected with the interaction between printing and tin box making. The final appearance of the printed surface depends upon a great many things. The surface of the metal itself is the first of these. The tin coating can vary in its reflectivity and there is, for example, a considerable difference between the surface of a tinplate sheet and the surface of an aluminium sheet. To achieve an identical appearance on both is almost impossible.

More frequently this change in gloss or reflectivity occurs because of the making-up process itself. For example consider a simple slip-lid tin. The body has been made by bending it into a circular shape. Its reflectivity has been wholly preserved and might even have been enhanced by the curvature of the metal. Fitting over this we have the vertical wall of the slip lid; it has been subjected to a drawing operation in a press tool which has redistributed the metal. Such working always tends to dull the appearance, if only slightly, and it generally tends to weaken the strong colours. The perfect match between lid and body required by the designer is not obtained, and probably cannot be obtained no matter what precautions are taken. The deterioration due to drawing is greatest on deep small-diameter components. One of the most demanding requirements is to match the drawn shoulder on a talcum powder tin with the body wall.

Registration between the printed design and the physical profile of the finished tinplate component is worth a word or two. Most of the time this creates no problem and, if a design appears misplaced in the lid of a tin, it is more likely to be due to an error in making up than in printing. However, designers often succumb to the temptation to surround their design with some sort of frame or margin. It would be ridiculous to debate the aesthetics of this in general terms, but examination of actual examples suggests that the trouble this causes on drawn lids and other press worked items is not justified by the objective sought. A design which is allowed to run across the lid surface and down the edge in one continuous pattern or colour will permit some inaccuracy in stamping the lids, even up to 3 mm on a large lid, before the error is noticeable. If the design on the top surface is surrounded by a sharp line in some contrasting colour, the slightest misplacement in stamping the lid is made glaringly obvious

Roller coating. Roller coating is not a printing operation in the sense that it allows the reproduction of an image. Roller coating provides the means to roll a continuous coating of solid colour on to the surface of a metal sheet. Many lithographed designs are printed on a roller-coated surface. Most often, the background 'colour' used is white.

The process is what its name suggests. Coating material is flooded over the surface of a rotating roller surfaced with synthetic rubber or some other elastic material. A sheet of metal is passed under the roller, against which it is held by another roller, and picks up a quantity of coating material. The application is followed by oven drying.

The technique requires less skill than printing. The machines run faster than printing machines, and the weight of coating applied is considerably greater than can be achieved by printing. It is the method used to apply protective lacquer to that side of the sheet which forms the inside of a can.

Dry offset printing

The meaning of 'offset' was explained under the previous heading. It was seen that lithography requires the use of water and grease in cooperation to confine the ink to the image areas. In dry offset printing, the water is eliminated and a printing plate with a raised image, like a letterpress plate, is used instead of the substantially flat printing plate used for lithography. Dry offset printing can be achieved on a lithographic printing machine with little more effort than the use of a letterpress plate and leaving out the water. Moreover the importance of dry offset printing has been accentuated by the development of the drawn and wall-ironed can.

Printing of the two-piece can. The metal of a DWI can undergo considerable deformation during the forming process and this means that these cans must be printed after they have been formed. The printing is carried out after the wall-ironed bodies have been trimmed, washed and dried. The first stage in the decoration is the same as for printing a flat sheet (the application of the white base coat). This is done by rotating the can against a rubber roller that is continuously fed with the white coating. The coated cans are then loaded on 'pegs' and pass through an oven to dry.

With offset lithography of flat sheets, colour to colour registration is achieved by working from the front edge of the sheet and one side of it, i.e. the machine has a way of detecting these two edges and placing the coloured image the correct distance from each. On a DWI cylinder, the perfectly smooth cylindrical surface provides no feature which could enable colour to colour registration to be achieved around the cylinder. Consequently each colour is printed in turn on to a rubber blanket from which all are offset simultaneously, on to the can surface.

To allow this, it is preferable to arrange matters so that no two colours overlap each other in the design. This means that multi-colour half-tone printing requiring dots of one colour to be printed over dots of another colour is more difficult to achieve.

The printing press has a number of separate printing units arranged at intervals around a large drum which carries a series of offset printing blankets on its perimeter surface. The printing units carry the components of the artwork, etc. on dry offset plates and as the drum revolves each image carrier is brought into contact with the blankets thus transferring the design to them. This design is then transferred to the can bodies as they pass the drum. Then, as was done with the base coat, the cans pass on pegs through an oven to dry, which takes about 5 s.

Cans may be printed with up to six colours in one pass. The quality of half-tone printing by the dry offset process has been greatly improved in recent years, largely by the use of better inks.

Silk screen printing

Silk screen printing uses the basic principles of stencilling. A fine screen of nylon polymer filaments or a silk mesh (whence the name) is mounted on a supporting frame and a stencil carrying the artwork is photographically produced as a mask on the mesh screen. The screens used may vary from 60 to 120 filaments per inch depending on the requirements; the lower range for bold effects and the upper range for fine detail of half-tone work.

Ink can pass through the screen except where the holes have been blocked by the stencil. This provides a system for reproducing a coloured image and for multi-colour work the ink must be dried between each pass. To print a can, for example, the can is rolled against the screen as the ink is forced through the mesh by a squeegee.

Drying is carried out in an infrared oven and inks can be cured in about 10 s at 100°C. It is of course much easier to print on metal for cans in the flat by lithography and make up the cans afterwards, but other considerations may make it necessary to print the container after production, particularly for small runs.

Paper labelling

Many food cans still carry paper labels (Figure 12.22). There are two main reasons. The first is that the thermal processing and automatic handling that a food can must suffer, can damage a printed decoration unless special precautions are taken in the design and management of the processing equipment.

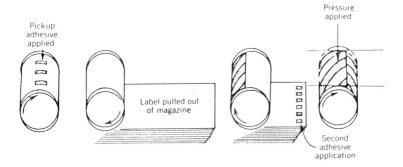

Figure 12.22 Stages in the labelling of cans. Adhesive is first applied to the container. As the container rotates over the label magazine, it picks up its own label, which is then rolled around it. A second adhesive application to the trailing edge of the labels, followed by pressure, completes the labelling operation.

Secondly, printed cans reduce the options. Each design may be used only for the product advertised on the outside and, if printed cans are used by a factory dealing with a variety of fruit and vegetable products, careful planning is necessary to ensure that the right cans are available at the right time and none need to be discarded as surplus.

Paper labels, on the other hand, are cheaper than cans and easier to store. Modern labelling machinery holds no terror for the plant manager and does a thoroughly effective job.

13
Aerosols (pressurized containers)

Historical development

The aerosol concept originated in 1923 when Eric Rotheim of Oslo, Norway, developed a wax spray for skis and other products using butane and vinyl chloride as propellants and brass containers fitted with needle valves. The feature that started this packaging system was the realization that a liquefied gas inside a pressure vessel can provide a constant pressure in that vessel even though it is being emptied.

Nothing further occurred until 1943, when Goodhue and Sullivan, working for the USDA, discovered that a 10-μm aerosol spray of insecticide did not disperse for at least 5 min and was effective against both flying and crawling insects. They patented the discovery, and the patent remains the basis of EPA requirements for pre-production registration of all aerosol pesticides.

Then, in 1947, Harry E. Peterson developed a super-strong beer can which could serve as an aerosol container. He also helped to create a relatively inexpensive valve that was simple to operate.

From such beginnings, a system has been developed by the introduction of new combinations of liquefied gases, different containers, sophistication of the valve mechanism, etc., into an accepted method of dispensing many products. Even over the comparatively short time that this method has been in existence, the pressurized container has become so important that it is essential to consider it as a way of doing a job.

Definitions

Aerosol: An integral ready-to-use package incorporating a valve and a product which is dispensed by pre-stored pressure in a controlled manner when the valve is operated.

Valve: A mechanical device, the operation of which permits the controlled emission of the product from the aerosol in a pre-determined manner.

Cap: A removable protective cover over the valve actuator, located in such a manner as to prevent accidental operation of the valve.

Propellant: A material which provides the power to eject the contents.

Types of pressurized packages

Three distinct classes of pressurized packaging exist. Two are self-pressurized; the chemical-propellant sprayer or aerosol; the non-chemical sprayer, of which there are several varieties; and the mechanical pump. The aerosol is by far the most common.

Aerosols are used to dispense a large number of products in containers commonly ranging from 3 ml to just under a litre. Liquid sprays are the most common but there are solid-liquid dispensers for special purposes and two trigger types valves for dispensing toothpaste. About 80–85% of aerosols produce sprays. They can also be used for foams, liquid streams, pastes, liquid-powder sprays and gas-powder sprays. Metering systems providing quantities from 0.05 to 5.0 g are also possible.

Worldwide in 1983 an estimated 6.5 billion retail units were sold.

Components

Figure 13.1 shows the essential parts of a pressurized package.

Caps

The cap performs two functions; it enhances the appearance of the container and protects the valve from accidental operation. Caps often have the orifice of the valve built-in together with either a locking device to prevent accidental operation or a shielded plunger. These 'actuator caps' have the advantage that the user does not have to remove them to operate the unit. While they make the item more attractive they are more expensive than the separate button and cap. In the early days of aerosols, all caps were of enamelled tinplate but nowadays they almost exclusively of plastic (PE or PP). Nearly all are the same diameter as the body of the container. Tamper-evident and child-resistant closure/caps for such products as insecticides and oven cleaners can be supplied.

Valves

The final orifice performs a vital function in the successful operation of the aerosol. It is incorporated into an actuator button and may take a variety of forms. It can consist of a single hole of closely controlled dimensions, in a plastic moulding; or of a multiplicity of holes leading to one final opening and, in such instances, often several plastic mouldings, or mouldings and

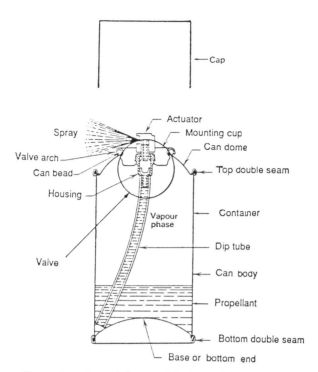

Figure 13.1 Aerosol dispenser (tinplate can) in operation.

machined parts, are locked together. Even a simple button actuator may still have a complicated maze of fine orifices and expansion chambers moulded into it. The button is mounted on a stem which is usually spring-loaded, so that a fixed rubber gasket closes over a hole in the stem when the unit is in the rest position (Figure 13.2).

Depressing the button uncovers the hole and connects the interior of the container to the atmosphere. The pressurized material inside is thus released. There are a number of ways of making this seal. For instance, some have a solid rubber diaphragm that is pushed away from the boss of the valve cup by the stem which has a slit in it. Others have a stem in two parts, but still have a hole in the gasket so that the product flows underneath the gasket to the stem, and thence to the button.

The housing (Figure 13.2) holds the spring and gasket in place and enables the cup boss to be deformed after assembling the valve so that the gasket is compressed. The housing may have a spigot for a dip tube, so that the container can be held upright whilst the valve is being operated. The spigot orifice is often specially sized to restrict the flow of product.

A vapour-phase tap valve is one in which vapour from the gas phase in the pack is introduced into the housing using a second hole. Here a stream of product and liquefied gas and a stream of vapour meet each other and alter

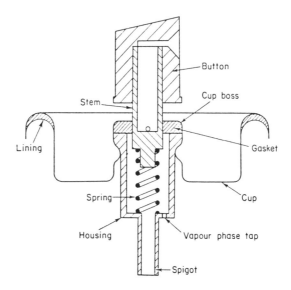

Figure 13.2   Parts of a typical valve.

the spray characteristics. The flow rates of these two, as determined by the relative sizes of the holes, must be carefully selected to ensure a satisfactory performance of the valve.

The cup can be made of tinplate or aluminium. The lining, which is used to make the seal between the valve and the container, may be of the 'flowed in' type as used for vacuum seals or consist merely of a rubber washer. The latter is usually used when either the product is very searching or the container is less uniform than the ideal.

The various components of the valve can be made from a number of different materials. For instance, the gasket can be soft or hard, resistant to some solvents and not to others. The spring can be of stainless steel or mild steel, and the housing of Delrin, nylon or metal.

The selection of a valve involves checking that the spray characteristics (spray rate, angle of discharge, particle size, distribution of particles, etc.) are suitable for the particular product, as well as seeing that it has a satisfactory operation throughout the anticipated life of the pack.

Containers

Metal, glass and plastic have been used for pressurized containers, and each has its place. Aluminium cans, while more expensive than tinplate cans, have the advantage of seamless construction, and hence the possibility of all-round decoration. They are manufactured by extrusion and can be of monoblock construction, in which instance the opening for the valve (which has been standardized) is made by a multiple forming operation. If the container is of two-piece construction, then the valve opening is manufactured as part of the forming process and merely trimmed. The end, which may be of tinplate or aluminium, is then double-seamed into place, using a sealing compound to ensure a satisfactory joint.

Glass and plastic containers are used, either where the product is too corrosive for metal, or where the product is expensive enough to justify the extra cost. With glass containers there is always a risk of shattering, and a general limitation on the size and internal pressure is necessary. To a certain extent this can be overcome by coating the glass with a suitable plastic material, or even by inserting the bottle into a steel, aluminium, or spirally-wound paperboard tube. This increases the cost, although the coating or the outer cover can be suitably decorated, thus saving the expense of a label. With plastic containers there are problems in that many plastics are permeable to some common ingredients of aerosol formulations, e.g. certain perfume oils migrate through them, and water vapour often has an appreciable transmission rate.

Steel. About 80–85% of aerosols produced in the United States (1984) had steel bodies (70% tinplate, 15% ECCS). Of the remainder aluminium accounted for about 13%, and glass for 2% with about 0.1% made in stainless steel and plastics; 80–85% in steel applies also to Canada, the United Kingdom, Australia and Mexico but many other countries use more aluminium bodies, particularly Italy where aluminium predominates and India, where because of import restrictions on tinplate, aluminium bodies constitute almost 100%.

Different thicknesses of steel are required for different parts of the container. Body plate is the lightest and is dependent on container diameter. End sections are heavier to withstand the internal pressure without deformation, domes being slightly stronger than bases.

Aluminium. Aluminium cans for aerosols first became available (in the United States) about 1948. They can be supplied for both very small and very large capacities because the impact-extrusion process used to make them is very versatile. Worldwide, the capacities vary from 10 ml to just under a litre and they have an advantage in requiring no side seam. The cans can be and are often internally coated to protect against aggressive contents and they can be decorated externally.

Glass. Glass aerosols are normally made from conventional soda lime glass and are mainly used for fragrances and pharmaceuticals. Many are plastic-coated to reduce the risk of breakage and fragmentation if they were to be accidentally dropped. All uncoated glass aerosols will have been surface toughened by a 'hot and cold end' processing during manufacture to reduce the effect of surface damage by scratching and bruising which lowers the strength. They are of small capacity (less than 30 ml) for the same reasons of safety. Small glass bottles are not only less likely to break but if they do burst in an impact, the fragments produced are smaller. Moreover, the fire hazard from the small quantity of any inflammable propellant is also minimized. Industry practice in the United States suggests that internal pressures should not exceed 40 psi in coated bottles up to 80 ml capacity and 30 psi for bottles between 80 and 120 ml. Above this 25 psi is suggested. The largest coated aerosol in the United States holds nominally 10 fluid ounces (approx. 300 ml).

Glass aerosols have fewer design possibilities than glass pump sprayers which not only resemble them in appearance but provide the main competition. As far as perfumes and fragrances are concerned the particle sizes of the sprays produced from both are similar and hence the pump sprayer has the edge. In the pharmaceutical field, however, particle sizes required in, e.g. a bronchodilator spray, are well below those obtainable with a pump sprayer.

Propellants

Various gases have been used, and in fact the soda-water siphon is an example of a pressurized pack. Compressed gases have the disadvantage that whilst they provide an internal pressure in the container, this tends to diminish as the container empties. This means that the spray characteristics alter throughout the life of the container and may become unacceptable. Proper selection of the valve, product carrier and propellant, can minimize this effect, but in general aerosols are not suitable for those materials which require a uniform spray pattern. The compressed gases which have been used are carbon dioxide, nitrous oxide and nitrogen. Table 13.1 lists properties of some of the common propellants.

It has been common practice to use a mixture of Propellant 11 and Propellant 12 and such a mixture has a vapour pressure of 240 kPa at 21°C. This is very suitable for use with tinplate containers which commonly have bursting strengths up to 1000 kPa. The use of the butanes as propellants is becoming widespread, and blends of these with propane are made in order to control the vapour pressure. There is some hazard with the use of propane and butane, particularly in the filling operation, and suitable precautions must be taken. However, once the pack has been sealed satisfactorily there is very little difference in the hazards between properly formulated packs with flammable and non-flammable propellants.

The selection of a propellant depends upon a number of factors such as:

1. Spray characteristics
2. Valve
3. Viscosity of product
4. Nature of product
5. Cost

Table 13.1 Properties of common propellants.

Common name	Chemical name	Chemical formula	Boiling-point (°C)	Density (g/ml at 21°C)	Vapour pressure[a] (kPa at 21°C)
Propellant 11	Trichlorofluoro-methane	CCl_3F	24	1.48	92
Propellant 12	Dichlorodifluoro-methane	CCl_2F_2	−29	1.32	585
Propellant 114	Dichlorotetra-fluoromethane	$CClF_2CCF_2$	4	1.47	190
Butane	n-butane	$CH_3CH_2CH_2CH_3$	−1	0.56	212
Butane	iso-butane	$(CH_3)_2CHCH_3$	−10	0.58	315
Propane	Propane	$CH_3CH_2CH_3$	−42	0.50	860

[a] $1 \text{ lb/in}^2 = 6.895 \text{ kPa} = 6.895 \text{ kN/m}^2 = 68.95 \text{ mbar}$.

The performance of the final pack is affected by several variables and, since these are not independent, the design must be carried out by experienced people.

Products

In theory, anything that can be made up into a liquid or paste form can be dispensed by this method. In practice, only those products which have some practical advantage in using this method of dispensing are so packed.

In general terms, it is important to ensure that the consistency (viscosity) is correctly adjusted to give satisfactory performance. Whilst it is possible to check that this has been achieved by suitable laboratory testing, in view of the wide range of valves and propellants available, it is essential that the formulator has sufficient experience to ensure that the most economical as well as the most advantageous combinations are used.

Having produced a satisfactory product, the physical characteristics must be recorded, and then a suitable series of test containers are packed. These containers are then stored and opened at intervals to check that their performance and the physical characteristics of the product are satisfactory. It is common practice to extend such tests over a period of months, although occasionally shorter times are used. Special techniques, usually available only in the larger laboratories, can reduce the intial screening for corrosivity to as little as one month.

The physical properties that the product must have and the behaviour of the finished pack should be noted. Typical factors requiring investigation might be:

1. The product:
   density
   viscosity
   pH

colour
low-temperature stability

2. The finished pack:
   moisture content
   spray characteristics
   residue
   internal pressure

Labels

No matter whether the label is printed onto the container body during manufacture or as an additional operation after filling, it should have a warning to the effect that:

> This container is pressurized. Keep away from heat, including direct sunlight. Do not puncture or incinerate, even when empty.

Filling

There are two methods of filling that can be employed: one of these, commonly called the *cold method*, involves the cooling of the product and the propellant below the boiling-point of the propellant. The second method *(pressure filling)* involves handling the product at ambient temperatures and filling it into an open container.

Cold filling (Figure 13.3)

It is essential that the product is cold stable, and that it does not become too viscous or precipitate at low temperatures (say −20°C). The system requires special filling equipment that must not only be capable of working at these temperatures but must also maintain the low temperature should the production line stop for any reason. Because of the capital cost involved and the large amount of energy required to heat the can and contents back to room temperature for testing, labelling and packing, the method has largely been superseded except for specific applications.

Pressure filling (Figure 13.4)

Alternative filling sequences are illustrated in Figure 13.4. After the product has been placed in the container, the air has to be removed; this is done either by replacing it with vapour propellant (purging) or by drawing a vacuum immediately prior to sealing the valve in place. The vapour purging method has the advantage of simplicity, but is more expensive to run since it

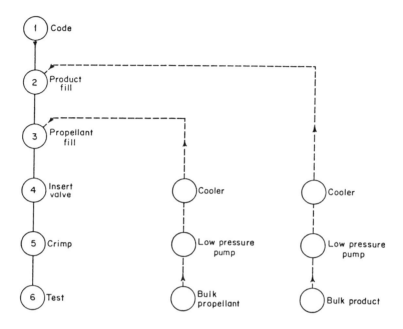

Figure 13.3   Cold-filling sequences.

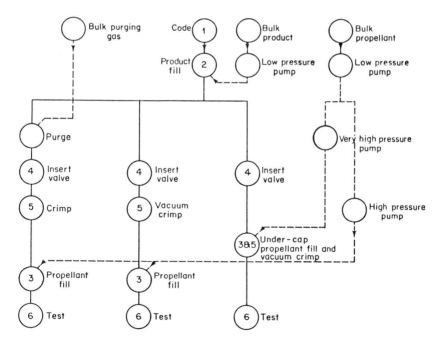

Figure 13.4   Possible pressure-filling sequences.

involves the dropping of several grams of Propellant 12 vapour into the can in order to displace the air before sealing. The exact amount required varies with head space volume and the product. The vacuum purging method requires a more complicated crimping machine and, unless this is a multi-head machine, it may limit the speed of production. It is more difficult to ensure that a satisfactory vacuum has been drawn, although the more sophisticated types have an automatic device for rejecting unsatisfactory cans.

The next machine makes a seal around the boss of the valve, or with certain valves around the valve stem, and then under high pressures (between 400 and 1000 lb/in^2) forces the correct amount of propellant into the container.

The amount of propellant shown in the space between the seals is lost when the head is removed, and it is important to keep this loss as low as possible. By suitably designing the head, it can be so arranged that the button of the valve is actually in place during the gassing operation.

Where the internal dimensions of the valve are restrictive to the flow of propellant, or the amount of propellant to be filled is very large, an 'under the cap' filling method may be used advantageously. In this method, the container with the valve in place but not sealed is presented to the filling head. The valve is lifted out of the way, and a vacuum drawn on the container. The correct amount of propellant is then metered in, the valve pushed back into place, and the crimping operation performed. The amount of propellant lost is rather larger than with the conventional pressure-filling method, but the extra cost may well be offset by the increased speed obtainable, on the advantage of filling with the button in place. The filling head is, of course, much more complicated. There are, for instance, three seals to be made: air/propellant, propellant/vacuum, vacuum/air, and the chances of failure are correspondingly greater. However, these machines have now been in use for a number of years, and the system of filling is well proven.

A variation of this system is to place the valve in position, and on one machine fill the product and propellant consecutively around the valve into the container, and then seal the valve to the container.

The specialized filling having been completed, it is now necessary for the containers to be tested to ensure the safety of the consumer. This is usually done in a hot water-bath, which should be large enough ensure that not only are the containers totally immersed, but that they are there long enough to raise their internal pressure to a suitable level. BS 3914: Part I: 1974 for aerosol dispenser specifies this, but for smaller sizes of metal cans, it is common practice to use hot water at 55°C and immerse for 3 min. The time must be increased for large containers, or containers which have low heat conductivities, such as plastic-coated glass or products which have low heat-transfer coefficients. The test not only exposes the containers to severe conditions and reveals any mechanical weaknesses, but it also detects gross over-filling and enables leaking cans to be rejected.

The cans are then dried, using hot-air blasts or sometimes infrared drying tunnels. Buttoning, spray testing, capping, labelling, and packing then follow, together with any special operations required for presentation of the final package. These operations are similar to those carried out on other types of package and will not be described in detail here.

Non-chemical sprayers

The aerosol industry worldwide is currently in a difficult period of development. The problems of depleting the ozone layer attributed in part to the release of chlorofluorohydrocarbons (CFCs) used as propellants overshadows the industry. In many instances particularly in the United States this led to a change over to hydrocarbon propellants such as butane which raises problems of flammability and increased fire hazard, particularly in warehouses. These problems have led to increased activity in producing other means of providing pressure inside the container.

Mechanical pump sprayers either of the push button or trigger type (Figure 13.5) have begun to compete in several areas. While they are most

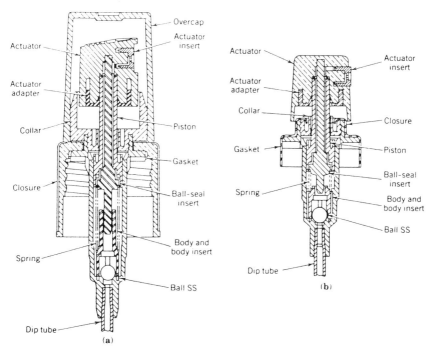

Figure 13.5 Model SL-200 (a) and SL-400 (b) Micromist pump sprayers. Reproduced with permission from John Wiley, New York.

costly, they can of course be refilled. Refills, moreover, can be larger than the original container and thus provide several fillings.

Pump sprayers, like the metered aerosol, provide a shot or dose of product. Once primed and full of product, a precise volume is dispensed for each actuation of the mechanism. When the button or trigger is actuated, a slight vacuum is created in the body of the valve which draws product up the polyethylene dip tube to fill the metering barrel. At the same time, the air pressure created by the actuation forces the product already filling the barrel though the orifice. A spring then restores the button or trigger to its original position.

Another alternative illustrated in Figure 13.6 is the *Exxel system*. This system consists of a thin-walled inner bottle made of lightweight durable

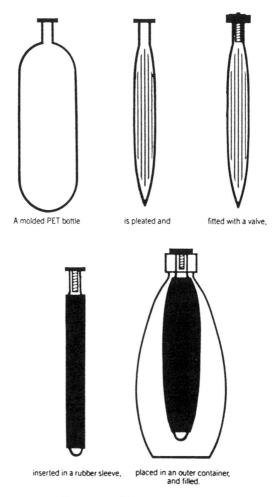

Figure 13.6   The Exxel process.

plastic, pleated, fitted with a valve and inserted into a natural rubber sleeve, which is then placed in a container and filled. The rubber sleeve expands as filling takes place and its tendency to return to normal size provides the power to dispense the contents. No propellant is needed and the viscosity of the liquid contents presents little problem.

The *Atmosol system* also avoids the use of liquid propellants by the use of a regulator using air, nitrogen or carbon dioxide. It can be used with fragrances, polishes and fresheners, etc. The novel regulator fits over the valve stem of a normal aerosol and produces an acceptable spray by providing automatic and accurate flow control until the can is emptied. Since it requires no change to existing containers, valves or dip tubes it can be incorporated readily into existing lines.

14
Metal and plastic collapsible tubes and aluminium foil containers

Collapsible tubes are flexible composite containers for the storage and dispensing of product formulations that usually have a pasty consistency. They may be made of metal or thermoplastic.

The metal tube has been manufactured since about 1885. They were originally made from tin or lead and lead remained the principal metal for tubes until the early 1930s, when aluminium alloys were used. The reason for the change from lead was that aluminium has better forming and handling properties, and the metal tube market now is about 90–95% aluminium, with the rest tin.

Plastics entered the field in the 1950s and increased the overall market for collapsible tubes by their use in cosmetics and other products requiring good barrier properties and an improved aesthetic appeal.

The laminated tube which became available about 1971 or 1972 has taken over a large part of the aluminium tube market for toothpaste in the United States.

These containers also have some markets in the food industry for products such as mustards, tomato pastes, etc.

General description

A collapsible tube without contents is essentially a cylindrical container with a shoulder, nozzle and closure at one end; the other end is open to allow filling with the product before this end is finally crimped to provide a completely sealed hygienic dispensing pack.

The size of the tube selected for any particular product will depend on the volume of product it is intended to hold. Charts giving capacity against dimensions are available from tube manufacturers. The nozzle and orifice sizes must be chosen to suit the dispensing properties of the product, and the approximate quantity required to be dispensed at a time. Dimensions of tubes range from a minimum of 9 mm diameter up to 76 mm; lengths corresponding to these diameters are dependent on the ultimate capacity of the closed tube and range from 40 mm to about 250 mm, but they can be longer if required. There is also a range of standard nozzle sizes and orifice diameters.

Methods of production

As most products packed in collapsible tubes require to be hygienically stored, a brief description of the method of production will serve to illustrate that the risk of contamination of the product by the container is negligible, always provided that the necessary features of the tube to provide maximum protection of the product have been thoroughly investigated prior to filling and marketing.

Collapsible aluminium tube production may be divided into two main parts: mechanical fabrication of the tube and treatment of the tube surfaces.

Mechanical fabrication

Mechanical fabrication of aluminium tubes involves the following stages of processing. Most of these are carried out automatically on a mass-produced flowline basis, with manual handling reduced to a minimum.

Aluminium of 99.7% purity is cast, rolled to a pre-determined thickness, and blanked to provide cylindrical pieces of metal of a defined diameter. These pieces of metal, generally referred to in the industry as *slugs*, are heat-treated to bring them to the desired metallurgical state for fabricating. The slugs are then lubricated and, by a process of impact extrusion, the lubricated slugs are converted in a specially designed press to the tube shape, with a shoulder and nozzle.

Treatment of the surfaces

Surface treatment of the tubes follows and includes the coating of the internal surface of the tube with a protective lacquer (if this feature is desired), followed by a high-temperature stoving of the lacquer. The tube is then coated over the whole of the outside wall with an enamel coating, followed by a moderately high-temperature stoving period. This external enamelled surface can then be printed with up to five colours as desired, on offset printing machines using thermally or UV-cured inks. Tubes coming off the processing line at this stage require only a cap or closure to be fitted to the nozzle before packing into suitable boxes or cartons for dispatch. Throughout the processing, inspection procedures ensure that a consistently clean and good quality container is produced.

Figure 14.1 outlines the first part of the process of making collapsible metal tubes. The impact (Figure 14.1a) cold works the metal slug, first to

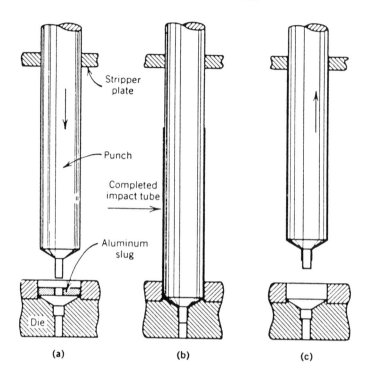

Figure 14.1   Metal tube production process. (a) Beginning of impact stroke. (b) Bottom of
stroke. (c) Beginning to strip. Courtesy of Herlan & Co.

form the nozzle and shoulder and then (Figure 14.1b) to force metal up the
surface of the plunger to form the tubular body. The metal is considerably
work-hardened in the process. The plunger then returns allowing the
completed tube to be removed (Figure 14.1c).

The tube is now trimmed to the correct length and the threads in the
nozzle are formed by rolling, cutting etc. The work hardened tube is then
annealed by heat treatment to recover the flexibility and dead fold
characteristics at about 590–600°C. This also removes the traces of lubricant
and sterilizes the tube. Any required product resistant lining is now sprayed
on, dried and the tube printed.

Tapered tubes were first used in Europe in the 1970s and allow the tubes to
be nested. The taper (to about 2°) allows them just to fit inside one another.
The coatings and inks must withstand this without cracking or flaking.
Makers and many packers prefer the tapered tube because packaging can be
accomplished for delivery and storage in much less space. Nesting can be
done by machine.

Collapsible plastic tubes

These are produced by a number of processes originating in the United States and in Europe. The more important ones are:

1. The Strahm method
2. The Downs method
3. The Magerle process

The first two start by the extrusion of a continuous thin-walled tube (0.35–0.46 mm thick depending on the diameter required). As the hot tubing comes out of the extruder it passes through a corona discharge to modify its surface for later printing. At the same time, the outer surface is cooled by cold water and the tube shrinks to the required diameter as it is drawn over a chilled forming mandrel. The 'sleeve' thus produced may be printed by dry offset at this stage or after the 'head' is applied, using thermally dried or UV cured inks. Some tubes for cosmetics, etc. may be decorated by hot stamping or silk screen to give better quality appearance. Barriers, e.g. two-component epoxy coatings, may be applied at the sleeve stage before heading.

The materials used to form the sleeve and the head of the tube must be compatible to give a firm bond. LDPE sleeves can be headed with either LDPE or HDPE but PP sleeves must be headed with the same material.

In the Strahm heading process, the top of the sleeve is trapped in an injection mould and the head is formed by injecting material into the mould cavity. The tube then cools and is transferred to other equipment which removes the sprue and fits the cap. They are then packed in trays for despatch.

In the Downs process, the heading is carried out in a very different manner. A continuous 3-mm thick strip of LDPE about 50 mm wide is heated so that it is quite soft. The sleeve on its mandrel enters a punch, and a disc cut from the hot LDPE strip and the leading edge of the sleeve bond together. While still soft, the disc plus sleeve pass to the next station where the head is moulded to its final shape.

The third process developed by Karl Magerle AG uses a small turret machine and injects a 'doughnut' of molten plastic into a female cavity before the mould closes. The head is formed by compression moulding. Although these machines have only one tool per station they incorporate capping on the same turret.

These three processes are the most important but there are several others. One, the Valer-Flax process, spin welds a premoulded head to the sleeve and a number blow a bottle and trim off the end. This produces a lot of scrap which must be fed back into the process to be economical.

Laminated tubes

Proctor & Gamble in the United States started to use a laminated tube of paper, foil and plastic for toothpaste in 1971. The barrier properties of this laminated tube were much better than the previously used plastic tube. The laminate is produced by an extrusion process. Large rolls of preprinted laminate (see Table 14.1 for typical composition) and then slit into the correct width for making tubes by seaming the long edges to produce sleeves. For this purpose the flat web is folded into a cylinder, the edges are slightly overlapped and then by heat and pressure some of the plastic layer is squeezed out around the edge of the foil and paper layers to seal them. The aluminium foil in the laminate allows the use of radio-frequency (rf) energy for induction heating in the seaming process. Since the foil is impermeable the seamed edges are the only places where water vapour could enter or escape.

The tubular sleeve is now cut to the correct length using printed registration marks. One method of completing the container is as follows. The head of the tube is a pre-moulded insert of polybutylene terephthalate (PBT). This is placed on the heading tool, the sleeve put in position and the head plastic locks the insert into place as it bonds to the sleeve. Several alternative methods of producing the laminate and sleeves have been developed as well as methods for completing the container. Inserts for the head are also made in a urea plastic.

Table 14.1   Structure of printed laminate.

OUTER SURFACE
    1. LDPE plus anti-static agent
    2. Printed white pigmented LDPE
    3. Paper
    4. LDPE
    5. Acrylic adhesive   Total thickness $\simeq$ 0.33 mm
    6. Aluminium foil
    7. Acrylic adhesive
    8. LDPE
INNER SURFACE

Types and styles of container

As explained in the introduction, collapsible tubes are essentially cylindrical containers. Types of tube are generally distinguished by either

1. The material used for fabrication (i.e. tin, lead, tin-lead alloy, tin-coated lead, or aluminium where metal tubes are concerned, and polythene or polyvinyl chloride where plastics are involved) or
2. The type of nozzle on the tube.

The first of these distinguishing characteristics is self-explanatory. The second is rather more involved and some further comment is required. Several types of nozzle may be used.

1. Conventional nozzles with more or less standard orifice sizes through which the product is dispensed. Typical of this type are the tubes used commonly for toothpaste, hand creams, and similar products.
2. Nozzles where the threaded portion has an extruded rigid 'cannula' of small diameter. These are used where the areas of application of the product are specifically defined. Eye ointment and veterinary cerate tubes are typical examples.
3. Nozzles where the orifice is covered by a thin membrane of metal which must be pierced before the product can be extruded. Such tubes provide a hermetically sealed package and are referred to as 'Membraseal' tubes.
4. 'Taper' or 'torpedo' nozzle tubes have no threaded portion at the nozzle area and require to be pierced with a pin before the product can be squeezed out. Certain adhesive tubes are typical examples.
5. Nozzle, cannula, and orifice features may be varied to suit particular requirements, provided these are within the practicabilities of the impact extrusion process on a mass-production basis.
6. Plastic nozzles attached to metal tubes. These are special innovations which have found increasing use (especially where abrasion between the tube closure (i.e. cap) and the conventional metal nozzle produces a blackening of the product). In other instances, a plastic elongated nozzle affixed to the tube is often used where the application of the pharmaceutical or veterinary product requires the tube cannula to be brought almost into contact with the area to be treated and there is some danger or likelihood of damaging the affected area by metal contact. Certain veterinary cerate tubes with cannulas are used as one-shot dispensers where the cerate is introduced into, for example, the udder or uterus of animals for treatment of mastitis or intra-uterine treatment.

   These nozzles are attached to the tube by mechanically crimping it to the shoulder, or by spinning metal round the moulded plastic nozzle, or by injection-moulding it directly over a metal nozzle of specific conformation.
7. Although tubes embodying various ideas for a captive closure were tried over a period, none completely solved the problem. First ideas were based on attaching the cap by some mechanical means to the metal nozzle of the tube. Although this solved the problem of cap loss, it did not remove the problem of abrasion between the metal nozzle and the cap, which results in discoloration of the product. It is possible to apply a plastic nozzle portion to a tube and to fix a captive closure to it without the necessity of having the nozzle threaded in any way. Such closures have still not gained general acceptance.

Sealing

Different materials require different methods of closing the ends (see Figure 14.2). Metal tubes normally have their open ends closed after filling by some type of crimping operation (Figure 14.2a). Plastic tubes are commonly sealed either by radiant heat or heated jaws (Figure 14.2b). High frequency heating is the preferred method for the laminated tubes but ultrasonic techniques can be used for both plastic and laminated containers. Contamination of the seal area can cause problems with all but ultrasonic sealing, but the last is a slower process.

Types of closure and methods of closing

Closures for collapsible tubes are generally made of plastic materials and both thermosetting and thermoplastic types are available with a variety of external shapes and internal threads for screwing tightly on to tube nozzles. Push-on varieties are also available.

With certain shapes of nozzle or cannula, elongated caps are made to cover both cannula and threaded portion, and seal the cannula at its top. Membrane seal-type tubes are usually pierced open with a 'spike' cap of polythene or hard plastic. There are also novel closures on the market which offer a captive feature.

Packing and dispatch

Collapsible tubes are generally packed in the United Kingdom into paperboard cartons or plywood boxes fitted with honeycomb-type divisions, so that individual tubes do not damage easily in transit, nor have their highly decorative artwork scratched. Tubes for the home market are

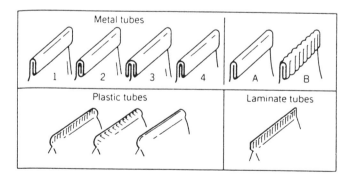

Figure 14.2 Metal, plastic and laminate tube closure types. (1) Single-fold, (2) double-fold, (3) saddle-fold, (4) double-fold, (A) plain, (B) crimped.

delivered by the manufacturer to the filler's premises in plywood or fibre cases, fitted with a flanged fibre lid to exclude dust. Palletized loads are delivered and the cases, which are returnable against a nominal charge, are collected for re-use at the tube works. Tubes are also exported packed in divisional paperboard cartons inside stout wooden overcases.

Aluminium foil containers

Introduction

Aluminium foil containers were first produced to provide low cost disposable pie plates and baking pans for bakers and then for disposable basins and dishes for use in hospitals. Today, foil containers are used for chilled and frozen foods, bakery products, baked meat pies and for much take-away food as well as for medical purposes. Foil for containers, unlike the thinner material used for wrapping, is often not coated or laminated but used plain. In recent times, however, a range of coatings has been developed to achieve product or brand identity, as well as protection. The decorative plain or patterned colours can be translucent as well as opaque and foil containers can also be printed. For some purposes internal white coatings to give a 'porcelain'-like appearance and/or scratch resistance can be provided. For food purposes all these coatings are non-toxic and odourless.

Manufacture

Foil containers are produced on combined mechanical and air presses which force light gauge foil into a female die to give a smooth appearance to the finished article. Figure 14.3 shows some typical types of container in common use.

Closure may be effected in a number of ways—heat-sealing a peelable film or laminate to the edges of a tray and by turning the edges of the container over a lid sheet of rigid paperboard laminate are two of the most common.

Use in microwave ovens

At one time it was considered dangerous to heat food in foil containers in microwave ovens but recent work has shown that more uniform heating is obtainable with these containers. The Aluminium Association, Washington and the Aluminium Foil Containers Association of Wisconsin in the United States commissioned a study of the effects of aluminium packaging materials on microwave oven performance which concluded:

Figure 14.3   Typical aluminium foil containers in common usage.

1. In most instances the food heating results were at least as good as that achieved with containers transparent to microwaves and in many instances it was more uniform.
2. Foil containers had no effect on the magnetron.
3. In just under 400 tests only once did arcing between the foil and the oven wall occur.

Another investigation by the National Research Council of Canada demonstrated that using foil containers in the microwave did not cause the magnetron to operate outside its allowable ratings. Other investigations have shown that only in some of the earliest microwave ovens (built before 1969) has any damage to magnetrons occurred.

15
Glass containers

Glass container manufacturing is a continuous high-speed automatic process. Each year glass container factories in the United Kingdom alone sell about 6000 million bottles and jars, for food (35%), beverages (45%), pharmaceuticals (9%), and toiletries and perfumery (4%). In value these sales constitute about 93% of all types of packaging, some competitive and others not competitive with glass.

Although there is no precise definition, a broad classification for glass containers is that those with narrow necks are known as *bottles* and those with wide necks as *jars*. As with other industries, the glass container industry over the years has acquired a number of terms appropriate only to its own processes and products.

Bottles account for the bulk. The bottle may assume a number of shapes, but the hallmark is a narrow neck which is definitely smaller than the body (Figure 15.1). This facilitates pouring and allows the attachment of suitable closures. Beverages and liquid foods such as salad dressing and vinegar are packaged in bottles. Jars, together with appropriate closures are wide mouth

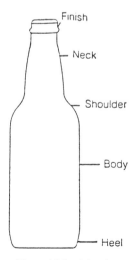

Figure 15.1 A bottle.

containers which afford easy access to the product. They are widely used for foods and cosmetics. Tumblers, wide-mouthed containers which resemble drinking glasses, are used to package jellies, preserves and spreads. Handled containers (Figure 15.2), large-size small-necked bottles equipped with a carrying handle, package beverages, institutional, industrial and household products.

Tubular containers

Amongst the smallest glass containers today are the ampoules and vials used in the pharmaceutical industry. The *ampoule* is a single-dose container manufactured fully automatically from glass tubing and used to contain liquid or powder injection medicines. The top of the stem is sealed after filling and then snapped off when the contents are required. *Friable ampoules*, which can be easily crushed and are used mainly for inhalants, are manufactured by hand from thin-walled tubing. *Vials* can be single or multi-dose containers, and are manufactured either from tubing or on conventional bottle-making machines. They are often sealed with closures that have rubber inserts, which can easily be pierced by the needle of a hypodermic syringe, but they are also designed for use with screw caps or cork stoppers.

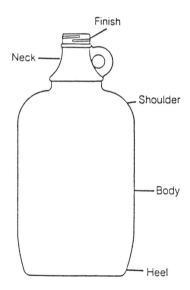

Figure 15.2   A handled container (a jug in the United States).

Glass container making

Forming the container

Container forming begins with the feeder which sizes, shapes and cuts individual gobs of glass (Figure 15.3). All glass containers are made in two stages, and therefore two sets of moulds are required. First, an initial blank or parison shape is formed and then transferred to the *blow mould* (Figure 15.4) where the final shape is blown. The parison can be blown or pressed. These stages are shown in Figure 15.4.

The Independent Section (IS) machine is now almost supreme in the industry for the manufacture of glass containers. It operates two basic processes. Narrow neck containers for liquid foods are made by the 'blow-and-blow' process. An accurately formed 'lump' of glass (known in the industry as a 'gob') is directed into a parison mould where a bubble is formed, at the same time as the moulding of the finish (that part of the bottle which will support the closure) is accomplished. The glass parison is inverted to the other side of the machine and the formation of the body of the bottle follows by means of a blowhead, blowmould and compressed air. The formed container is then transferred to a conveyor which transports it to the annealing operation.

Wide-mouth containers, which are used for semi-liquid and solid foods, are made by the 'press-and-blow' process (Figure 15.5a, b and c). The accurately formed gob of glass is shaped into a parison and the finish is moulded by means of the upward action of a driven plunger. The formed parison is then transferred for blowmoulding as with the 'blow-and-blow' process.

The accuracy of the final mould will determine the ultimate shape of the

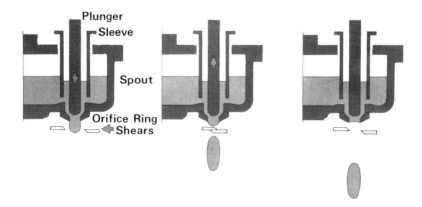

Figure 15.3 Gob forming. The feeder consists of a plunger which extrudes the glass that flows along the forehearth. A pre-arranged weight of glass is cut off to form the gob.

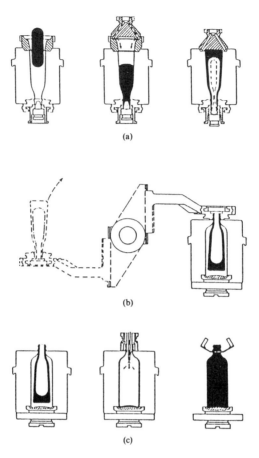

Figure 15.4 Functional diagram of parison and container forming by the blow-and-blow process. (a) Blowing parison. The first stage of a glass container in which the parison is blown. (1) Gob drops into the parison mould, (2) settle blow to form the finish, (3) counter blow to complete the parison. (b) Transfer from blank mould to blow mould. (c) Final blowing. The second stage of the manufacturing of a glass container in which the parison is blown.

glass container, but of equal importance is the distribution of glass and wall thickness. It is in determining this that the parison plays an important part. The shape of the parison, and the temperature of both the glass and the mould, will affect the distribution of the glass when the final shape is blown. The initial temperature of the glass is about 1000°C and of the mould about 500°C, and the aim is to keep the temperature as constant as possible. If the mould cools, the glass will not flow easily; if it is overheated, the glass will stick to the surface.

The moulds are made from fine-grained cast iron (Figures 15.5 and 15.6) with the working surfaces highly polished. The subsequent application of lubricants containing graphite forms a very fine carbon film. The heat of

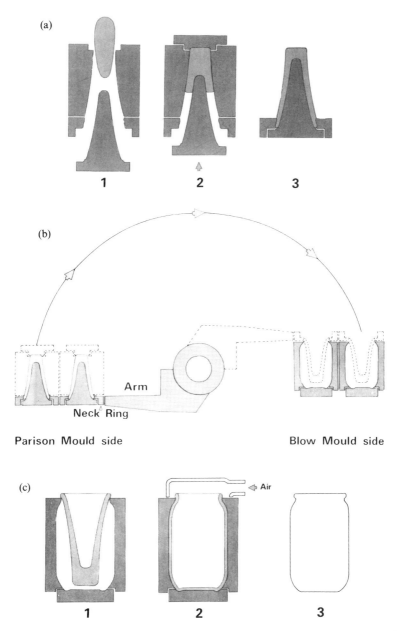

Figure 15.5 The press-and-blow process. (a) Parison pressing. The first stage of the manufacturing of a glass container in which the parison is pressed. (1) Gob drops into parison mould, (2) plunger starts to press parison, (3) parison completed. (b) Transfer mechanism. The gob, having dropped into a parison mould has been pressed into its initial or parison shape. It is then transferred to the blow mould for completion. (c) Final blowing. The second stage of the manufacturing of a glass container in which the parison is pressed. (1) Parison in blow mould, (2) jar blown to shape, (3) finished jar.

forming causes oxidation, which has to be cleaned off. This, together with fair wear and tear, means that the mould cavity will gradually enlarge. Since tolerances on the body diameter and the finish are important, it is customary to machine a new mould to slightly below the mean specified dimensions of the container. The life of a set of moulds will depend ultimately upon the agreed tolerances.

The IS machine can be adapted according to the volume of production planned and the container size for a particular range of bottles. It can consist of a number of sections and each section can be equipped with moulds for handling one to four gobs.

Two machines can be operated in tandem; thus two six-section machines in tandem can produce an output equivalent to that of a twelve-section machine, a technique which has been developed during the last five years to meet large demands for one particular bottle design.

Because the machine consists of independent sections separately timed, it is possible to make bottles of different shapes on the same machine, provided they can be formed from a gob of common weight. By changing mould equipment the IS machine is capable of making a wide range of bottle shapes and sizes, using either the 'blow-and-blow' or the 'press-and-blow' processes, and performs a most versatile manufacturing operation.

The manufacture of a bottle depends on the tuning of a heat flow process which takes place between the glass and the cast iron moulds which form the bottle. Hence the need for close control of glass viscosity. The thermal conditions are adjusted to ensure the stability of the bottle as it emerges from the blowmould at maximum speed. Cooling is therefore of fundamental importance.

Vertiflow is the trade name given to a new process for cooling moulds. Air is ducted through the body of the mould iron. Thermal imaging is used to calculate the distribution of ducting to effect optimum cooling. Compressors used for the traditional methods of cooling are being replaced by fans which are cheaper to purchase as well as operate.

Accurate machine timing within an IS machine has been achieved by means of a rotating pianola type of drum associated with each section. Each mechanism is air operated; studs supported on the drum operate air valves by means of a series of latches and levers. Thus, each mechanism is correctly 'programmed' (in a mechanical sense) for its operation. Provided the machine is in a high state of maintenance, the mechanisms operate in ordered sequence.

This system has now been replaced with much simpler, magnetically operated air valves for each mechanism, which can be adjusted from a console situated adjacent to the machine. Programmes can be built into the console for setting up all sections at the same time, which is more rapid than the adjustments which were required for each drum.

Electronic timing is now replacing drum timing as a means of improving efficiency of operation, reducing job changeover times and introducing higher levels of safety for machine operators.

As production speeds have increased, so the cycle time on an individual section available for the formation of a bottle has reduced. A recent development in Germany has added a turret to the blow-mould area; this supports a number of individual moulds, each of which receives a parison for final blowing. The turret indexes with the machine cycle, thus increasing the glass contact time and the heat extraction without altering the overall production speed of the section. The RIS (Rotating Independent Section) development, which is the name given to this modification to the IS machine, has increased production speeds whilst at the same time improving the control of bottle dimensions, but its capital cost is high which restricts its use to high volume markets.

Semi-automatic process

The process that has been described is automatic, and most glass containers are made in this way. But automatic high-speed processes in any industry, if they are to be economic, depend on long runs. The automatic glass container process, with moulds costing in the region of £10 000–15 000 is no exception.

In the semi-automatic process, the glass is gathered and the moulds are operated by hand, with the result that, although production speeds are lower, so too are the costs of moulds. The semi-automatic process can cater, therefore, for special orders, trial runs, and those markets where small quantities are required. Figure 15.6 shows the sequence of operations.

Glass containers must be annealed, inspected and packaged to enable them to be transported to and handled through customers' filling lines. They must also be able to perform well in the retail environment when they are filled with the product.

Annealing

When the finished bottle leaves the final mould (Figure 15.7) its temperature is in the region of 450°C. If it were left to cool on its own, a differential rate of contraction across the glass would set up sufficient strain to make the bottle unstable. This is because glass is a poor conductor of heat. Because of the heat retained within the bottle, the internal surface of the bottle wall will cool less rapidly than the outer surface, and the body of the glass between the surfaces even more slowly. The aim, therefore, is to ensure that the rate of cooling, and hence any contraction, is as even as possible throughout the body of the glass. To do this, bottles are fed from the forming machines to an annealing *lehr*, which is a long tunnel, where they are heated to a temperature of about 600°C (the annealing temperature used depends on the geometry of the container) and gradually cooled (Figure 15.8).

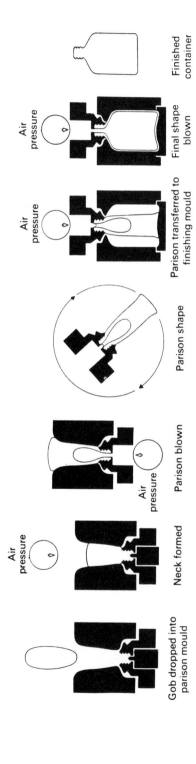

Figure 15.6 A semi-automatic blowing system. The gob is gathered by hand on an iron and the correct amount of glass is dropped into a preliminary mould. Compressed air is introduced to form the neck of the article. The embryo shape (parison) is then transferred to the finishing mould in which the final shape is blown.

Figure 15.7   The bottles emerge from the blow mould and are conveyed to the lehr. Courtesy
of Glass Manufacturers' Federation.

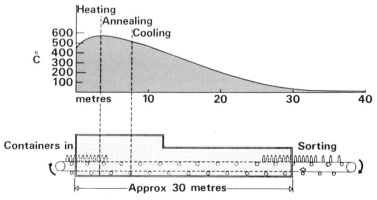

Figure 15.8   The annealing process.

In order to cope with the ouput from larger IS machines and the space required for inspection equipment within existing buildings, annealing lehrs have had to become wider and shorter in length. Improved design and operation have reduced the costs of annealing.

Inspection

When glass containers emerge from the annealing lehr, each one is inspected for faults (Figure 15.9) such as *striae* or *cord*, *seeds*, or *stones*; *checks*, *smears* or *crizzles* which are cracks on the surface of the glass; or *chokes*, constrictions in the bore of the neck and other forming defects. The bottles and jars are then packed in outer containers, or bulk-palletized and dispatched to the warehouse.

Twenty years ago, a visitor to a glass factory would have witnessed several employees in the inspection department examining bottles by hand as they were conveyed away from the annealing oven or lehr. As size and output of the IS machines increased, so the number of employees had to be expanded. Now, in a modern factory, these repetitive uninteresting tasks have been replaced by a range of purpose-built machines. The cost savings have been significant because for each person actually seen at the lehr, there are also three additional employees, one who has finished work and is on his way home, one in bed asleep and a third getting ready to come to take over at the end of the shift. Twenty-four hour, seven-day week operation dictates the

Figure 15.9   The glass containers emerge from the annealing lehr (see Figure 15.8) and pass on to be inspected for faults. Courtesy of Rockware Glass.

necessity for this 'four-crew three-shift' type of employment. Some factories have managed to eliminate all labour in the sorting room; some still retain a final inspector examining bottles as they pass in front of an illuminated screen, but the trend is towards complete automation.

A typical modern inspection line consists of: a squeeze tester, for removing any obviously cracked bottles; 'image grabbing' equipment for comparing the bottle with an inclusion-free stored image and rejecting bottles which are contaminated; bore gauging, for checking the entry dimensions of the finish, thus ensuring compatibility with the customer's filling tubes; check detection, for close examination by photoelectric techniques of sensitive areas of the bottle (such as the finish) for cracks and 'crizzles'; wall thickness measurement, where bottles are rotated against horizontal sensor bands which measure wall thickness at the point of contact. Glass distribution can be assessed by increasing the number of sensor bands in relation to the height of the bottle. This device has been most helpful for improving the quality and consistency of bottles used for the packaging of carbonated beverages.

Quality control

In ensuring that a run of containers will meet the specification, a wide range of factors has to be considered. It is important, therefore, that a constant check be kept on every part of the process where variability is likely to occur. In a highly productive process, where two days may elapse between the feeding of raw materials to the furnace and the emergence of finished containers, a considerable number of faulty containers could be made if there were not a recognized procedure.

Already an indication has been given of the factors likely to affect the quality of a glass container. The manufacturing process is continuous and, from the point of view of control over quality, it can never be put in reverse, i.e. an error at any one stage of the process cannot be corrected later. Apart from the obvious checks on the quality of raw materials, and the glass mould condition, temperatures and rate of annealing, sample bottles are taken immediately they leave the finishing mould, and are checked for weight and dimensions. A general summary of these controls is shown in Figure 15.10. Many of these are routine, necessary for the efficiency of the process as a whole.

At the inspection stage, a more rigorous check is made to ensure that the containers meet the specification. The number involved prevents a complete check on every bottle and jar. Not only would such a procedure add considerably to the cost, but it is unnecessary. However, for some specific characteristics, e.g. height, diameter, neck bore, cracks and crizzles, each container may be subjected to an electronically operated inspection.

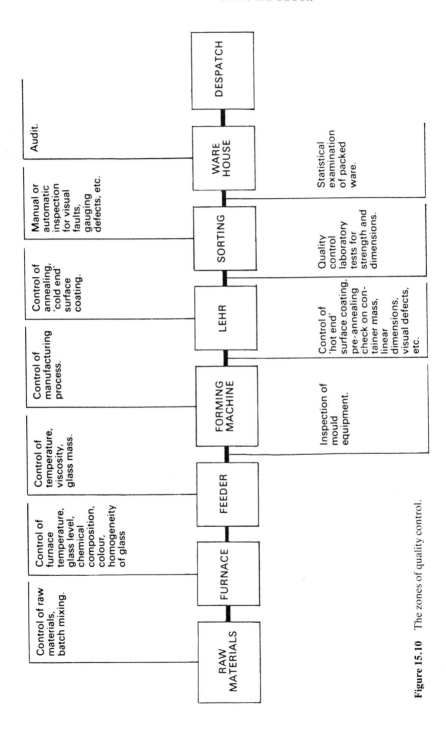

Figure 15.10 The zones of quality control.

Detailed inspection for other factors is done on a random sample selected at regular intervals from the production line.

In drawing up a quality-control procedure, defects are classified into three categories; minor, major and critical. Minor defects are those which will not affect the utility of the container, but which may look unsightly. In this category come seeds, striae, prominent mould marks, and the like. Major defects are those which could, but do not necessarily, cause the container to give an unsatisfactory performance. Fine cracks or checks could give trouble if the container were to meet considerable tensile stress, but such stress may not be encountered during the life of the container. Critical defects are those which render the container unusable. These are rare and, given normal control over production, such containers should not get to the annealing lehr.

The aims of quality inspection are to maintain the reputation of the glass manufacturer, and to ensure that the customer receives containers of the standard required. For minor and major defects, an acceptable limit for the percentage of defectives is set, and this determines whether or not the batch of containers from which the sample has been taken is acceptable. For major defects, this level is more severe than for minor defects and in the unlikely case of critical defects, the limit is zero. The over-riding aim throughout is that the levels should be set from a knowledge of what is fit for purpose.

In order that its customers may study in more detail the quality-control procedures in current use, the glass container industry has published guides.

Acceptance sampling

Not all customers will want the same *quality* of bottle. The product, how it is filled and capped, the method by which it is to be sold, will each in its own way determine what is, and is not, important about the glass container.

When a customer for one reason or another wishes to carry out additional checks on quality, an acceptance sampling scheme is agreed. This clearly is a matter for negotiation and may well affect the cost of the bottles.

Tests and test procedures

The most critical of the dimensions on a glass container are those relating to the height, the body diameter (or relevant dimensions if the container is not cylindrical) and the finish. These are important to the packer, especially on high-speed lines where the filling heads, the star wheels and the capping machinery can cause damage if there is too great a variation in these dimensions between one container and the next.

Reference has already been made to the variation in the dimensions of a set of moulds during its working life, and the tolerances that can be reliably maintained by glass manufacturers have been set out in detail.

The cavity of the moulds will not, of course, wear evenly over its surfaces during the working life. The differential wear on surface causes *ovality*, which is usually expressed as the difference between the axes. In practice, ovality is kept to a minimum, particularly in those instances where bottles are conveyed on the sides or rolled for labelling.

Verticality is the relationship of the centre of the finish of a bottle to the centre of the base. This is an important feature, particularly in tall bottles where an excessive eccentricity could cause difficulty in locating the filling head.

To check the contour of the finish, a shadowgraph is used in which the enlarged outline of the finish is projected against the specified contour.

Capacity

The introduction of legislation calling for an exact definition of the weight or quantity of goods offered for sale has emphasized the need for accurate 'capacities'. In recent years improved control over production has resulted in close control over the capacity of glass containers. Nevertheless, since capacity depends upon the volume of the mould, the volume of glass in the mould, and the accuracy with which the glass reproduces the mould shape, variations are bound to occur. The existence of these has led to the adoption of the *bulk test* as the normal method of control. In this, the capacities of a given sample of bottles, selected at random, are measured, and statistical calculations indicate whether the mean capacity, the sample and the spread of individual measurements satisfy predetermined requirements.

The legislation in respect of measurement of containers places the responsibility of supplying a container of specified capacity on the glass supplier provided the bottler has filled to the level indicated on the container. A reversed epsilon (ɜ) embossed, usually on the insweep at the heel, indicates that the container has been made to measuring container tolerances.

When glass containers are to be filled with hot products, or heated during processing, a test for resistance to *thermal shock* has to be carried out. The procedure is to immerse the bottles in hot water for a given time, and then to plunge them into cold water. This is designed to test the effect of thermal shock, and the important points are the close control of the temperatures, the time the containers are immersed in hot water, and the transfer time. The details of the tests will vary according to the nature of the bottle and the severity of the shock it is likely to meet on the packaging line. All glass containers, once they are packed and closed, will be subjected to variations in internal pressure, although these will be minimal in many instances. Variations are likely to be more severe when the containers hold carbonated liquids. Where internal pressure is known to be an important feature, bottles are regularly subjected to pressure tests.

Specifications

The specification should set down in precise terms the details of the container the packer wants and which the glass manufacturer can supply. Ideally, it specifies those requirements that are vital to the efficient handling of the container. In practice, there is always a danger that unimportant, or relatively unimportant, details will be included.

The specification should, therefore, deal only with essentials: the height, diameter and capacity, for example, and the weight and colour of the glass, and if resistance to pressure or thermal shock is important, then details of the tests should be set down. Dimensions that are critical should carry tolerances, and these should be realistic; otherwise the container will carry unnecessary additional cost.

In short, the specification is the basis of the contract to which all parties will work, and to which dispute, should it arise, can be referred. It must, therefore, be acceptable and practicable. The terms used in a specification are set out in Figure 15.11.

Packaging

The packaging of bottles for warehousing at the glass factory and subsequent despatch to the bottler has also been automated. Hand loading of cartons and crates has been almost completely superseded by bulk palletization, involving the assembly of layers of bottles on a wooden pallet, each layer being separated by means of a tray or layer board and the whole being held together with shrink film. Following the inspection operations, bottles are collated to coincide with a frame which supports inflatable rubber tubes. The lowering of the frame allows the tubes to be placed between the bottles and near the finish. Inflating the tubes allows them to grip the bottles, which are then elevated away for stacking on a layer board supported by a partially assembled pallet. The Schaberger principle has become almost universal for palletizing bottles, although some wide-mouthed bottles are handled in a different way involving a sweep-off motion.

Glass manufacturers are more and more recognizing the importance of the packaging needed to deliver their product. Rockware Glass in the United Kingdom for example spend in excess of £8 million on packaging materials to ensure correct delivery of glass containers to their customers. They have recently implemented 'TRANSPAC', a software system developed to accurately specify, pictorially in 3D and in normal projection,

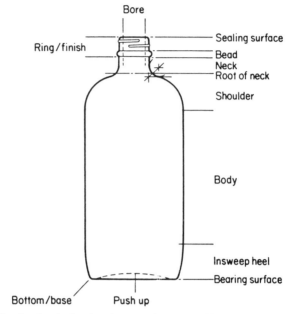

Figure 15.11   The glass bottle showing glass container terms. The terms *ring* and *bottom* have in general been used in this terminology. The term *finish* is sometimes used to denote the *ring*, and the terms *base* and *punt* to denote the bottom.

the packaging needed, i.e. pallets, crates, cases, etc. All the required materials for construction are listed and specified—costs are also available— to be fed into the now computerized product costing system. TRANSPAC is interfaced with the CAD systems for bottle design and hence can be used to tackle other problems of delivery and marketing.

Design

Aesthetic as well as technical considerations play an important part in the development of new designs. The range of designs is very wide. Nevertheless, no matter how exotic the designer's ideas, or how compelling the marketing manager's argument, in the final analysis the container has to be made on a machine, it has to hold a given product, and it must be capable of being filled, closed, transported and used with ease by a consumer. Each design, therefore, is a compromise, the ultimate aim being to achieve the most effective total package in terms of customer satisfaction.

It is important, therefore, that packers should consult all those concerned with marketing of the product at a very early stage. Furthermore, the glass container manufacturer, and the manufacturers of closures and filling machinery, must be given precise details of what is required, otherwise something less than optimum efficiency will be achieved.

Stress analysis

Finite element analysis is a computerized technique which enables an engineer critically and accurately to examine the stresses which exist in a civil engineering structure at the design stage provided he has the basic strength constants of the materials involved.

Such techniques are now being introduced at the design stage to predict the internal pressure resistance (IPR) and head load resistance (HLR) of carbonated beverage bottles, thus saving effort and time in making and sampling bottles to prove a design prior to bulk production.

Optimization of design (to give maximum strength against dynamic stressing), as on a filling line, and even more exciting, the flow of glass in the mould will be a fact before long. These could all lead to reduction in lead times for delivery of new designs to customers.

Since glass containers are blown from a ductile thermo-elastic material, the ideal basic shape is a sphere. This is the shape of many early bottles, whose only departure from the spherical shape was the addition of a neck for filling and decanting the product, and a pushed-in base to enable the bottle to stand upright. Production men favour glass container designs which approach the shape of a sphere. With these shapes, glass distribution is easier to control and production speeds can be increased. Unfortunately, the market demands almost the complete opposite! In the consumer's eye, height represents quantity. Bottlers attempt to modify their designs to give maximum height in order to obtain a sales advantage.

For most practical purposes the cylindrical shape is the best. It can be manufactured relatively easily; bears a close relationship to the sphere, which gives maximum strength; and because it is stable, it lends itself to high-speed filling and automatic handling. A departure from this shape, although adopted in almost every design, brings certain disadvantages. The following is a rough guide to the strength of various shapes relative to the cylinder.

1. Cylindrical cross-section   1
2. Elliptical cross-section   0.5
3. Square cross-section (with well-rounded corners)   0.25
4. Square cross-section (with sharp corners)   0.10

There are, of course, many bottles on the market which have a square cross-section with sharp corners, and their strength is adequate. The guide is for bottles with a comparable distribution of glass. An adjustment can be made by varying the amount of glass in order to increase the strength of the container. This ensures that the distribution of glass over the surface is adequate to resist any shock the container is likely to meet in use.

Since the natural shape of blown glass is spherical, there is a resistance to flow into rectilinear shape. A square corner is thus likely to have less glass in it. This weakness is accentuated by the fact that, during handling, corners are more likely to be knocked. This problem has to be faced in every glass container, since they have to stand vertically, and it is overcome by 'insweeping' the side wall to join the base (Figure 15.12). Ideally, any change of shape, from the base to the sidewalls, from the sidewalls through the shoulder to the neck, for example, should flow smoothly. The glass manufacturer will then achieve a good distribution of glass, which is of the utmost importance. It is important, for instance, that the walls of the container should be slightly thicker at likely points of impact. If a lightweight container is required, then the design should be such that the distribution of glass is well under control and approaches that of the ideal. Handling methods and high speeds in packing plants impose their own discipline on the designer. Simpler shapes can be handled at lower costs, and care in matters such as the position of the centre of gravity can enable the filling line to run faster. The label is not a matter to be left to the last moment, since the bottle designer must know the area needed, and whether a recessed panel is required to protect the label from scuffing.

Shape can also be used to represent product and customer image. Certain products such as salad cream and tomato ketchup have been packed in particular bottle shapes from which it is now difficult to move away, for fear of losing customer recognition. Particular brands of whisky are packed in shaped bottles which have been maintained over the years in spite of strong pressure from production departments to replace them with a cylindrical bottle to improve line handling. In the United Kingdom, shape has been marketed by the glass industry as a main attribute of glass packaging for the purposes of emphasizing brand identity.

Designing glass containers using traditional drawing board methods was a long and tedious process extending over several days and involving many repetitive calculations and visits to the customer. The application of desk-top computers has reduced the time taken to minutes rather than hours and days by performing the many repetitive calculations at high speed. The appearance of automatic drafting machines also accelerated the design process by producing engineering drawings of consistent quality without the

Figure 15.12   The 'thin spot' produced by a right-angled junction (left) and the remedy, the 'inswept junction' (right).

aid of the draughtsman. A customer and his assistant team (a design house representative, a line handling engineer and a marketing manager) are now able to gather round the computer screen with a glass container product designer and agree on a design, thus avoiding time consuming visits by the designer to the customer's premises often to be told that his work is unacceptable or in need of further modification. Relieved of the repetitive calculation and drafting work, the designer is able to focus his attention on the customer's requirements and match them to the production capability of the IS machine.

Developments in both software and hardware have enabled designers to extend this technique to the development of shaped and handled containers. Translucency and multi-point illumination are introducing a greater degree of realism to the image on the computer screen, enabling a customer to view his product more precisely (Figure 15.13) and commit himself to a new concept at an early stage. In the near future it is hoped to introduce a further version of the software which will permit the label to be added to the bottle.

There is no doubt that this area of development is employing some of the most sophisticated modern technology and as such it is one of the most

Figure 15.13 Computer-aided design (CAD). The cachet bottle is typical—because of its shape, it could only be designed in 3D on a computer screen. The mould joint is on the shoulder.

exciting developments in the industry. Work is now proceeding to link the design of moulds to this facility; this will be closely followed by the development of CAM techniques for mould manufacture at reduced cost and with faster production time and greater accuracy.

Decoration

Glass containers are decorated to meet the needs of marketing expensive up-market products e.g. cosmetics and spirits. Techniques available in the United Kingdom are:

1. Acid etching: HF is used to produce a silky etched effect
2. Spray coating
3. Ace decoration: silk screen painting using coloured enamel inks followed by firing near the annealing temperature of the glass
4. Colorama: already described in Chapter 6
5. Nu-tec coating: polyurethane spray coating to produce a range of coloured bottles

The subsequent usage of the bottle, and how it is to be marketed, are also of importance; the designer should be provided with these facts. Most product markets use non-returnable bottles but, where the delivery of filled and the collection of empty bottles is under the effective control of the packer, as in the case of the brewing and dairy industries, a multi-trip bottle is an economical pack (Figure 15.14). Where distribution is such that collection is difficult or costly, the non-returnable bottle is more effective and can be lighter in weight.

Marketing through self-service stores suggests the need to display merchandise to the consumer and this in turn imposes conditions on the shape of the container and the label. The ability to stack jars is another factor which calls for stable containers and attention to the closure.

Developments

One of the most interesting features in glass-container manufacturing is the varied nature of the developments that are occupying the time and energy of engineers and glass technologists.

In recent years much research has been brought to bear on the strength of glass. Success in this field of research is, of course, of importance to the manufacturers of many glass products other than containers. The fact is that in use the tensile strength of glass products is far below its theoretical value. For example, the theoretical tensile strength of glass is estimated to be well in excess of 70 000 kg/cm^2 (1 000 000 lb/in^2), and in laboratory experiments

Figure 15.14 Example of lightweighting: the progressive reduction in the UK (1 pint) milk bottle since World War II. Courtesy of Rockware Glass.

values of 14 000 kg/cm^2 (200 000 lb/in^2) have been achieved. In practice, however, a bottle will have a tensile strength ranging from 700 to 1400 kg/cm^2 (10 000–20 000 lb/in^2). For normal usage this is sufficient and, although a bottle breaks occasionally, this is sufficiently infrequent to cause little bother. Nevertheless, with each increase in filling speeds, a hold-up on the filling line becomes more costly, and each reduction in breakage makes the glass container more economic.

Reduction of surface damage

It has been known for many years that the loss in tensile strength is due to *microcracks* in the surface of glass which arise as the result of, or develop after, the manufacturing process. In the container field, therefore, research has tended to concentrate on the less costly method of treating the surface in such a way that microcracks are less likely to develop or, alternatively, to create a more stable surface layer. Silicones and other chemicals have been used commercially to give greater lubricity, but in some instances these have led to difficulty in handling. A more recent development in the United Kingdom has been to spray a liquid metal organic compound on to the surface of the glass whilst it is still hot. The main effects of this are to reduce the development of microcracks, and at the same time to increase the lubricity of the surface. A reaction between the compound and the glass takes place, such that some penetration of the surface occurs, inhibiting the development of microflaws. The addition of an organic lubricant when the bottles are cold increases the resistance to damage during handling.

It has been pointed out that much can be done with existing technology to preserve the inherent strength of glass container surfaces. A prime requirement is homogeneous, thermally well-conditioned glass gobs. The bottles need to be made on well-maintained, electronically timed machines and the moulds lubricated (swabbed) with dry solid lubricants. Alternatively mould materials which diminish or avoid the need for swabbing are used. If swabbing is necessary then the use of robots is being tried. Narrow-neck bottles have also been made lighter without loss of performance by adopting press-and-blow techniques developed from conventional press-and-blow by increasing the material duty of the individual components. The narrow neck press-and-blow process (NNPB), pioneered by the Heye Glass Company of Germany, gives improved glass distribution in the formed bottle, which increases bottle strength. Improved cooling has been an important element in achieving a satisfactory performance with this process. Lightweight beer bottles with a weight of 155 g and a capacity of 330 ml are now in full production using this technique. Many other designs have been similarly converted and this work will continue in the future. 'Lightweighting' and improvements in production efficiency are development programmes of the utmost importance to the future of the industry.

Other workers have stressed the importance of avoiding surface damage in container forming and processing as by far the cheapest way of producing high strength containers. A review of progress to date in surface treatment of containers, stresses the advantage to be derived, in all types of treatment, from reducing damage to the glass surface before the coating is applied.

Current approaches to conserving or improving container strength

The treatments currently applied to glass containers to conserve strength may be differentiated as chemical (phosphate or nitrate), physical (hot-end and cold-end coatings), physico-chemical and external (plastic sleeve or prelabelling). Of these the physical methods are by far the most frequently used. Usually a combined hot-end and cold-end system is employed, the consensus view these days being that a hot-end treatment alone can be counter-productive.

Surface treatment

Much greater emphasis is being given to this area of operation. Thickness measurements quoted in coating thickness units (CTUs) are being specified for hot-end treatment in order to avoid torque-release problems associated with closures and cap rusting, specifically related to crown closures. Also, slip angles measured by the use of standard (AGR) tilt tables are being used for specifying cold-end treatment. Both are essential to assist the adequate conveying of glass containers along high speed filling lines and to reduce/ avoid breakage during transportation of bulk pallets. More and more, customers are imposing quantitative specifications for surface treatment on glass suppliers.

During recent years container manufacturers worldwide have initiated several cooperative research and development projects aimed at improving strength and thus resisting the inroads being made by alternative materials in their traditional markets.

Experiments have been made with a variety of coating materials and application methods, some of which appear to have distinct advantages over the conventional hot-end/cold-end treatments.

There is now much greater awareness of the need to minimize surface damage during the manufacturing process, especially at the earliest stages of forming, and progress is being made by research directed towards devising practical methods for preventing surface damage.

The cold-end systems available are too numerous for any sensible discussion of relative merit, the choice in any individual case being influenced by so many factors other than lubricating efficiency. The toxicity characteristics, ease of labelling, durability in the case of returnables, resistance to washing, particularly with caustic, method of application and cost, may all prove to be more important for any particular manufacturer

There has also been some interest in single-layer coatings, applied cold but having the same effect as a combined hot- and cold-end coating. Typically such experimental systems were based on dilute phosphoric acid and were abandoned because of the difficulties of application and effect on appearance, but current work on resins reacted at low temperature with acrylic resins may prove more promising.

In Japan, a process has been developed where the surface of the glass is strengthened by a chemical strengthening treatment (CST); a one litre carbonated beverage bottle capable of withstanding an internal pressure of $200 \, lb/in^2$ has been made at a weight of 203 g using this process. Without CST the bottle weight would have to be at a minimum weight of 470 g with sleeving to ensure an acceptable service performance. The capital cost of the process outweighs the advantage obtainable from light weighting in the European and UK industry.

Plastic coatings, especially for carbonated beverage bottles, have been used for many years. At first PVC, and later polyethylene, were used but several incidental problems were encountered. More recently, a whole host of patents, covering a wide range of possible plastic coating materials, has appeared on the scene, including ionomer resins.

Pre-labelling consists of covering a substantial area of the bottle with a wrapper extending all round the perimeter of the body. There are three main types, commonly known as 'sleeves', 'Plastishield' and 'all-rounder' labels.

The combination of bottle and protective covering, trade-named Plastishield, developed by Owens-Illinois in the United States in 1971, uses extruded polyester film shrink-wrapped round the bottle. In the process developed by the Fuji Seal company in Japan, transverse mono-axially orientated PVC envelops the bottle and then is shrunk on to the bottle surface by heating. Polypropylene and polyethylene may also be used. The 'all-rounder' method involves the use of paper laminated with polypropylene fixed to the surface of the bottle by the hot-melt method.

16

Moulded pulp, expanded polystyrene, moulded and thermoformed plastic containers

Moulded pulp containers

Moulded pulp provides a lightweight form of packaging; a container with dimensions 100×100×150 mm can weigh between 25 g and 100 g depending on thickness and density. Such containers and forms are able to absorb shock without transmitting it to the packed item because although the material itself is only slightly resilient, the shock is dissipated by distortion and crushing at the point of impact.

Containers and forms can be made from any of the pulps generally used in making paper and board, from pure bleached sulphate pulp right through to pulps made entirely from waste material. The type of pulp used will depend on the articles to be packaged. Thus foods such as fresh meat or fish would require the best quality bleached sulphate and/or sulphite pulp while corner protectors for furniture might be made from an all-waste pulp of the right strength.

The density and, as a result, the thickness of the material can be varied by leaving it as first formed to give a softer and less rigid structure or by pressing it to provide a harder and more rigid material. In this way, thicknesses of 0.75–6.4 mm can be obtained, the density varying from about 0.2 g/cm^3 to 1.0 g/cm^3.

The forming process used in the production of moulded pulp containers is, in many respects, similar to paper-making but with the important difference that, whereas paper is formed on a travelling wire screen moulded pulp containers are made in a mould fitted with a screen, formed to the shape and profile of the container to be produced.

Manufacture

There are two distinct moulding processes employed for the production of pulp containers. These are known as *pressure injection-* and *suction-moulding*, respectively. Although these two processes produce articles

having different characteristics, the preparation of raw material is basically
similar for both, and follows the same lines as that used in the paper-making
industry.

The water-borne pulp is either forced under pressure into and through
a perforated mould or is sucked through by applying a vacuum on
the far side of a perforated mould, so that, in both methods, the
fibres are deposited on to the mould. Whichever method is used the
perforated mould approximates the shape of the item to be packed and
produces a seamless moulded product of the correct consistency and
strength.

Before an article can be produced by any type of moulding process; it is
first of all necessary to make a suitable mould. In the pressure injection
process, one mould is the minimum. If very large quantities of mouldings are
to be produced, multiple pocket moulds are employed; but the process is
also well adapted to the production of relatively small quantities of not less
than 5000 of any one article. It will be appreciated that the cost of the mould
has to be spread over the number of articles produced, and hence the greater
the quantity, the lower the tooling charge per unit.

Pressure injection process. In the pressure injection process the mouldings
are produced on semi-automatic machines. Each machine is fitted with a
mould normally having six sections of which five are movable. A measured
amount of pulp and water mixture is admitted into the mould, and the article
is formed by blowing in air at a pressure of approximately 4 kg/cm^2 (50 lb/
in^2) and at a temperature of approximately 480°C. The moulded articles
leave the mould containing 45–50% of moisture, and are subjected to an
after-drying operation which also sterilizes the mouldings. Pulp mouldings
produced by the pressure injection process combine great strength and
rigidity with light weight. The thickness of the walls is normally
approximately 2.5 mm, but this may be increased or decreased at will. It is
possible to produce mouldings of almost any shape, and this point is
illustrated by the fact that large quantities of novelty articles such as toys,
animals, hats, etc., are produced by this method.

Mouldings can also be produced with special characteristics, such as
waterproof or wet strength, by adding suitable materials to the water-pulp
mixture.

Suction process. In the suction process the pulp mixture is deposited and
formed to the shape of the mould by application of a partial vacuum to one
side of the mould screen. As compared with the pressure injection process,
the formed article as it leaves the mould contains a very much higher
percentage of moisture, generally of the order of 85%. This residual
moisture has to be removed by an after-drying process. For certain
applications an after-pressing process, using heated dies, gives a stiffer and

smoother article which can more readily be colour-printed with the customer's name, trademark, etc.

The suction process is normally operated by fully automatic moulding machines which require a minimum of two forming moulds and one transfer mould. It is general, however, that several moulds are used, anything up to 24 per machine, dependent on machine size and article to be manufactured. The tool costs involved are therefore relatively high, and the process is normally employed only for the production of articles required in very large numbers. Moreover, by virtue of the design of the machines, the process is mainly used for the production of open-top tray-type containers.

Irrespective of the forming method employed, the product may then be pressed to the desired thickness and density between hot tools, one of which would have a means of dispersing the steam generated during the operation. Alternatively, the moulded product may be dried in an oven and may then be pressed between tools which do not have to deal with steam even if they are hot.

A degree of water resistance is generally incorporated at the moulding stage using wax/rosin or synthetic polymers and containers and forms can be rendered impervious to water and resistant to water vapour by lining with plastic films and by impregnation.

The material can be coloured by adding dye to the pulp furnish or by spraying after manufacture. It can also be printed. Inscriptions or designs may be produced either embossed or debossed, without any extra operation. Similarly, recessed panels can be provided for labelling. The pH value of moulded pulp normally lies between 5 and 9 and can be adjusted to meet specific requirements. The principle competitive materials are:

1. Expanded polystyrene thermoformed sheet and mouldings
2. Pads and fittings made from single face corrugated board
3. Fittings in single and double wall corrugated board

Moulded pulp packaging has a relatively low fire risk compared with expanded plastics, and can be produced in far more complicated profiles than any of the corrugated boards. Compared with expanded plastic, moulded pulp forms produce virtually no static problems.

Typical uses

The following are examples of typical current usage of moulded pulp packaging (Figure 16.1).

1. Egg trays and cartons
2. Bottle and jar sleeves and trays
3. Trays for fruit, vegetables, meat, fish and other packaged foods
4. Containers and trays for electronic components, electrical instruments and switchgear

5. Protectors and inserts for a variety of goods including fluorescent light tubes, picture frames, door locks and car components
6. Protectors for the corners and edges of furniture, radiators, windows (with or without glazing) and domestic appliances (white goods)
7. Domestic sanitary ware and ceramic items

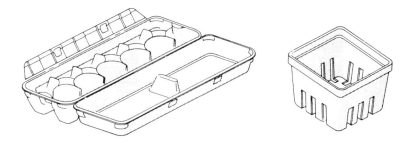

Figure 16.1   Some typical uses for pulp packaging.

Choice of manufacturing process

The choice of moulding process for any given article is largely determined by the considerations set out above. Some articles can be made to best advantage by pressure injection and others by the suction process.

Pressure-injection moulded-pulp articles are particularly well suited to the packing of glass bottles and jars. Examples of successful use are the packing of whisky for export. This form of packing consists of a pulp sleeve moulded to the profile of the bottles which are packed head to tail in a wood or fibreboard case. This method enables a considerable amount of space to be saved.

Another successful type of pulp container is designed to hold and protect Winchester quart and other bottles of various capacities from 30 ml to 5 litres, used by the essence, perfumery and similar trades. These containers have a square or rectangular cross-section and the bottles are supported on specially designed panels which prevent the transmission of shock to the bottle.

Pressure-moulded containers are also used in very large quantities for packing and protecting electrical and engineering components, for the protection of highly finished machined parts, and for inter-departmental transit during manufacture.

Manufacturers' names, decorative designs, trademarks and other features can be embossed on the article at no extra cost. The containers may also be labelled, coloured by dyeing the pulp, or finished in any desired style by means of a spray gun or other decorative process.

Other types of moulded-pulp articles are also produced in the form of novelties, such as animals, dolls and other complicated shapes. Further use for the process is the manufacture of specially shaped collecting boxes for charitable societies. Well-known forms of moulded-pulp articles made by the suction-moulding process are egg flats, food trays, and many other forms of tray-shaped articles for packing fruit and other commodities.

In summing up, for specialized shapes requiring a high degree of protection and a low tooling cost in relation to quantity, the pressure-moulding process is most suitable, but for large quantities of mouldings with low unit cost, the suction process would be selected.

Moulded plastic containers

Definitions

Moulded in the context used here, may be defined as shaped, under the influence of heat and pressure, either in or around an appropriate form.

Plastics are not so easily defined, but the BSI Glossary of Terms used in the plastics industry (BS 1755:1951) defines them as a wide group of solid composite materials which are largely organic, usually based on synthetic resins or upon modified polymers of natural origin and possessing appreciable mechanical strength. At a suitable stage in their manufacture, most plastics can be cast, moulded or polymerized directly to shape. Some plastics are rubber-like, while some chemically modified forms of rubber are considered to be plastics.

Large containers such as plastic drums are dealt with in another chapter, so that the emphasis here will be on the smaller containers associated with retail use. Excluding those made from plastic film, there are three major forms of plastic packages: blow moulded bottles and jars; thermoformed packages, such as trays, tubs, cups and inserts; and injection-moulded containers and components such as boxes, closures and lids.

Methods of production

Blow-moulding. The basic techniques of plastic blow-moulding were derived from those used in the production of glass bottles. In each instance, air is forced under pressure into a sealed molten mass of material which is surrounded at the right distance by a cooled mould of the required shape. The pressure of the air causes the molten mass to move out to the mould walls where it cools on contact. Finally, the mould is opened and the moulded article ejected.

The first attempt to blow-mould hollow plastic objects, over a century ago, was with two sheets of cellulose nitrate clamped between two mould halves. This was very unsuccessful. In the early 1930s, cellulose acetate and polystyrene were tried by Owens-Illinois, using glass blowing techniques. Unfortunately, neither of these materials offered any advantages over glass bottles. The development of low density polyethylene in the mid-1940s provided this advantage. 'Squeezability' gave the plastic bottle a feature that glass could not match. With the development of high density polyethylene in the late 1950s, many of the problems of low density polyethylene were avoided.

The acceptance of high density polyethylene for packaging bleach, detergent, household chemicals and milk in the 1960s and 1970s, and the acceptance of poly(ethylene terephthalate) for packaging carbonated beverages in the 1980s has expanded the blow-moulding process. It is now one of the high growth technologies of today. The principal materials used today are polyethylene, mainly HDPE, and oriented polyethylene terephthalate, PETP, but several others are also in use PVC, PS, AN, and PP. Table 16.1 gives typical properties and uses.

An important point to remember with all plastic containers is that they lack the stiffness and stability that is found in their glass analogues. Additionally they are not so inert towards foods as is glass and must be selected so as to be compatible with the particular food concerned and not to contain components that could leach into the food and cause taint. Stiffness and stability can be compensated to some extent by selective design but often this must also be taken into account by employing load bearing constructions in the secondary packaging (i.e. fitments in the shipping case).

In the moulding of plastic bottles, the blowing processes in commercial use can be divided into two main classes, namely, *injection blowing* and *extrusion blowing*.

Injection blowing is the process that most closely resembles the blowing of glass bottles. The plastic material is moulded round a blowing stick, inside a more or less conventional injection moulding machine. This gives a thick-walled tube called a *parison*. The parison and blowing stick are transferred to another (blowing) mould where compressed air is passed down the blowing stick. The injection-moulded parison is still molten at this stage, and is thus blown out to the shape of the second mould. The sequence is shown in Figure 16.2.

Injection blow-moulding is used to a great extent for pharmaceutical and cosmetics bottles because the bottles are frequently small, precise neck finishes are important, and the process is more efficient than the extrusion blow-moulding alternative. The resins most commonly used are high density polyethylene (HDPE), polypropylene (PP), and polystyrene (PS).

In the early days of the injection blow-moulding process it was hand-operated, i.e. the blowing stick and parison were removed from the injection-moulding machine and transferred to the blowing mould by hand. Later developmnets included the provision of blowing moulds on either side of each injection mould, the parisons being transferred automatically into one or other of the blowing moulds alternately. Another development has been the use of multi-cavity injection and blowing moulds. This process is generally used for small bottles, usually less than 500 ml capacity. The process gives extremely accurate weight control and neck-finish detail. On the other hand, proportions are limited and containers with handles are not practical. Tooling costs are also relatively high.

Extrusion blowing was a later development, but now accounts for a greater percentage of blow mouldings produced today. It is used for bottles larger than about 200 ml capacity. Tanks larger than 4500 litres weighing over 110 kg have been blow-moulded. Compared to injection blow moulding, tooling costs are less, and part proportions are not so limited. Containers with handles and offset necks are commonplace. Flash must be trimmed from each part and recycled. Operator skill is more crucial to the control of quality (see Table 16.2).

Instead of the injection-moulded parison of the first process, we have a continuously extruded tube. At intervals a pre-determined length of tube is trapped between the two halves of a split mould. Both ends of the tube are sealed as the mould closes, and the trapped portion is inflated by compressed air introduced via a blowing pin. Once again, the inflated plastic is cooled by

Table 16.1 Properties of plastic bottle materials.

Properties	Polyethylene		Polyester (oriented) (PET)	PETG Copolyester	PVC	Polystyrene	Acrylic multi-polymer	Polypropylene	
	Low density	High density						Regular	Oriented
Resin density	0.91–0.925	0.94–0.965	1.35–1.40	1.27	1.35	1.0–1.1	1.09–1.14	0.89–0.91	0.90
Clarity	Hazy but transparent	Hazy but translucent	Clear	Clear	Clear	Clear	Clear	Hazy transparent	Clear
Barrier to water vapour	Good	Very good	Moderate	Moderate	Moderate	Poor	Poor	Very good	Very good
oxygen	Poor	Poor	Good	Good	Good	Poor	Good	Poor	Poor
CO_2	Poor	Poor	Good	Good	Good	Poor	Moderate	Moderate	Moderate
Resistance to acids	Fair to very good	Fair to very good	Fair to good	Fair	Good to very good	Fair to good	Poor to good	Fair to very good	Fair to very good
alcohol	Fair to very good	Good	Good	Good	Good to very good	Fair	Good	Good	Good
alkalis	Good to very good	Good to very good	Poor to fair	Poor to fair	Good to very good	Good	Poor to fair	Very good	Very good

mineral oil	Poor	Fair	Good	Good	Good	Fair	Good	Fair	Fair
solvents	Poor to fair	Poor to good	Good	Poor to good	Poor to good	Poor	Poor	Poor to good	Poor to good
Resistance to heat	Fair	Fair to good	Poor to fair	Poor to fair	Poor to fair	Fair	Fair	Good	Good
Temperature at which finished product distorts (°C)	71–104	71–121	38–71	60–71	60–65	93–104	82–90	121–126	121–126
Resistance to cold	Very good	Very good	Good	Good	Fair	Poor	Poor	Poor to fair	Very good
light (UV)	Fair	Fair	Good	Fair	Poor to good	Fair to poor	Good	Fair to good	Fair to good
Stiffness	Low	Moderate	Moderate to high	Moderate to high	Moderate to high	Moderate to high	Moderate to high	Moderate to high	Moderate to high
Resistance to impact	Excellent	Good to very good	Good to excellent	Poor to fair	Fair to good	Poor to good	Poor to good	Poor to good	Very good
Food uses	Mustard, edible oils	Milk, chocolate syrup	Carbonated beverages, liquour, edible oils	Foods	Edible oils, vinegar	Vitamins, spices	Foods	Syrups, juices	

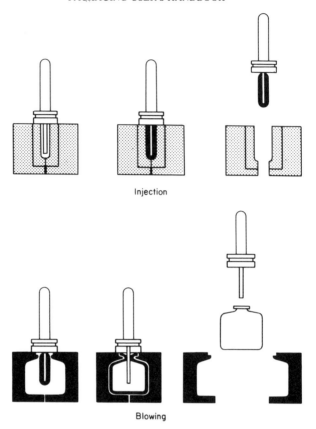

Injection

Blowing

Figure 16.2 Basis of injection blowing.

Table 16.2 Comparison of injection versus extrusion blow moulding.

Injection blow moulding	Extrusion blow moulding
1. Used for small sizes, typically <473 ml	1. Used for larger sizes, typically >237 ml.
2. Best process for general purpose polystyrene; no flash to recycle, no pinch-off scars, no post-mold trimming	2. Best process for poly(vinyl chloride)
3. Injection-moulded neck provides more accurate neck-finish dimensions and permits special shapes for complicated safety and tamper-evident closures	3. Much fewer limitations on proportions permitting extreme dimensional ratios; long and narrow, flat and wide, double-walled, offset necks, moulded in handles, odd shapes
	4. Low cost tooling often made of aluminium; ideal for short run or long run production
5. Accurate and repeatable weight control	5. Adjustable weight control ideal for prototyping
6. Excellent surface finish or texture	

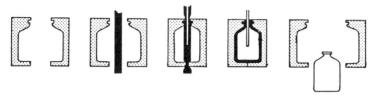

Figure 16.3 Basis of extrusion blowing.

the cold mould walls, the mould is opened and the bottle ejected. The sequence of operations is shown in Figure 16.3.

A variation on this is to install the blowing pin at the base of the mould. The molten plastic tube is then extruded vertically downwards so as to fall over the pin. The two halves of the mould then close around it as before, sealing it at both ends. This enables moulding to be speeded up, since the mould plus blowing pin can be moved away from under the extruder, allowing a second mould to be put in its place. Extrusion of a tube for a second bottle then takes place while the first bottle is being inflated, cooled and ejected.

This technique can be extended by the use of rotary moulds, or by having a series of matched mould halves mounted on two endless belts, so that they first close on the extruded tube, travel with it during blowing and cooling, then diverge, thus ejecting the bottle. Another way to speed up production is to fit the extruder with multiple die-heads, and feed the various extruded tubes to a multi-cavity mould.

One disadvantage of extrusion blow-moulding is that when a length of plastic tubing of even wall thickness is blown into a bottle shape, the material is thinned more at the extremities of the mould than elsewhere. This is important because it is at these points (the base corners and the shoulders) that extra strength is needed in the bottle. Machines are now available that will extrude a tube of variable wall thickness, such that extra material is available where it is most needed.

One method of doing this is known under the descriptive name of *dancing mandrel*. The mandrel forms the inner part of a circular die gap, and in the dancing mandrel extruder it is conical, and so produces a variable die gap as it is moved up and down thus giving thicker or thinner parts to the wall.

Stretch blow-moulding. Stretch blow-moulding uses either an injection-moulded, an extruded-tube, or an extrusion blow-moulded preform. It uses resins with molecular structures that can be biaxially oriented. Stretch blow moulding is generally used for bottles between 450 ml and 1.8 litres in capacity. It can be used for smaller bottles, however, and some bottles as large as 24 litres have been moulded. The process by orientation enhances the stiffness, impact, and barrier performance of bottles making a reduction in weight or a lower cost possible. The technique is, however, limited to simple bottle shapes.

The major stretch-blow resin is PET, and the principal application is for carbonated beverage bottles. Three other resins are stretch blown: PVC, PP and AN. The process is based on the crystallization behaviour of the resin. A preform is brought to the right temperature and then rapidly stretched and cooled in both directions. For best results the conditioning, stretching and orienting should take place at just above the glass transition temperature (T_g) where the resin can be stretched without the risk of crystallization (see Figures 16.4, 16.5 and 16.6).

The advantage of the process is the improvement in bottle impact strength, transparency, surface gloss, stiffness and barrier properties. This in turn permits lighter, cheaper bottles, and sometimes the use of resins that would normally not be suitable.

When the stages of parison production, stretching and blowing take place in the same machine we have the so-called one-stage or in-line process

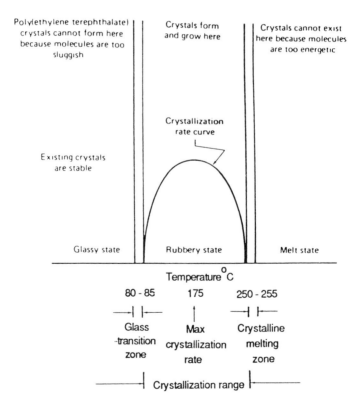

Figure 16.4 Diagram of crystallization behaviour. Courtesy of Society of Plastics Engineers.

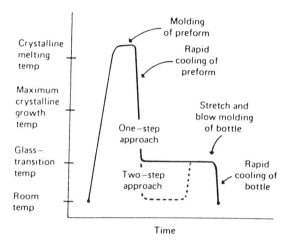

Figure 16.5   Basis of the stretch blow process. Courtesy of J.S. Schaul.

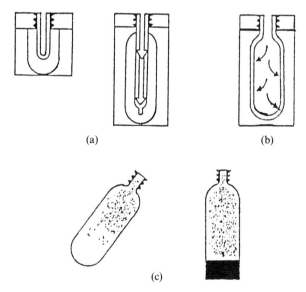

(a)   (b)

(c)

Figure 16.6   Basics of stretch blow-moulding: (a) A 'test-tube' shaped preform is made by extrusion or injection moulding. It is then heated and placed in the bottle-shaped mould. It is then stretched to twice its length; (b) Compressed air then blows out the lengthened parison to fill the mould thus stretching the PET in the other direction; (c) The bottle is then ejected and fitted with a base to permit it to stand. It is now biaxially orientated.

(Figure 16.7). In the two-step process, parison production is done separately from parison stretching and blowing. The machines are called 'two-stage', or 'reheat-blow' machines. The main advantage of the one-step approach is energy savings. The parison is rapidly cooled to the stretch temperature. In

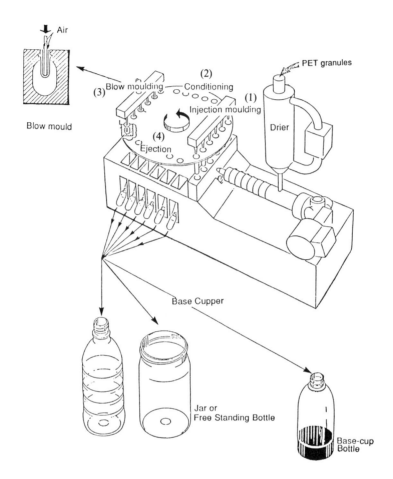

Figure 16.7 Integrated one-stage stretch blow-moulding process. PET granules are fed through a drier into the extruder and then in (1) an injection moulded test-tube shaped parison is blown, in (2) the parison is conditioned and passes to (3) where the parison is stretched and blown to shape. They then pass to (4) where they are ejected either as free-standing jars or bottles or pass to a base cupper to be fitted to a base.

the two-stage process, the parison is cooled right down to room temperature and is reheated to the stretch temperature (Figure 16.5). On the other hand, the two-step approach can be more efficient; a minor breakdown in one stage does not halt the other, and preforms can be made in one place and bottles blown elsewhere. Both processes use either injection-moulded, extruded, or extrusion blow-moulded parisons.

Injection-moulded parisons are virtually the same as those used in injection blow-moulding. With the two-step process (Figure 16.8), the parison is injection-moulded, sorted and later placed in an oven for temperature conditioning and blow-moulding. A rod is most often used

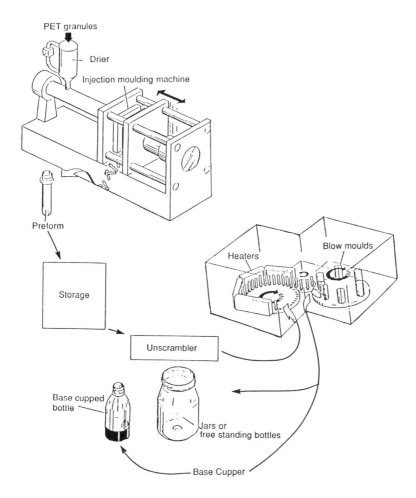

Figure 16.8   Two-stage stretch blow-moulding process. Here the test-tube shaped parisons are produced as in the single-stage process but are stored until required. They then pass via an unscrambler into a heating station where they are raised to the right temperature for stretch-blow moulding as before.

inside the parison, in combination with high air pressure, to complete the stretch (Figure 16.9). Injection stretch blowing is most often used for PET resin.

With the extruded parison, a one-step approach is used. A tube is extruded and fed directly into an oven for conditioning. After conditioning, the tube is cut into parison lengths. Mechanical fingers grab both ends and stretch the parison. The two halves of the mould close and air pressure expands the stretched parison to the walls of the mould. In the two-step approach, the extruded tube is cooled and cut to length. The cut tubes are

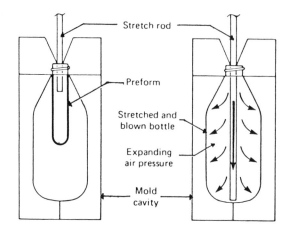

Figure 16.9 A temperature-conditioned preform is inserted into the blow-mould cavity and is then rapidly stretched. A rod is often used to stretch the preform in the axial direction with air pressure to stretch the preform in the radial direction.

placed in an oven for conditioning as required. This stretch blow-moulding method is most often used for PP and PVC.

The *extrusion blow-moulded parison* is shaped and temperature conditioned in a preform cavity in the same way a bottle is extrusion blow-moulded. From this preform cavity, the parison is transferred to the bottle cavity where the expansion takes place. The one-step process is more usual but a two-step approach is also practised. PVC is most often stretch blow-moulded using this method.

Multi-layer blow-moulding. The important requirements of many bottles are reasonable cost, strength, clarity, product compatibility, and gas barrier properties. Polyethylene and polypropylene, for example, are relatively low cost resins, suitable for food contact, excellent water vapour barriers but poor oxygen barriers. They are not, therefore, suited for packaging oxygen-sensitive foods requiring long shelf-life. Poly(ethylene-vinyl alcohol) (EVOH) on the other hand, is a high cost material with excellent oxygen barrier properties but it is sensitive to water. A thin layer of the latter sandwiched between two layers of the former can provide both barrier needs.

All of the basic blow-moulding process methods have been used with multi-layer blow-moulding, additional extruders being used for each resin.

A common problem of multi-layer materials is that the different layers do not adhere to each other. An adhesive layer is required to create the bond. As a result, three or more extruders are required. With the HDPE/EVOH/HDPE example decribed, five layers are actually required: HDPE/adhesive/EVOH/adhesive/HDPE.

Aseptic blow-moulding. A more recent development is the extension of the form-fill-seal principle to blown bottles. There are quite a number of ways in which this can be done, but the general principle for all is much the same. The blowing tube is integral with the extrusion die, i.e. air is blown from the top. Concentric with the blowing tube is a filling tube. The bottle is blown as usual, then the liquid to be packed is metered into the bottle. The liquid helps to cool the bottle and so reduces the cooling cycle. A further concentric tube is also fitted to allow the air displaced during the filling operation to escape. When the bottle is full, the neck is heat-sealed and the filled bottle ejected from the mould. The process often involves a series of split moulds, the matching halves of which are mounted on two endless chains, positioned vertically beneath the filling/blowing head. The bottles are heat-sealed while still inside the moulds, and are ejected when the mould halves diverge again at the base of the machine. One advantage of the form-fill-seal bottle-blowing approach is that aseptic filling is possible; another is the elimination of empty bottle storage and transport.

Injection-moulding. In the injection-moulding process the material is softened in a heated cylinder, then forced under high pressure into a closed mould. After the material has hardened (by cooling) the mould is opened and the moulding is removed by ejector pins or compressed air.

The essential elements of an injection-moulding machine are shown diagrammatically in Figures 16.10 and 16.11.

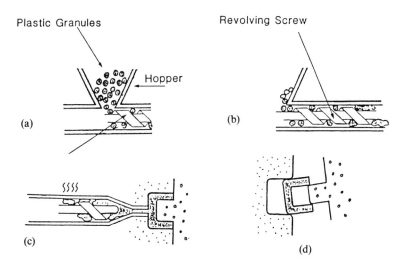

Figure 16.10 Injection-moulding: (a) Plastic granules are fed into the hopper of the extruder; (b) The revolving screw moves the granules towards the heated zone of the extruder; (c) The heater raises the temperature of the molten plastic to the right level and forces (injects) it into the cold, closed mould where it solidifies and cools; (d) The mould then opens and the container is removed.

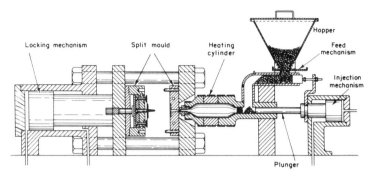

Figure 16.11 Diagrammatic representation of an injection-moulding machine.

The complete cycle of operations is as follows:

1. The mould is closed and a locking force is applied. This force must be large enough to prevent the two halves of the mould being pushed apart when molten plastic is forced in.
2. The plunger moves forward, taking a fresh charge of granules with it into the heating zone of the cylinder. At the same time it displaces already molten material left there during the previous cycle. This molten material flows out of the nozzle, through the 'sprue' opening in the die, and into 'runners' which terminate in a 'gate', leading to the mould cavity itself.
3. The pressure on the plunger is maintained during the period while the material in the mould cools and contracts. If no pressure were applied during this period, the contraction during cooling would cause depressions or 'sink marks' to occur on the surface.
4. The plunger returns to the fully retracted position.
5. The mould opens and the moulding is removed.

The cycle is then repeated. Continuity is maintained by coupling the feed mechanism to the plunger. Fresh material is then deposited during the return stroke of the plunger, ready to be pushed forward during the next cycle.

Modern injection-moulding machines have a screw feed instead of a plunger. The screw aids plasticization and so reduces the cycle time. Some definitions may be helpful at this stage. A *gate* is a point of entry into the mould cavity. Gates can be of various sizes and shapes, depending on the viscosity of molten plastics, and the size and shape of the mould cavity. *Runners* are the channels that feed the molten material to the mould, when there is more than one gate, and they connect with the 'sprue'. The *sprue* is the path from the external entry of the molten material into the die, to the gate (if there is only one) or to the runners. One reason for having more than one gate is that the mould is a multi-cavity one. This is often the case when small items are moulded. Another reason is the necessity to have more than one point of entry into a large mould cavity, to give a more even flow of material into the cavity.

The following points about injection moulding should be noted:

1. High production rates can be achieved, particularly with multi-cavity moulds
2. Reverse tapers and pronounced undercuts cannot be moulded by this method, because the finished article could not be removed from the mould.
3. Mould costs are high, because of their massive construction (necessary to withstand the high pressures involved in moulding). Short runs are, therefore, uneconomic.

Thermoforming from sheet. In this process a plastic sheet is softened by heat then forced either into or over a mould. Figure 16.12 illustrates the steps in this operation. Heating, forming, cooling and sealing are the most important steps in producing the thermoformed (deep-drawn) package. Thermoforming the film is especially important. Figure 16.13 contains a summary of the possible forming methods, of which the most commonly used are negative vacuum forming, with or without plug assist and negative compressed air forming, with or without plug assist. In these methods, the heated film is formed in a mould (negative forming). In vacuum forming, the force is provided by the difference in pressure between the evacuated mould and the atmosphere of 1 bar (1 kg/cm^2). In compressed air forming, forming pressures of 6–8 bars are common.

The simplest *vacuum forming* has the plastic sheet clamped over a box containing the mould. The sheet is heated by electric panel heaters. The air in the box is withdrawn through the holes in the mould, thus creating a vacuum between the sheet and the mould. The atmospheric pressure of the air on the sheet forces it on to the mould, where it cools sufficiently to retain its shape when removed from the mould (Figure 16.14).

Pressure forming is very similar to the vacuum-forming method, but air pressure is applied from above to push the softened sheet on to the mould. One important difference is that the pressure that can be applied during forming can be greater than atmospheric pressure, so that better mould definition can be obtained. The most uniform wall thickness distribution is produced by means of compressed air forming with plug assist. Compressed air forming without plug assist also results in more uniform wall thickness than with vacuum forming. However, the machinery required for this

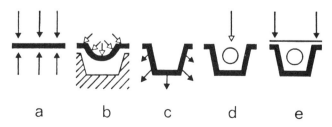

Figure 16.12 The steps in making thermoforms. (a) Heating, (b) forming, (c) cooling, (d) filling, (e) sealing.

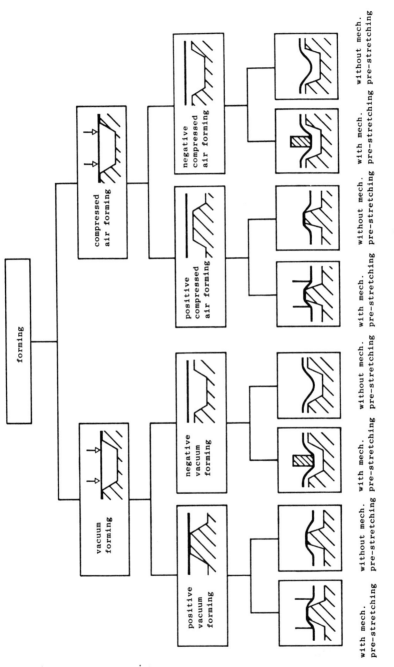

Figure 16.13 The possible ways of thermoforming.

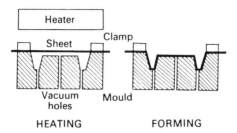

Figure 16.14   Vacuum forming into a male mould.

method is more complicated—and more expensive—than comparable machines for vacuum forming.

Forming with matched moulds is particularly valuable when complex shapes are required. As the name implies, the heated sheet is pressed into shape by trapping it between matched male and female moulds.

Mould costs for vacuum and pressure forming can be low, although they obviously vary with the complexity of the mould, the length of run and the degree of automation. For short runs and prototype moulding, wood and plaster moulds are often used, while for longer runs aluminium or filled epoxy resin moulds are popular. Matched mould forming is more expensive, because two moulds are required and dimensions are more critical.

There are six main materials used for thermoforming:

1. *Rigid PVC film* (polyvinyl chloride). This is produced primarily by calendering. It has good thermoforming properties, and is generally processed clear, coloured and printed. It can be sealed by HF, ultrasonic, radiation, heat impulse and heat contact methods.
2. *Polystyrene film*. This is produced almost exclusively by extrusion. It has good thermoforming properties, similar to rigid PVC film. Clear stretched polystyrene is brittle, but non-transparent copolymers have good impact strength. Polystyrene cannot be sealed by means of HF but is well suited for heat contact sealing.
3. *Polypropylene film*. This is produced by extrusion and generally employed in a laminate with other films, e.g. polyamide.
4. *PVC-PVDC* (polyvinyl chloride-polyvinylidene chloride) The polyvinylidene chloride is applied as a dispersion or emulsion and dried.
5. *PVC-PVF* (polyvinyl chloride-polyvinyl fluorides). The two films are laminated together. The laminate is currently employed for products which are extremely hygroscopic to be sold in areas with tropical climates.
6. *PVC-PE-PVDC* (polyvinyl chloride-polyethylene-polyvinylidene chloride). This triple laminate has been on the market for several years.

The requirements for processing laminated films (PVC-PVDC, PVC-PFC, PVC-PE-PVDC), and for producing more complicated shapes, are basically different. As a result of the required high forming force in the partially limited forming temperature range, only the compressed air forming method, either with or without plug assist, can be employed. Mechanical pre-stretching is always required for complicated shapes. Compressed air forming alone does not provide uniform distribution of the material in the deep-drawn cavity.

Moulding of expanded polystyrene bead. The moulding of expandable polystyrene beads is different from other processes and deserves special mention. It is carried out in three stages, the first of which is a pre-expansion using steam. This softens the polystyrene and increases the internal pressure in the beads (due to the liquefied gas, usually pentane, contained therein). The degree of expansion, and consequently the density, is controlled by the time of heating and the temperature. The pre-expanded beads are then allowed to cool. At this stage there is a partial vacuum inside each bead and they are, therefore, weak and easily collapsible. The beads are left for about 12 to 24 hours to mature, during which period air permeates into the beads, which are then ready for the moulding process.

In the moulding operation, the mould is completely filled with pre-expanded matured beads, then closed and heated by injecting with steam. The steam softens the polystyrene and causes any remaining pentane, and the air that entered the beads during maturing, to expand. Since the beads are confined in the mould, they simply merge together and fill all voids in the mould. Cooling then hardens the moulded block.

Types and styles of container

Moulded plastic containers have no special terminology and are usually named after traditional containers of similar shape. Thus, the plastic containers are referred to as bottles, jars, pots, tubes, jerricans, punnets, etc., but within these broad classifications there are ranges of size, colour and style.

There is also much overlap between the types of containers and the methods of manufacture. For example, bottles can be made by all three types of blow-moulding and by injection-moulding (in two parts subsequently joined), while thin-walled tubs can be made by injection-moulding, vacuum forming and blow-moulding.

Blow-moulded containers. The blow-moulding process is principally used to produce bottles, jerricans, carboys and other containers where the neck diameter is small compared with the overall diameter of the container.

The materials used for blow-moulded containers include polyethylene (both low and high density), PET, PVC, polypropylene, polyacetals, polycarbonate and cellulose acetate. By far the most common of these is polyethylene. Up to about 1961, low-density polyethylene dominated the bottle-blowing market in the United Kingdom, but large numbers of high-density polyethylene bottles and stretch blown PET bottles are also produced, as well as PVC and polystyrene bottles. The choice of material is governed by cost, ease of processing, chemical resistance, permeability to water vapour and gases, and whether or not a transparent or opaque, flexible or rigid container is required. Typical of the speed with which developments in packaging can occur is the rapid growth of PET bottle for carbonated drinks.

Injection-moulded containers. Injection-moulded containers include jars, tubs, bottles (moulded in two halves subsequently welded together), phials and boxes. The main polymer used in the field of injection-moulded containers is polystyrene, but the use of polypropylene is growing rapidly. Both polystyrene and polypropylene are available in special grades where the containers are to be used in the packaging of foodstuffs.

Polystyrene containers can be produced both with a very wide range of colour effects and in clear grades to allow full visibility of the contents (Figure 16.15a and b). Polypropylene containers can also be obtained in a range of colour effects, but the clarity of polystyrene cannot be matched in the clear grades.

High-impact grades of polystyrene are also used, particularly for those thin-walled pots where general-purpose polystyrene would be too brittle. High-impact polystyrenes are also non-transparent, and the choice between polypropylene and high-impact polystyrene is usually resolved by consideration of the required chemical resistance, permeability and impact strength. In the absence of any technical advantage, the issue will probably depend upon cost. The prices of any raw materials on a volume basis are usually fairly close together, and vary from time to time. Hence, it is impossible to give a once-for-all answer to the relative economics of these two materials.

Bottles, or jars with necks narrower than the maximum internal wall-to-wall diameter of the containers, cannot be injection-moulded in one piece, because of the difficulty of removing the male half of the mould. On the other hand, injection-moulding gives perfect control of wall thickness, which is important when the container has to withstand internal pressure. The problem has been solved by injection-moulding the bottle in two parts, which are subsequently joined together, either by an adhesive, by solvent welding (particularly useful in the case of polystyrene, for which simple solvents such as ketones are available) or by spin welding. In the last process, one part of the bottle is held stationary in a clamp, while the other is rotated at high speed in a magnetic clutch. The frictional heat which is generated causes the polymer at the two surfaces in contact to melt. At this point the rotation is rapidly halted by the magnetic clutch, and the polymer allowed to cool. The seam produced by this method is said to be as strong or stronger than the rest of the bottle.

One of the most important developments in the field of injection-moulding is the production of a container and lid in one piece, using polypropylene.

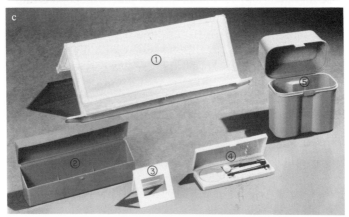

Figure 16.15   (a) A selection of food packs and sample tasting cups made in a special heat resistant high molecular weight grade polystyrene. (b) Dispenser packs for pharmaceuticals moulded in polystyrene. (c) Hinged components in polypropylene. (1) Filter unit for a dishwasher, (2) box for photographic slides, (3) photographic slide mount, (4) cosmetic case, (5) water softener test kit. Courtesy of Shell Chemical Co. Ltd.

This polymer has the property of forming a practically indestructible hinge when once a thin section has been flexed. Repeated flexing of such a hinge fails to cause breakage, even after as many as a million cycles. The moulding of a container with an integral hinge cuts assembly costs, and of course it cannot corrode or stick, as may sometimes happen with a metal hinge (Figure 16.15c).

Thermoformed containers. Tubs, trays and box inserts are the commonest containers formed by this method, particularly where very thin walls are required such that it would be difficult for a polymer to flow between the mould walls in an injection-moulding.

Polystyrene, cellulose acetate, PVC, polypropylene and high-density polyethylene have all been used for thermoformed containers. The latest material in this field is foamed polystyrene sheet, which can be thermoformed to give trays and containers with built-in cushioning properties.

Expanded polystyrene containers. Expanded polystyrene mouldings are widely used as contour-spreading packages for items such as typewriters, cameras, microscopes, hi-fi equipment and similar complex-shaped and shock-damageable products. More recently, higher density material has been used for moulding boxes for the carriage of 5 to 6 kg of fresh fruits and vegetables, such as cherries, grapes and tomatoes. Fittings for corrugated outer cases are also moulded, to be used for the packaging of television sets and similar items.

Types of closure

Screw caps. Screw caps are injection-moulded from polyethylene, polystyrene or polypropylene, or compression-moulded from phenol-formaldehyde and urea-formaldehyde. They are commonly used on plastic bottles or jars, and conventional capping equipment is usually suitable for applying them.

Polypropylene caps are particularly valuable as closures for cosmetic preparations, because of their design possibilities. Polypropylene has good resilience, so that mouldings having slight undercuts can be 'jumped off' the mould core without damage to the moulding. Decorative inserts, such as imitation jewels, can then be pressed into these moulded-in undercuts to give caps with highly effective sales appeal. The resilience of polypropylene also makes the design of linerless closures possible.

Plug fittings. Plug fittings are normally injection-moulded from low-density polyethylene, since its softness and flexibility enables it to give a good seal, even against hard smooth surfaces, such as the walls of a polystyrene tube. The plug itself is often ribbed to give even better sealing.

An interesting example of the design possibilities inherent in the use of plastics is the plug closure which incorporates flexible prongs on the underside of the plug. The use of this closure for tablet tubes eliminates the necessity for a wad of cotton wool on top of the tablets to prevent their movement, with consequent risk of breakage during transport. Again no special equipment is necessary for closing.

Push-on covers. Push-on covers are normally used for injection-moulded plastics pots or jars, and for some types of vacuum-formed containers, e.g. beaker-shaped containers with a curled rim. As well as plastic push-on covers, paperboard ones are still used in some instances. Before applying the push-on cover, a foil diaphragm is often crimped over the top of the pot or jar. This serves to give extra protection or a tamper-evident seal. In thicker gauges, aluminium foil is sometimes used as the only closure, e.g. in the case of yoghurt containers. Flexible push-on covers (in low-density polyethylene) can also be used for bottles.

Heat-sealed covers. Heat-sealed covers are often used for closing vacuum-formed containers of the tray type, or the deep-drawn pyramid type for fruit drinks. The cover may be flat or recessed to give a shallow plug-type fitting to the container, with a consequent increase in rigidity and strength. Sealing is carried out with a heated jig. With certain types of equipment it is possible to vacuum-form plastic sheet continuously from the reel, fill the depressions so formed, then cover them with another plastic sheet fed from a separate reel. After heat-sealing the cover on, the containers are cut out and trimmed.

Miscellaneous closures. The possibilities for design inherent in plastic moulding have led to many special types of closures, such as combined plug and snap-on covers, and plug or screw-caps which incorporate means of dispensing the contents in droplet form or as a jet or spray. Such closures are usually fitted to 'squeeze' bottle designs. Another interesting design feature which is often incorporated in plastic closures is the integral moulding of a nozzle and cap to give a captive closure. The assembly is fitted as a plug in the bottle neck. The same result has also been achieved by the use of a snap-on-action retaining ring.

Uses of moulded plastic containers

Detergents. The first large market for plastic bottles was for washing-up liquids. The great majority of liquid household detergents are now packed in these bottles, the most common material being low-density polyethylene. This is used because it combines chemical inertness, low weight and resistance to breakage, with a flexibility that allows the production of 'squeeze' bottles, thus giving the convenience of easy dispensing. The bottles can be attractively printed to give them sales appeal. In the United States, liquid detergents are commonly packed in high-density polyethylene. These bottles are more rigid than those in low-density polyethylene and were brought on to the US market as direct replacements for cartons. If required, high-density polyethylene can be used for the manufacture of 'squeeze' bottles by reducing the wall thickness and designing the bottle to improve 'snap-back'. One such bottle used for car shampoo has an elliptical cross-sectional shape to give the required 'snap-back'.

More recently, PVC bottles have been used in large numbers for the packaging of detergents, particularly when the detergent has an attractive colour. The clarity and sparkle of a PVC is often effective in such instances in increasing the sales of the product.

Foodstuffs. A problem of importance in the use of plastics for the packaging of foodstuffs is that of tainting, but special grades of several plastics have long been used for packaging a wide range of food products. Polystyrene, for example, is used in the form of pots or small jars in the packaging of butter, jam and cheese, while in bottle form it has been successfully used to give a 'new look' to the retailing of vinegar (Figure 16.16). The vinegar bottle is a good example of an injection-moulded two-piece container.

Polystyrene containers are both light in weight and chemically inert. Polystyrene can also be used in crystal-clear grades if visibility of the contents is required. Vacuum-formed trays or punnets with heat-sealed covers are made from polystyrene and PVC, and used for a variety of foodstuffs. Both materials have good rigidity, so that very thin-walled one-use containers can be produced.

A toughened grade of polystyrene (containing a small percentage of a synthetic rubber) is used to produce deep-vacuum-formed containers for fruit juices, while special high-flow grades are used to give thin-wall injection-moulded tubs for such products as cream and yoghurt. The squeeze-action dispensing possibilities of polyethylene have also been exploited in the foodstuffs field, with the production of containers shaped and coloured like lemons and containing pure lemon juice, plastic 'hot dogs' (containing mustard) and many variations on the same theme.

One of the most important developments of recent years in foodstuffs packaging is the use of rigid PVC blow-moulded bottles for cooking oils, wine and fruit-juice concentrates. PVC is an ideal material for many foods (especially oils) because of its chemical resistance, clarity and cheapness, but it had always been difficult to blow into bottles because thermal degradation occurs at temperatures very little above that needed to melt the material and give adequate flow properties. Advances both in blowing equipment and in compounding PVC have since made it possible to produce bottles of extremely good quality.

Figure 16.16   Vinegar bottles: injection-moulded two-piece containers in polystyrene. Courtesy of Shell Chemical Co. Ltd.

The development of the PET bottle

Practically every product purchased today is packaged in some way or other. The general acceptance of plastic bottles for liquids attests to their fulfilment of such customer demands as safety, convenience, lightness and ease of handling. Typical liquid products packaged in plastic bottles include household and industrial chemicals, foods and beverages, toiletries, cosmetics, medicines and health care products.

The important factors needing consideration when selecting a material for a particular application are:

1. Will the product be cold or hot filled?
2. What gas or water barrier properties are required?
3. Is high tensile strength needed?
4. Is high clarity desired?
5. Is impact strength critical?
6. What closures will be used?
7. Are squeeze bottle characteristics needed?
8. What regulations must the bottle comply with?

Beverages

The soft drinks industry was the primary driving force behind the development of the PET carbonated soft drink bottle. Many small, heavy and breakable glass bottles were used for soft drinks and in the larger sizes the weight became excessive.

The concept of 'strength through orientation' in PET originated in the 1950s when it was found that by drawing polyester polymer in its longitudinal direction its tensile properties were greatly improved. Thus 'uniaxial orientation' was started and the polyester fibre industry began. Today well-known polyester fibres include 'Terylene' and 'Dacron'.

In the 1960s the technology of 'biaxial orientation', i.e. stretching in one direction and then again at right angles to the original stretch, was developed. This produces a polyester film with excellent mechanical and gas permeability properties now used in food packaging and many other market areas such as audio and video tapes.

In the United States DuPont had been researching into the possibility of producing a high strength bottle from biaxially oriented polymeric material. In the late 1960s the breakthrough was finally made.

A United States patent covering the PET bottle was issued to DuPont on May 15th, 1973. Patents were also granted in many countries around the world, including the United Kingdom.

DuPont and Pepsi Cola began to work together and in 1974 PET bottles were produced on a pilot scale. They were used in test markets in New York State. The tests involved more than 1 million cases and the results clearly established the use of the PET bottle. Pepsi Cola introduced a commercial 2 litre PET bottle in March 1977, and were quickly followed by Coca Cola. The PET bottle became standard for the 'family size' market in the United States.

The United Kingdom was the first country in Europe to adopt the PET bottle on any scale. By the beginning of 1980, most bottlers of internationally marketed drinks had either moved into PET or had plans to do so. Other beverages quickly followed until now beer, cider, wine, spirits and mineral waters are all in PET bottles.

Having developed a glass-clear, lightweight plastic bottle, many people had the idea of using them to replace the heavyweight breakable wide-mouth glass jars used for sweets, etc. With the development of PET bottles for beverages like Coca Cola this seemed possible but these are pressurized bottles and are used differently from the wide-mouth jars used for mayonnaise, spices, toffees or coffee.

Attempts to use the proven process for PET beverage bottles for jars as well, showed that although technically it was possible to make very nice jars, they could not compete with glass on price, as they were simply too expensive due to the thick wall needed.

About four years ago in Japan, Katashi Aoki, founder of the Nissei Group, discovered that the use of a very simple, disc-shaped preform, allowed the production of thin-walled wide-mouth containers with excellent strength, gloss and wall distribution, at an affordable price (Fig 16.17).

Enhanced PET containers

Oxygen sensitive products. For certain products the gas barrier properties of PET are inadequate. Beer, for example, is very sensitive to oxygen and a small amount permeating through the wall of a normal PET bottle would cause an off-taste. To improve the oxygen resistance a very thin layer of a high gas barrier material may be coated on the outside of the bottle or be co-injected at the preform moulding stage to form a PET/barrier material/ PET sandwich. It may also be blended with the PET prior to injection-moulding of the preform.

Hot-filled products. Many food products are filled hot into their containers. This poses a problem for the standard PET bottle. The blown side wall of a PET container retains a 'memory' of the moulded preform from which it was made and even at room temperture will shrink very slightly over a period of time. At high temperatures the shrinkage is so much greater that a normal PET bottle cannot be used above 60°C.

A heat-setting technique can be applied to PET containers to raise their maximum working temperature. The container is held inside a blow-mould under pressure at high temperatures until any strains frozen into the wall during the stretch blow-moulding process have been relaxed. This enhances the performance of the bottle sidewall without affecting the properties of the neck which remains in the 'as moulded' amorphous state and will soften heated above 72°C. This could distort the neck and ruin the integrity of the closure. Filling temperatures up to 85°C are possible but for higher temperatures, further improvement is needed.

A multi-layer structure provided by co-injection of a thermally stable polymer as a sandwich between two layers of PET can be used. The middle layer prevents distortion during and after the filling operation. An alternative is to crystallize the neck material. Crystalline PET, which is opaque white, will not soften at temperatures above the glass/rubber transition.

Cosmetics. The wide range of colour effects possible with polystyrene has led to its extended use in packaging of cosmetics where 'sales appeal' is extremely important. In addition to being chemically inert to many cosmetic products, it has a pleasant touch and is light in weight. Typical uses are for

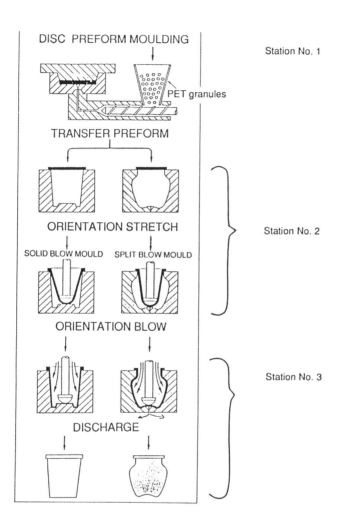

Figure 16.17 Stretch blow-moulding of thin-walled wide-mouth jars. Station No. 1: In the first station of the fully encapsulated machine, a disc-shaped preform is injection moulded from the PET raw material granulate. Station No. 2: Still warm, but with its top section kept in cooled clamping rings, the preform is transferred by a rotary table to its second station where with the heat that is still in the core of the material and not in the skin the preform is optimally mechanically stretched to a conical 'trumpet' giving the orientation in one axis (longitudinal). Station No. 3: Air is explosively introduced, forming the trumpet to its final jar shape and bringing about orientation in a hoop direction. Only the centre part of the bottom and the top section are non-oriented. The orientation is in 2 directions at 90° and produces a very strong and glossy-clear container which is cooled and ejected from the machine.

face-powder boxes, cream jars, holders for sticks of deodorant and shaving soap, compacts for face-powder boxes and holders for lipsticks. Later developments are blow-moulded containers for talcum powders using polystyrene.

Polypropylene is also finding uses in cosmetics containers because of its good gloss, coupled with the possibilities for moulding containers with hinged lids in one peice. Powder compacts and eye-shadow cases are typical examples. One use of polypropylene in cosmetics packaging is a blow-moulded polypropylene aerosol container for perfume. It is extremely difficult to break and, even if breakage does occur, the container only splits to allow the gradual release of its contents.

Both low- and high-density polyethylene are used for packaging shampoos, hand-creams, talcum powders, sun-tan preparations and deodorants. The absence of rusting when kept in a moist bathroom atmosphere is an advantage when compared with tins, while resistance to breakage gives an advantage over glass containers.

Pharmaceuticals. Polystyrene is used successfully for the packaging of tablets in injection-moulded phials. Unbreakable, but with the visibility of glass, they have the added advantage over glass of being easier to print.

Where greater protection against moisture vapour is required, high-density polyethylene tubes can replace those of polystyrene. Another advantage of high-density polyethylene is its higher softening point, which enables it to stand up to sterilization temperature.

Low-density polyethylene is also used for a wide variety of pharmaceutical packs, ranging from hydrocortisone to foot powder. A squeeze action makes it particularly suitable for dispensing powders, and for eye-drops and nasal sprays. Toughened polystyrene has been used for prescription containers, for pills, tablets and ointments.

Household products. Resistance to corrosion and to breakage have been the most important factors leading to the use of plastic containers for packaging liquid household products. Examples include rust removers, window cleaners, oven cleaners, writing inks and fly sprays. One oven cleaner pack makes full use of the potential of plastics to give a novel dispenser and applicator. High-density polyethylene is used for lavatory cleaners in both liquid and powder form. Liquid bleaches are also packaged in high-density polyethylene bottles, thus giving an unbreakable container for a potentially hazardous product.

Plastic containers are also being used for scouring powders in place of the spirally wound composite containers. They are made either by vacuum-forming high-impact polystyrene in two halves, subsequently spin-welded together, or by blow-moulding high-density polyethylene.

Miscellaneous. An interesting example of the packaging of a corrosive product is the use of polyethylene bottles for hydrofluoric acid. This acid attacks glass and was formerly packed in either guttapercha or wax bottles. Polyethylene bottles are now used, giving a safer pack and one in which the liquid level is visible. Warning notices and identification can be printed directly on the bottle, so that there is no possibility of losing the label. Both low-density and high-density polyethylene are used for the production of blow-moulded jerricans. These are used as water containers for camping and many aqueous solutions, while those made in high density polyethylene can also be used for the short-term storage of paraffin. The normal design of a jerrican is not suitable for petrol, but specially designed containers have been approved for such use in Germany. The main difference lies in an increased wall thickness of the jerrican designed for petrol.

Use of newer materials for containers

Polyacetals are more expensive than the materials already mentioned, and are used mainly as alternatives to light metals, such as aluminium, in engineering applications. The high strength of the polyacetals has led to their use for aerosol containers, and usage can be expected to increase.

Polycarbonate is again a rather expensive material, but is of interest in packaging because of its combination of high impact strength, high softening-point and good clarity. It can be blown into bottles and has been used in Japan for the manufacture of soda-water siphons.

Polymethylpentene is more widely known under the name TPX (trade name of Imperial Chemical Industries Ltd.) It has extremely good clarity and a high softening point but is a poor barrier to gases and moisture vapour. Its packaging uses would appear to be mainly for coating and laminating, rather than in blow-moulded or injection-moulded containers.

Ionomer is basically a modified low-density polyethylene with ionic bonds between the polymer chains. These ionic links give it a high melt strength (very suitable for extrusion coating) and also modify its crystallinity, consequently improving clarity. Bottles can be produced, therefore, with the high impact resistance of low-density polyethylene, but with significantly improved clarity.

Modified acrylics: Straight acrylics such as polymethyl methacrylate (Perspex, widely used in World War II for aircraft windshields and radomes) have many excellent properties, including extremely high clarity, but have rather high melt viscosities which make it difficult to blow them into bottles or extrude them into film. Modified acrylics have been developed to overcome these processing defects, without losing the valuable end-use performance characteristics of unmodified materials. One of the earliest modified acrylics was developed in the United States under the name XT polymer. It has been used to produce bottles having a high resistance to oils and greases, together with good contact clarity (although the bottles are somewhat hazy when empty). XT polymer has also been extruded into sheet, which can subsequently be thermoformed into trays and tubs. Products packaged in this material include medicinal paraffin oil and peanut butter.

More recent examples are Lopac (trade name of a material manufactured by Monsanto Chemical Company) and Barex (trade name of a material manufactured by Vistron Division of Standard Oil of Ohio). Both were developed in the United States. The main monomer from which Lopac is produced is methacrylonitrile, with small percentages of styrene and methyl styrene. Lopac is a hard, rather brittle polymer, but it has excellent barrier properties, good resistance to cold flow or creep, and is very clear. Barex consists mainly of acrylonitrile, copolymerized with methyl acrylate, together with a small percentage of a butadiene/acrylonitrile rubber. This is also a clear polymer with good creep resistance and excellent barrier properties. It has, in addition, good impact strength. Both Lopac and Barex are being developed, mainly as bottle-blowing materials for the packaging of carbonated soft drinks.

Developments in container manufacture

One of the disadvantages of normal thermoforming techniques is that two separate heating processs are involved, one when the granules are heated during extrusion of the sheet and the second when the sheet is heated prior to forming.

Much work has been carried out on the cold forming of sheet and some success has been achieved. Present cold forming techniques, however, have not yet reached the high degree of sophistication attained in thermoforming methods, and the cost of equivalent-sized containers is still normally greater by cold forming. A saving is shown, though, when the comparison is made for printed containers. Cold forming then becomes potentially cheaper, because printing of the flat sheet, prior to forming, is possible. This is quicker and cheaper than rotary printing of formed containers. Printing in the flat can be carried out, because cold forming subjects the sheet to less movement by stretching than does thermoforming. Most of the work on cold forming has been carried out on acrylonitrile/butadiene/styrene copolymer (ABS).

A process which does involve some heating of the sheet is *stretch forming*. A heated disc of the plastic sheet is placed over a split mould, the disc fitting into a recess at the top of the mould. The disc is then clamped into position and a forming punch descends, stretch forming the disc into the shape of the mould. The method is suitable for polypropylene, which is at a disadvantage in normal thermoforming processes, because of the difficulty of controlling it at the high temperatures involved and also because of its longer cooling time. Stretch forming has another advantage, since the resultant containers have good transparency, which cannot be achieved by most normal methods of processing polypropylene.

Stretch forming involves the use of special equipment, but it is also possible to convert polypropylene into transparent containers by more or less conventional pressure-forming equipment, using sheet heated to a temperature below its crystalline melting-point. Containers with diameter to depth ratios of 2:1 have been produced in this manner.

Developments such as these could increase the use of polypropylene in areas where its other properties give it an advantage over polystyrene and PVC, which are the commonly used transparent plastics. These other properties of polypropylene include a higher softening point, resistance to oils and greases, and toughness.

Another new development is melt-to-mould forming in which melted plastic is extruded directly onto a drum-type apparatus containing female moulds on its surface, producing thermoform-type containers without requiring production and reheating of the sheet.

17
Closures and dispensing devices for glass and plastic containers

It has been said that 'a bottle or jar is only as good as its closure'; but this is only part of the story, since a good seal is the outcome of a cooperative effort between the finish of the bottle or jar and its closure. The main duties that properly marrying closures and bottle finishes are required to perform are as follows:

1. The closure must keep the bottle or jar sealed until the contents are required for use. This usually means that the contents cannot escape, and external environments cannot enter the container. The degree of tightness required by the seal is, however, dependent upon the nature of the product packed; many products, such as non-hygroscopic powders and tablets, do not need a completely hermetic seal.

2. It must be possible to open the bottle or jar without difficulty and re-seal it properly with ease when only part of the contents is used at a time. Alternatively, the closure can be provided with a dispensing device, such as an orifice or spout, which is operated without removing the closure; or the closure may be fitted with a pierceable wad as is used for blood transfusion and injectable fluids.

3. The closure must neither affect nor be affected by the contents of the bottle or jar; it should be inert to any climatic conditions to which it may be exposed, and may need to withstand conditioning or processing treatment, such as pasteurization or sterilization, to which the pack is subjected.

4. The closure may need to provide a pilfer-proof device to show whether it has been removed prior to use, thus giving the consumer assurance that no-one has tampered with the contents. This is important when there might be a temptation to 'water down' the contents, as with potable spirits; or to inspect the contents out of curiosity, as with certain food packs on supermarket shelves when, by so doing, the sterility of the product is destroyed.

5. The closure may need to provide a tamper-evident membrane adhered to the mouth of the container which must be torn to gain access to the contents. Such membranes are usually also designed to keep out moisture and/or oxygen, so as to preserve the freshness of the product. Membrane seals are mostly used on jars of instant coffee and dried milk powders.

6. Special-purpose closures may need to retain a vacuum or internal pressure within the container.

7. Not only must the closure perform its mechanical and protective functions correctly, but it must often blend with the graphics of the container, enhancing the appearance, and adding to the sales appeal of the pack.

The products packed will be liquids, creams, pastes, powders, granules or other solids. they will vary from completely innocuous to highly corrosive, from very penetrating liquids (e.g. oils, detergents, solvents and the like) to solids which have little or no penetrating properties. Even so, some solids (e.g. fine chemicals) are hygroscopic or even deliquescent, and these require particularly good water-vapour-tight seals.

Other products that are sensitive to the ingress of moisture include many pharmaceutical tablets, both plain and sugar-coated, as well as effervescent salt mixtures in both tablet and powder form. Many medicines (such as cough mixtures) contain small percentages of chloroform which must not escape. Nail varnish and nail varnish removers also contain volatile solvents.

Products such as bleaches, containing peroxides or hypochlorites, develop internal pressures on standing, and it is important to be able to relieve those pressures without risk of leakage when taken home in shoppers' baskets. This requires the use of a liner that vents when the pressure in the container reaches a predetermined level.

Some liquid products, of which whisky is an example, have high coefficients of expansion, and containers may develop high internal pressures on exposure to tropical temperatures. It is essential that the closures on such containers are able to contain the developed internal pressures without leaking.

Carbonated beverages must be sealed with pressure-tight closures, and the vacuum in vacuum packs must be preserved until the jars are opened for use.

The general requirements for a good seal

A seal is usually made by causing a resilient material to press against the sealing edge or rim on the top of the finish of the container. The pressure must be evenly distributed and maintained to ensure a uniform seal around the whole of the edge of the cushioning material in contact with the rim.

The resilient material is frequently a wad cut or stamped out of a composition-cork or pulpboard sheet; usually protected by a facing material against interaction with the contents. The combination of a wad and a facing material is called a *liner*. Hence, caps have to be fitted with liners in order to make good seals on bottles.

Liners consist of materials that combine the resilient properties of the wad with the protective properties of the facing material. Such liners can be discs of solid rubber, PVC, polyethylene, EVA, etc., or they can consist of purpose-designed fitments moulded from these or other similar resilient thermoplastics. Liners also can be flowed-in to form wads, sealing rings, or gaskets. A flowed-in liner results from injecting a PVC plastisol, or a natural or synthetic rubber colloidal dispersion, into a metal cap and causing the compound to set by stoving.

In addition to the requirements for the cap and liner, the third but no less important aspect is the container finish, which must mate completely with its closure and liner. Fundamentally, the sealing edges or rims on the finishes of both glass and plastic containers must be seamless, smooth, even, and free from roughness or defects. Perfection in this respect is rarely attained in practice, but very good commercial finishes are now consistently obtainable from production runs, and the resilience of the liners then compensates for the small variations that still occur.

The mechanics of a good seal

To make a good seal, the liner is pressed against the sealing rim on the finish of the container with sufficient pressure, which must be maintained during the shelf-life of the pack. The higher this pressure (within reason) the more effective will be the seal, but it would obviously be self-defeating to increase the pressure to the point at which the cap can either break or deform, the bottle finish become chipped (glass) or deformed (plastics), or the liner break down by splitting or collapsing.

The tightness with which a cap is screwed on to a container is known as the *tightening torque*. Part of this torque is used in overcoming the frictional forces encountered in screwing on the cap, and the remainder is converted into a direct top-sealing pressure on the liner.

With rolled-on, crimped-on, and pressed-on caps, the effectiveness of the seal depends upon the direct pressure exerted on the top of the cap during the capping operation.

The sealing pressure ensures that the liner makes good contact with the sealing rim on the finish of the bottle or jar. The greater the area of the sealing surface on the rim of the bottle finish, the greater will be the area over which the load exerted by the closure is spread, and the less effective will be the seal for a given cap-tightening torque. It follows, therefore, that to obtain a good seal without having to use unduly high cap-tightening torques, the width of the sealing edge must be kept as narrow as possible, consistent with not damaging the liner or its facing. There are thus practical difficulties in maximizing the effective sealing pressure for low cap-tightening torques and, in practice, a narrow rounded sealing edge is particularly suitable for the finishes of glass bottles and jars, but there are also many completely satisfactory applications using flat sealing edges.

Plastic bottles and jars are usually made with flat sealing edges, the best seals being obtained when these are kept reasonably narrow.

A further point to remember, when designing a good closure system, is that the impression produced in the liner by the sealing rim of the bottle finish must not be too near to the outside perimeter of the liner. If this occurs, it can cause the edge of the liner to collapse, particularly if the wad is of composition-cork or similar easily compressible material.

Both composition-cork and pulpboard wads make comparable seals up to diameters of about 40 mm. Above this, composition-cork wads are better able to take up any slight waviness that can occur in the sealing rims of bottles but, because of flexibility, composition-cork wads can be difficult to retain in large caps (say above 48 mm), unless the liners are glued-in. Nowadays rim waviness is a much lesser problem with glass containers, and it can be completely controlled on plastic containers.

The importance of thread engagement and thread pitch

Thread engagement is the number of turns given to a cap from the point of first engagement between the start of the thread in the cap with the start of the thread on the bottle finish to the point when the sealing edge on top of the finish of the bottle makes contact with the liner in the cap. In order that the liner should be pressed down uniformly around the whole of its circumference on the sealing edge of the bottle, at least one full turn of thread engagement is required. The greater the thread engagement, the better the cap stays on, and the more effective is a given cap-tightening torque in keeping it there.

Thread pitch is measured in terms of the number of turns of thread per unit of traverse. The pitch determines the slope or steepness of the thread. The lower the number of thread turns per centimetre, the steeper will be the slope of the threads, and the more rapidly will the caps screw on and off the bottles. Also the steeper the slope of the thread, the deeper the caps will have to be to achieve a given thread engagement.

It is thus clear that the performance of a screw cap depends both on the thread engagement and the thread pitch.

There is no difficulty in producing plastic bottle finishes with a finer thread pitch than can be achieved satisfactorily with glass. Thus, while it is possible to design a shallower skirted cap having the requisite minimum of one full turn of thread engagement, plastic bottle finishes tend to follow the pattern of glass bottle finishes, if only for the reason that the same caps can then be used on both glass and plastic bottles and jars interchangeably. With plastics, of course, many speciality designs are available, where the cap and bottle thread forms are specific to each other.

Applying the correct tightening torque to screw caps

Having made sure that both the cap and the bottle finish requirements for a good seal have been satisfied, it remains to ensure that the caps are applied properly to the bottles. To do this we need a yardstick to measure the effectiveness of cap application.

This can be measured by means of a *torque tester*. To determine the cap application torque, the bottle is clamped on the spring-loaded table of the torque tester, the cap is screwed on to the bottle, and the torque applied is read directly from the scale. In practice, the torque tester cannot be placed under the capping head of a capping machine, and the torque applied by a capping machine must be determined indirectly by measuring the *immediate loosening torques* of the caps applied by the machine. The torque at which the caps should unscrew is first obtained by applying several dozen caps to the bottles, by hand, to the required tightening torque using the torque tester. The cap loosening torque for each is then measured. Having determined the range of values of the cap loosening torque for the required cap tightening torque, it is then only necessary to adjust the capping machine until the loosening torques of the caps applied by the machine fall within the same range of values as those of the caps applied by hand.

The cap tightening torque varies with the diameter of the cap, and each torque tester is supplied with a chart giving the minimum and maximum recommended cap tightening torques for the range of cap diameters in normal use. There are nevertheless occasions when the maximum recommended cap tightening torque may be exceeded, provided the caps are sufficiently robust. With thin plastics or aluminium caps, it is generally not advisable to exceed the recommended torque.

The importance of vacuity or ullage

Vacuity or ullage can be described as the amount by which a bottle falls short of being brimful with liquid. It is the volume of the air space in the head of the container above the level of the contents. It is also known as the *head space* (or *expansion space*) and is expressed as a percentage of the liquid contents of the bottle at the time of sealing.

If the liquid contents of a bottle expand as the temperature rises, then the higher the coefficient of expansion of the liquid, the greater the provision that must be made for vacuity. For example, whisky and gin have high coefficients of expansion, and a 2% vacuity will result in an internal pressure of the order of 500 kN/m^2 after a rise in temperature of about 22°C above the filling temperature. With a vacuity of 5% the internal pressure would only reach about 140 kN/m^2 after a rise in temperature of 35°C above the filling temperature. In practice, vacuities between 3% and 5% are used for spirits (such as whisky and gin) depending upon the type of closure used.

It is clear that the temperature of the product at the time of filling is important. If the product is at 5°C when filled, and no adjustment is made to the height of fill, effective vacuity will be lost by the time the contents of the bottles reach room temperature. This means that a vacuity normally allowing a rise of, say, 30°C above a product filling temperature of 16°C would, if the same vacuity were used at a filling temperature of 22°C, only provide for a rise of 16°C above the ambient temperature of 16°C before leaking would occur.

A vacuity of 10% has to be provided for products like peroxides and hypochlorites which are liable to give off gases. In practice, these products are frequently sealed with closures that vent to relieve the excess gas pressure that develops.

The above considerations apply to all rigid bottles. Flexible and distensible plastic bottles could accept lower vacuities since the effective pressure developed against the insides of the caps is reduced as such bottles distend under internal pressure. But, if they distend too much, they can become unstable and fall over or burst.

Types of closures

When classified according to their primary function, all closures fall into one of four main groups:

1. Those that provide a normal seal: these are primarily intended to keep the containers sealed and, although not specifically designed to resist pressure, will cope with normal rises in internal pressure resulting from increases in external temperature, subject to the use of proper vacuities (Figure 17.1).
2. Those that ensure a vacuum-tight seal: closures that are designed to make an air-tight seal when (at least) a partial vacuum has been developed in the head space of the container as a result of the sealing or processing conditions to which the packs are submitted—the maintenance of vacuum being necessary for the proper preservation of the contents.
3. Those that are designed to contain high internal pressures of the order of 700 kN/m^2: these closures (Figure 17.2) are commonly used for sealing pasteurized beers and carbonated beverages.
4. Those that have a venting feature: closures that allow venting (Figures 17.3, 17.4) when the internal gas pressure reaches a pre-determined level, excluding vacuum seals (where venting takes place during the processing of packs that are required to develop at least a partial vacuum as they cool to room temperature).

Some overlap can occur between these groups. Certain pressure-tight and vacuum-tight seals can, for example, serve for both purposes or be used as a normal seal. Similarly, some normal seals can withstand quite high internal pressures.

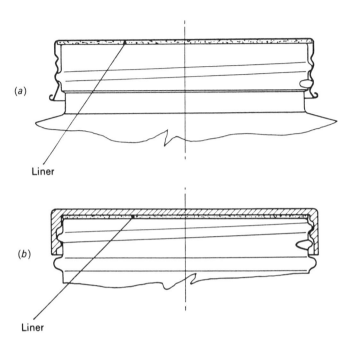

Figure 17.1 Conventional pre-threaded screw caps. Cross-sections of (a) a metal screw cap and (b) a plastic screw cap applied to the finish of a glass or plastic container.

Figure 17.2 Crowns: shallow fluted-skirt crimp-on tinplate closures with plastisol liners. These are the best pressure-tight seals but require an opener.

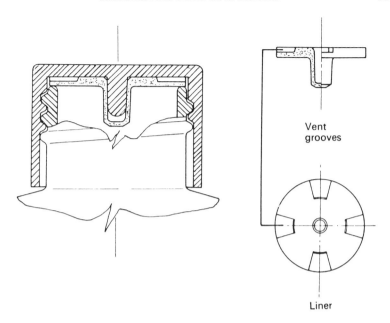

Figure 17.3 Venting closures developed originally for sealing bleaches (hypochlorites) and peroxides. They are designed to relieve excess pressures. A venting closure is basically a standard external screw closure fitted with a liner that vents. The diagram shows an example of the grooved venting liner. The liner is moulded in suitable resilient material, such as low-density polyethylene and is provided with grooves that arch when excess pressure develops in the bottle, thus allowing the gases that produced this pressure to escape.

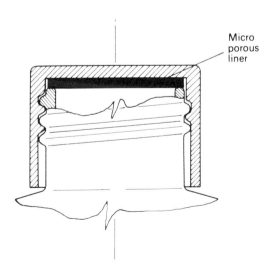

Figure 17.4 A closure fitted with a microporous liner, such as microporous PVC, which allows gases to escape via the micropores when excess pressure develops in the bottle.

Selection of closures

The main aspects to be taken into consideration when selecting a closure within groups 1, 2 and 3 are as follows.

Normal seals (Table 17.1)

The basic requirements for a good seal are:

1. The liner must be compatible with the contents.
2. The liner must be compressed evenly over the whole surface of the sealing edge of the container. With screw caps, a minimum of one full turn of thread engagement is required for maximum security of the seal but, in practice, a minimum thread engagement of three-quarters of a turn is usually satisfactory for general-purpose applications.

Table 17.1   Normal seals.

Method of performance	General classification	Materials in which made
Pre-threaded screw on, screw off, generally single-start threads	Compression-moulded plastics	Thermosets
	Injection-moulded plastics	Thermoplastics
	Vacuum-formed plastics	Thermoplastic sheet
	Metal	Tinplate Hi-Top plate Aluminium Aluminium alloy
Two-piece pre-threaded screw on, screw off	Metal	Tinplate Aluminium Aluminium alloy
Lug type screw on, twist off	Metal	Generally tinplate
Roll on or spin on, screw off	Metal	Aluminium Aluminium alloy
Press on, prise off	Metal	Tinplate
Crimp on, prise off	Metal	Aluminium
Lever type crimp on, lever off (now obsolete)	Metal	Generally tinplate
Crimp on, screw off	Metal	Tinplate Aluminium
Push in, pull out	Driven corks	Cork
	Stoppers	Corks with either plastic tops or wooden tops inside decorated aluminium covers
	Stoppers with ribbed or finned hollow plugs	All thermoplastics, generally polyethylene
Push on, pull off	Snap on Push on	Thermoplastics, generally polyethylene

3. The sealing edge on the rim of the container should be seamless, smooth and reasonably narrow (about 0.6–1.2 mm wide) so as to ensure maximum pressure against the liner or gasket for the recommended cap tightening torque. For plastic containers, it is important to avoid sealing rims that are so narrow and sharp that they cut into the liner.
4. Sufficient vacuity must be allowed to provide for the expansion characteristics of the product. Even more must be provided for products such as ether, which develop high internal pressures for only moderate increases in ambient temperature.

Vacuum-tight seals (Table 17.2)

All closures used in vacuum-tight seals are made of metal, and, in practice, they are all fitted either with flowed-in liners, or gaskets, or with rubber rings. The reasons for this are as follows.

In any closure (metal or plastic) fitted with a conventional liner (such as a faced pulpboard or composition-cork disc), the vacuum in the container pulls the liner down on to the finish of the bottle. With a metal cap, fitted with such a conventional liner, the pull of the vacuum is likely to pull the liner out of the closure. With a plastic closure, the liner is usually glued-in, and the closure will then be so difficult to unscrew that the liner will shear before the vacuum is broken. Moreover, vacuum packs generally involve some form of processing or steam injection, and moisture entering the closure would seriously affect any liner based on a pulpboard or composition-cork wad.

The sealing medium must, therefore, be a flowed-in compound containing a slip additive to make the closure easy to unscrew or twist off. Flowed-in compounds do not adhere to plastic closures, which in any event will not

Table 17.2 Vacuum seals.

Method of performance	General classification	Materials in which made
Screw on, twist off – lug caps	Metal	Almost invariably tinplate, but aluminium and its alloys are possible
Press on, prise off	Metal	Tinplate
Press on, twist off	Metal	Tinplate
Two-piece screw on, screw off	Metal	Tinplate Aluminium Aluminium alloy
Two-piece roll on, screw off	Metal	Aluminium Aluminium alloy
Crimp on, prise off Crimp on, screw off	Metal	Tinplate Aluminium Aluminium alloy

withstand the temperatures at which flowed-in compounds have to be stoved. Hence, only metal caps are used for vacuum-tight closures.

A vacuum-tight closure is thus either a screw-off or a twist-off metal closure lined with a flowed-in compound containing a slip additive, or a prise-off metal closure lined with a flowed-in compound without a slip additive, or fitted with a rubber sealing ring. A prise-off closure is lifted directly off the finish of the bottle to break the vacuum and no twisting is required.

Pressure-tight seals (Table 17.3)

Pressure-tight seals are generally required to contain pressures from about 350 kN/m^2 to over 1000 kN/m^2. This means that consideration must be given to the strength of the closure in relation to the total force that will be exerted against it. For a closure about 25 mm diameter on the inside, the internal surface area is about 2.5 cm^2 and the force exerted against it, when the bottle is pressurized at 350 kN/m^2 will be 164 N or 500 N if the bottle were pressurized at 1000 kN/m^2. But, if the internal closure diameter is 50 mm, its internal surface area is about 10 cm^2 and the force then exerted against the inside of the closure is 670 N at a pressure of 350 kN/m^2 and 2000 N at a pressure of 1000 kN/m^2.

Thus, the larger the diameter of the cap, the thicker must be the plate from which the cap is made, or it will not withstand the internal forces that will be directed against it. But the thicker the plate, the less economical the

Table 17.3 Pressure seals.

Method of performance	General classification	Materials in which made
Screw in (internal screw), screw out (now almost obsolete)	Moulded plastics	Cold-moulding (ebonite) Thermosets Thermoplastics
Screw on, screw off	Moulded plastics ('Estseal') (now obsolete) Metal ('Lo Tork')	Thermosets Thermoplastics Tinplate
Crimp on, lever off Crimp on, screw off	Metal	Tinplate Aluminium Aluminium alloy
Roll on or spin on, screw off	Metal ('Eurospin' and 'Flavorlok')	Aluminium Aluminium alloy
Crimp on, pull off or rip off	Metal	Tinplate Aluminium Aluminium alloy
Crimp on or roll on with tab-locked band, undo band	Metal (Phoenix) (gradually going out of use)	Tinplate

cap becomes. Hence, in practice, the maximum diameter of commercial closures to withstand high internal pressures is likely to be about 40 mm.

The best pressure-retaining seal is the tinplate crown (crimp-on; lever off, Figure 17.2) but the spun-on aluminium closure has good performance, particularly when fitted with a flowed-in liner that is made to hug the top and outside edges of the sealing rim (Figure 17.5). Another closure, particularly in the larger diameters, that makes an excellent pressure-tight seal is the Eurocap which makes possible the replacement of cans by glass jars without altering the processing conditions. It is the only relatively large-diameter closure that will withstand autoclaving without counter pressure.

An internal screw provides the best pressure retention for plastic closures, but it is now expensive, and cumbersome. To contain pressures of the order

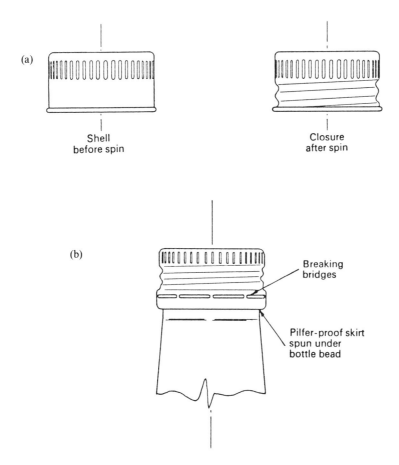

(a)

Shell
before spin

Closure
after spin

(b)

Breaking
bridges

Pilfer-proof skirt
spun under
bottle bead

Figure 17.5 (a) Roll-on closure. (b) Roll-on or spin-on tamper-evident cap applied to the finish of a glass container.

of 400 kN/m^2 or 500 kN/m^2, an external screw plastic closure must have ample thread engagement, and the finish of the container a downward and outward tapering sealing edge which makes a more effective seal with the liner than a narrow conventional flat or rounded rim.

Secondary functions of closures

The secondary functions of closures, whether for use on glass or plastic containers, all have particular objectives such as inviolability, non-refillability, measuring and pouring facilities, child-resistance, sales appeal or two or more of these functions in combination. Additionally, advantage can be taken of the flexibility of some plastics to combine a squeezing action with a cap feature for dispensing small quantities of liquids at a time, or to provide a spray from a spray nozzle incorporated in the closure system.

The main secondary functions of modern closures for both glass and plastics containers are as follows.

Inviolability

A closure must demonstrate that unlawful access to the contents or their unwitting exposure to the atmosphere has occurred. Strictly, no closure is inviolable, since it must be possible for the consumer to gain access to the contents of the container, but tell-tale features can be introduced into the closure system that show up immediately the seal has been broken.

Pilfer- and tamper-evident. Here a band is joined to the skirt of the closure by means of frangible bridges that fracture when the closure is unscrewed, leaving a tell-tale ring round the bottom of the neck of the bottle. These caps are normally represented by the so called roll-on pilfer-proof (ROPP) aluminium closure (Figure 17.5b) that is spun on to the bottle finish, the tell-tale pilfer-proof ring being tucked under a bead on the neck of the bottle during the capping operation.

Plastic pilfer-proof closures (Figure 17.6) are designed to snap over a security bead on the necks of glass or plastic bottles so that the closures cannot be opened until the band connecting the skirt to the pilfer-proof ring is torn away.

Tamper-evident packaging. Ever since the days of barter there have been thieves and pilferers and traders have had to counteract their depradations. This was fairly simple when goods were exchanged for other goods or for money on a face to face basis, because the buyer could examine the goods

(a)

Primary seal   Hinge open   Snap shut

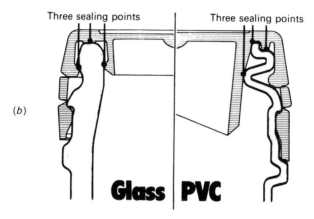

(b)

Three sealing points   Three sealing points

Glass | PVC

Figure 17.6   Plastic pilfer-proof closures are designed to snap over a security head on the neck of the container, so that they cannot be opened until the band connecting the skirt to the pilfer-proof ring is torn away. (a) Polythene Jaycap, currently used on both glass and rigid plastic containers. (b) The sealing features of a Jaycap on both glass and rigid plastic bottles.

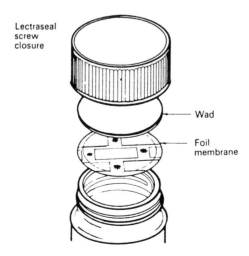

Lectraseal
screw
closure

Wad

Foil
membrane

Figure 17.7   The pre-threaded tamper-evident closure (metal or plastic) with a membrane seal.

beforehand without either the intervention of a third party or a period of display and/or transport outside the care of the seller. So there is nothing new in products being stolen or otherwise interfered with to permit undiscovered theft to take place. Indeed, from early times the seller of many foods, for example, adulterated them himself for commercial advantage. Flour adulterated with chalk, sugar with sand and milk with water were commonplace right up to the end of the 18th century. In fact, one of the major reasons for food manufacturers to package their products in retail size-branded packages was to prevent such adulteration occurring. Even today adulteraton for extra profit is still far too common in many of the developing countries; and legislation does not prevent it.

Terms like tamper-proof, tamper-resistant and tamper-evident have been used to describe many kinds of closure and the meanings of these must be understood. There is confusion over the use of the right term, many advertisements refer to tamper-proof devices although as pointed out by many experts, it is not attainable under conceivable retail circumstances. The difference between pilfering to taste a product or to steal without detection in a retail store and removing one or more packages to deliberately and maliciously contaminate the product for revenge, sadistic amusement or blackmail as has occurred in the recent past, lies in the time and resource factors available to the criminal concerned. In the former case, the pilferer has limited time and can only use a simple instrument. In the latter instance, time and equipment can be as required to achieve the objective since the tampering is done unobserved and the packages are then returned to the retail shelf apparently untouched. So 'tamper-proof' is unattainable; the determined criminal can always achieve his objective.

What of tamper resistance and tamper evidence? This brings us to the consideration of security of packaging as a whole. It is, of course, only one of the several essential characteristics of any package and must be considered in the overall context of what the package is required to do. This includes containment, protection, communication and convenience as well as closure integrity and tamper resistance/evidence. Security packaging does not guarantee to prevent tampering but should make it difficult to accomplish without rendering the product unusable and/or making it evident that the package has been tampered with. In 1982, the Proprietary Association of Great Britain (PAGB), which represents the makers of medicines available over the counter to the general public, issued a set of guidelines on security packaging following many requests for a definition of 'tamper-evident'. It reads as follows: 'Security packaging . . . a container or package having an indicator or barrier to entry which, if breached or missing, can reasonably be expected to provide visible evidence that the said container or package has been tampered with.'

There are many ways of creating a degree of tamper evidence and Table 17.4 lists those accepted by the FDA in the United States.

Table 17.4 Current tamper-evident packaging methods.

Film-wrappers	Must be transparent with distinctive design or print for each product; must be cut or torn to gain access and be a tight fit (e.g. by heat shrinking); sealed overlapping end flaps are only acceptable if they cannot be opened and resealed without leaving visible evidence.
Blister or strip packs	The backing material must not be separated from the blister without visible evidence and each individual compartment must be broken, cut or torn to gain access
Bubble packs	Requirements as for Blisters
Heat-sealed bands or wrappers	Must be placed over the union of the primary closure (cap, lid, etc.) and the container; must have a distinctive design, (pattern, name, etc.); must be shrunk on by heat (wet shrinking is not acceptable); must be cut or torn to gain access; must not be worked off without leaving visible evidence, and a perforated tear strip enhances tamper evidence)
Pouches	The endseals must not peel or separate without visible damage and the pouch must be cut or torn to gain access
Inner seals	Whether made of paper, plastic film, PS foam, foil or combinations, they must have a distinctive design (pattern, name, trademark, logo or picture) and must be torn or broken to gain access, heat induction seals are considered superior to adhesive seals (see Figure 17.7)
Tape seals	These are only acceptable if they incorporate some unique feature that makes it apparent that they have been removed and reapplied; if used on cartons they must be applied over all flaps
Breakable caps	The cap or part of it must be broken in order to open the container and remove the product; it must not be possible to replace it in its original state (Figure 17.5)
Sealed metal tubes, plastic (blind end) heat sealed tubes	These must have both ends sealed; the mouth or blind end has to be punctured to gain access to the product; crimped ends are only acceptable if they cannot be unfolded and refolded without visible damage
Sealed paperboard cartons	Current glue and flap cartons are not acceptable but the FDA recognises that technology may in the future achieve an acceptable result
Aerosols	These are considered to be inherently tamper resistant; direct printing on the container is preferable to a paper label
Cans and composite containers	Top and bottom of a composite must be joined to the walls so that they cannot be pulled apart and replaced; again direct printing, not labelling is required

Non-refillability

Non-refillable closure systems are designed to prevent or at least strongly deter bottles from part or complete refilling with spurious liquors by unscrupulous handlers. These closure systems usually consist of devices which are firmly locked on to the finishes of the bottles, and which contain valves and weights that only allow flow in one direction, namely out of the bottle.

Dispensing, measuring and pouring devices

These aids control the rate and manner of flow, or the method and/or amount of dispensing (Figures 17.8–17.11) so that the product is used to the best effect and convenience (see Figures 17.12–17.14).

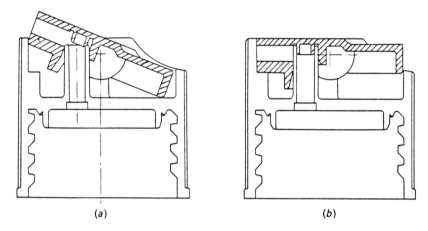

         (a)   (b)

Figure 17.8 Toggle dispenser closures in (a) the open and (b) the closed positions. The closure makes a plug seal in the neck of a rigid plastic container and the contents of the bottle are dispensed after depressing the disc so as to open the spout.

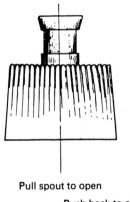

Pull spout to open

Push back to close

Figure 17.9 The Tip-Top dispenser makes a plug seal in the neck of a rigid plastic container. The contents of the bottle are dispensed by pulling the spout out and the bottle is sealed by pushing the spout in again.

Figure 17.10 The flip top is a low-cost one-piece fixed snap-on polythene closure with dispensing spout and captive re-seal push-on cap. It is widely used for sealing detergents and similar products.

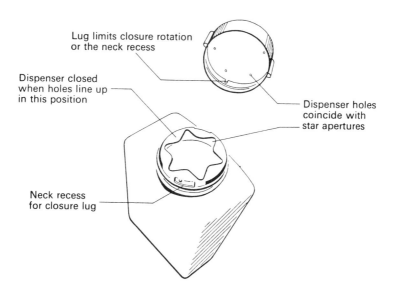

Figure 17.11 A one-piece talc dispenser snap-on closure.

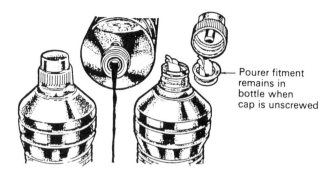

Pourer fitment remains in bottle when cap is unscrewed

Figure 17.12   The Rohill oil-pourer closure.

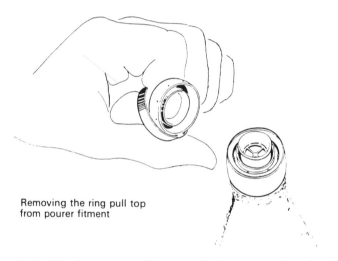

Removing the ring pull top from pourer fitment

Figure 17.13   The ring pourer cap. The top portion is a tamper-evident ring closure.

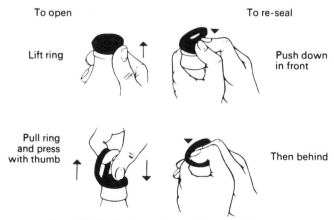

To open   To re-seal

Lift ring   Push down in front

Pull ring and press with thumb   Then behind

Figure 17.14   Opening and re-sealing a ring closure.

Child resistance

Child-resistant closures are designed to make it difficult for young children to gain access to the contents of bottles and jars while enabling adults to do so without difficulty (Figure 17.15).

Press and turn. The cap is removed by applying a downward force while the closure is rotated (e.g. 'Clic-loc') (Figures 17.16, 17.17). Closures which require the application of an upward force while it is rotated (e.g. 'Ringuard') may be placed in this general class, involving vertical and rotational forces.

Squeeze and turn. Force is applied to the sides of the closure while rotation is supplied (e.g. 'Squeeze-lok').

Combination lock. The cap incorporates a freely rotating component, which has to be held immobile before the closure can be removed (e.g. 'Hold'n Twist').

Press and lift. Downward or side pressure has to be used, either to relax the grip of the cap on the container neck, or to release a portion of it which can then be grasped for lifting off (e.g. 'Pop-Lok') (Figure 17.18).

Key closures. Two-piece closures, one part of which has to be removed, and

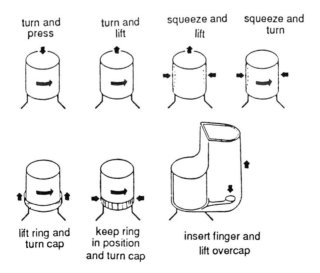

Figure 17.15   A child-resistant closure requires two coordinated actions.

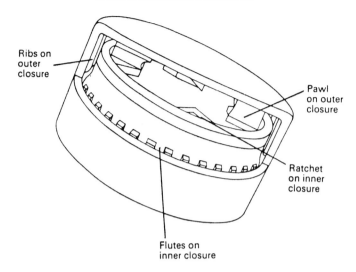

Ribs on
outer
closure

Pawl
on outer
closure

Ratchet
on inner
closure

Flutes on
inner closure

Figure 17.16   The Clic-Lok closure, a child-resistant closure suitable for use on standard glass bottles. This is a press down and turn closure with the added feature of an audible alarm when an attempt is made to unscrew it by a normal turning action.

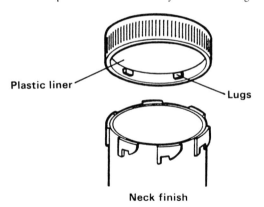

Plastic liner

Lugs

Neck finish

Figure 17.17   The Screw-Loc press and turn closure which is more suitable for plastic than glass containers because of the special finish required to mate with it.

Comes off plastic
tube when recessed
tab is raised.

Figure 17.18   The Pop-Lok closure fits tightly on the container and cannot be removed by pulling the sides. Pressing the correct spot on the top of the cap releases a tab which is easily gripped to pull off the cap.

so could be stored separately. This removable part is then used as a key to open up the actual closure portion of the cap (e.g. 'PCC' Figure 17.19).

Double cap closures. Two-piece closure with the over-cap having CRC function requiring insertion of a finger at least 6 cm long to release. The over-cap can be obtained separately and can be attached to closures of certain dimensions (e.g. 'Kidlok').

Sales appeal

There is virtually no limit to the versatility of craftsmanship in design, colour and decoration of plastic closures. There is much less scope in the design of metal caps—being limited by the constrictions on their method of manufacture—but they can be decorated in many colours and in an almost endless range of attractive printed designs.

The manufacture of closures

Closures fall into two main classes:

1. Metal caps
2. Moulded caps including those made from thermosetting and thermoplastic materials.

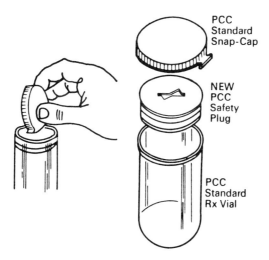

Figure 17.19 The PCC Safe-Key cap consists of an outer snap cap and an inner flush-fitting plug seal. The tab of the snap cap is provided with lateral projections that fit into a moulded-in keyhold opening in the top of the plug. The user first removes the snap cap and then inserts the projecting tab ends into the keyhole to pull out the plug.

Metal caps

Metal caps are stamped out of sheets of either tinplate, aluminium or aluminium alloy, generally in a thickness of about 0.25 mm, depending upon the use for which the caps are required. The sheets are first coated by a roller coating machine with lacquers or enamels compatible with the product packed and resistant to any processing or climatic conditions the caps are called upon to withstand; they are frequently printed on their tops and/or their skirts with decorated designs. All lacquers, enamels and printed coatings are stoved immediately after application by conveying the coated sheets through an oven. This not only develops the hardness in the coating essential to withstand scuffing during subsequent tooling and handling operations, but also the required resistance to the product with which the coating may come into contact, as well as the processing treatment it may have to withstand.

For screw caps, the decorated sheets are fed to presses that stamp out the cap shells which are then automatically fed into the threadforming, beading and knurling machines. The liners are automatically inserted into the finished caps but, in some instances (for example, when there is a possibility of damaging the edges of composition-cork wads), the insertion can take place after the cap shells have been knurled, but before the threads are formed. Other types of metal caps, such as lug caps (Figure 17.20) and crowns, will either be complete after pressing or will require one or two further tooling operations according to their design. Crowns and practically all lug caps are fitted with liners after they are otherwise complete, and the type of liner used in these caps is now almost universally a flowed-in plastisol

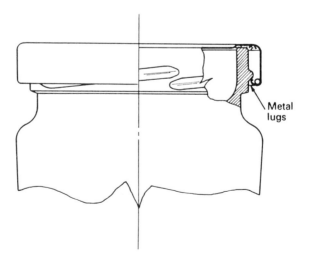

Figure 17.20   Lug cap applied to the finish of a glass jar.

gasket or sealing ring, although there is still some demand for flowed-in water-based compounds. There are also two-part metal caps in which each part is fabricated separately.

In the case of spin-on or roll-on closures, the manufacture of the cap shells is complete after the drawing operation, apart from any bead or pilfer-proof feature that may be required. The threads are developed in these closures when the cap shells are rolled or spun on to the threaded finishes of the bottles. The liners are fitted or flowed-in (depending upon the type of liner to be used) immediately after the caps are complete in their shell form.

Moulded caps made from thermosetting plastics

These caps are also known as *compression moulded* since they are formed in presses that compress by means of heat and pressure the moulding powder into the cap moulds.

Three types of moulding powders are generally used for thermoset screw closures: UF is a paper-filled urea-formaldehyde resin; WF is a woodflour-filled urea-formaldehyde resin; PF is a woodflour-filled phenol-formaldehyde resin.

The manufacture of thermoset closures is based on the principle that urea and formaldehyde on the one hand, and phenol and formaldehyde on the other, will combine together under heat and pressure to form hard, insoluble and chemically resistant polymeric materials.

The cap moulder either uses a moulding powder as received, or preforms it by initial compression into tablets (or pellets) according to the requirements of the moulding press. These are fed into the cavities of the moulding press, and the caps are removed by unscrewing as soon as the curing cycle is complete. The whole process is generally completely automatic.

In essence a mould consists of a steel cavity which determines the outside configuration of the cap, and a threaded pin which forms the threads on the inside of the cap. The cavity is loaded with a measured volume of powder or with a pre-determined size of pellet. Both cavities and pins are heated to a temperature determined by the nature of the powder and the moulding conditions (generally of the order of 145°C for UF and WF powders and 160°C for phenolic powders) so that, when the pin is pressed into the cavity under a pressure of the order of 15–20 MN/m^2 the moulding powder flows into the form and shape of the finished cap and the resin sets. Both pins and cavities are highly polished and also chromium plated, so as to resist corrosion and produce the best finish possible on the caps.

All thermoset caps have a light moulding flash where the excess material exudes between the cavity and the pin during the moulding process. This flash is removed either by rumbling the caps in a revolving barrel of wire mesh, or by passing them through a deflashing machine which automatically grinds the flash off the ends of the skirts. The caps are then polished, either in a barrel with rags or by hand on rotating polishing mops. They can also be given decorative finishes by spraying, hot leaf stamping, metallizing, etc. Finally, the caps are automatically fitted with liners, which are almost always glued-in, before dispatch.

Moulded caps made from thermoplastic plasics

These caps are also known as *injection moulded* since they are formed in presses that inject the hot fluid plastics material into the moulds. There is an increasing variety of thermoplastic materials from which caps can be made. The most commonly used are polystyrene, polystyrene blends and copolymers, high- and low-density polyethylene and polypropylene.

Unlike thermoset powders supplied to cap manufacturers in partly polymerized form, all thermoplastics are completely polymerized when delivered to the cap moulders. All that the latter have to do is to heat the thermoplastic material to temperature and inject it under pressure into the moulds, which are then cooled to cause the thermoplastic material to set. The moulds are opened for the removal of the caps, and the process then starts all over again. As with thermosets, the whole moulding process is generally completely automatic.

Although the process is very simple, the tools for the injection moulding of thermoplastic materials are considerably more expensive than those required for the thermosets. This is because injection moulding tools consist of multi-impressions fed by runners from a central injection point, and the moulds also have to be made so that they can be water-cooled.

Once moulded, thermoplastic caps are complete and require no further finishing. Generally, too, thermoplastic caps have to be fitted with liners, but there are applications for thermoplastic closures without liners (Figures 17.21, 17.22).

Thermoplastics re-soften when heated and can, therefore, be recovered by grinding into granules for re-use. Thermosets, in contrast, are not reversible and do not re-soften when heated; reject thermoset mouldings cannot, therefore, be recovered and re-used.

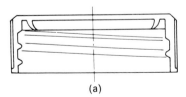

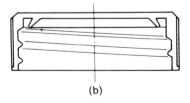

(a)   (b)

Figure 17.21   A linerless closure is rigid enough to function as a closure yet resilient enough to make a seal on the finish of the container without requiring any form of gasket.

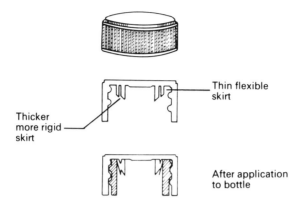

Figure 17.22 The VIMA wadless closure.

Choosing the right type of closure

From the point of view of normal performance there is little to choose between the various materials used in the manufacture of conventional closures, and the choice is governed mainly by cost and appearance. However, aluminium will deform if caps are screwed on too tightly; tinplate is the most rigid of the metal caps, and the aluminium alloys are intermediate between tinplate and ordinary aluminium. But tinplate can rust at its edges, and both aluminium and aluminium alloys are free from this defect.

The strengths of moulded caps depend mostly upon the thickness of the sections in which they are made, caps in normal sections being comparable with tinplate caps. The material used does, nevertheless, have an influence: the phenolics (PF) being stronger than the urea-formaldehydes (UF and WF), and the medium- and high-impact polystyrenes being stronger than normal polystyrenes. Caps in either high-density polyethylene or polypropylene are among the strongest, but screw caps in low-density polyethylene are a little too flexible and liable to jump the threads, particularly when applied to glass bottle finishes. Low-density polyethylene is consequently much more suitable for snap-on closures.

The outside appearance of metal caps is essentially functional, but moulded caps (whether thermosets or thermoplastics) lend themselves to an infinite variety of shapes. The external configurations and designs of plastic closures can be made to suit packs ranging from the most exotic for cosmetics to the purely functional pharmaceutical caps. Whilst the decoration of metal caps is, with a few exceptions, carried out on the metal sheets from which the caps are fabricated, moulded caps are decorated after manufacture. The principal methods consist of filling in designs with coloured paints, silk screen printing, hot foil stamping, decalcomania, the spraying of the outsides of the caps with gold bronze and other lacquers, or in metallizing them with a vacuum-deposited aluminium coating protected by a clear or coloured lacquer to produce gold and other highly reflective finishes. Even electroplating can be used to achieve particularly fine and lasting effects.

The fit of the liners in the caps is another feature that is determined by the material in which the caps are made. All metal screw caps have a recess, generally a knurl, in which the liner is retained, and the liners are consequently not glued into metal caps. Since moulded thermoset caps cannot be made with the same depth of recess as metal caps, the liners are almost invariably glued into compression-moulded caps. It should be added immediately that moulded thermoset caps often have shallow liner-retaining recesses, but these never retain the liners as well as the knurls of metal caps. In the past, moulded thermoset caps have been fitted with oversize liners retained in the caps as a result of an interference fit. This is not good practice, since it generally causes the liners to bow, and frequently to remain on the finishes of the bottles when the caps are removed. Thermoplastic caps can be injection-moulded, using collapsible core-pins that enable them to be produced with sufficiently deep recesses to retain the liners in position without having to glue them.

Liners

In its most general form, a liner consists of a cushioning material (i.e. the wad) with a facing material. The facing material isolates the contents of the bottle from contact with the wad, and is chosen principally for its resistance to the product packed. Its resilience or lack of resilience (i.e. hardness) is also of importance, particularly for products that seep easily for one reason or another.

The wad material will generally be either composition-cork or pulpboard, and the facing will be either a coated paper, a paper faced with a plastic film, a plain metal foil, a lacquered metal foil, or a metal foil faced with a plastic film or coated with a layer of a wax mixture.

Composition-cork consists of granules of cork bonded together with either a gelatine-type glue or a synthetic resin into blocks that are then sliced into the sheets from which the wads are cut. Glycerine is used as a plasticizer or humectant for the gelatine-glue-bonded cork, and diethylene glycol for the resin-bonded cork. The function of the plasticizer is to make the cork composition more pliable and resilient. The glue-bonded cork will support mould growth under damp conditions, and consequently resin-bonded composition-cork is always used whenever there is any risk that such growth can occur.

Pulpboard is produced from an 85/15 mixture of mechanical and chemical woodpulp. It is generally free from mould growth troubles, providing a suitable facing material is used to prevent contact with products such as emulsions and creams, which are susceptible to moulds.

Composition-cork is more resilient than pulpboard and is consequently better able to take up unevennesses in the sealing edges of the bottle finishes. Nowadays, however, the quality of both glass and plastic bottle finishes has improved to such a degree that pulpboard is replacing composition-cork more and more as a wad material. Also, composition-cork wads develop a greater permanent set (i.e. they lose their resilience) than do pulpboard wads, and this generally results in lower cap-loosening torques after the bottles have been in stock for some time.

There are a number of materials used as complete liners that require no facings, since they possess in themselves both the resilience of the wad and the resistance of the facing to the products packed. Such materials are rubber, PVC, polyethylene and EVA. Additional resilience can also be achieved by using plastics in expanded form. In this group also are included moulded liners in materials such as polyethylene and polypropylene, the shapes of these liners generally being such as to present an arch or the base of a groove against the sealing edge of the bottle, thereby increasing the effective resilience of the material used. Yet again, a moulded liner in one of the above-mentioned thermoplastic materials, particularly polyethylene, can be in the form of a disc made captive to a peg in a moulded cap and, if the cap also had a cored-out dome, the liner is made to seal against the edge at the entrance to the bore of the bottle, and a particularly good seal results. On the other hand, the cored-out shape can be moulded in a thermoplastic material and can then be fitted into a cap.

Both *natural* and *synthetic rubbers* are available in almost any degree of resilience or rigidity that may be required for particular applications. Natural rubber, however, has the disadvantage of a high coefficient of friction that causes caps to spring back after tightening. This can be overcome by wetting the rims of the bottles or the wads before applying the closure. Two of the main advantages of rubber are that it will withstand high-temperature processing and that it is self-sealing after piercing with a hypodermic needle. It is, therefore, extensively used in pathological and blood transfusion closures. However, its increasingly high cost encourages users to seek alternative materials.

Plastisols and *water-based (flowed-in) compounds* are also liners that require no facings, since they possess both resilience and inertness. Plastisols are dispersions of PVC resins in plasticizers. They are flowed into metal caps either in the form of annular sealing rings or overall lining discs. They are fluxed at a high temperature relatively rapidly to set the plastisols into resilient sealing-rings or discs. Water-based compounds are colloidal dispersions of either natural or synthetic rubber in water. They are flowed

into metal caps in exactly the same way as plastisols, but need prolonged
stoving to drive off the water. Flowed-in compounds are almost invariably
used in vacuum closures as well, for all food processing applications, and are
replacing the synthetic rubber (usually nitrile rubber) rings that were
popular with a number of vacuum closures.

Compatibility between closures and the contents of containers

The product must in no way become tainted or affected physically or
chemically by any of the closure materials with which it may come into
contact. Nor must the product itself have any effect on the closures.

This compatibility must also include the sealing compound in the closure,
namely, the wad, liner, or sealing ring. Pulpboard on its own has a degree of
porosity; it is sensitive to moisture, and absorbs aqueous products readily.
Gelatine-bonded composition-cork is sensitive to moisture and disintegrates
in the presence of aqueous products; resin-bonded composition-cork is
better in this respect, but it is still liable to taint aqueous products and will
also disintegrate in them, although more slowly.

Consequently, pulpboard and composition-cork wads, even when waxed,
are only used in closures for sealing dry moisture-insensitive products. In
other instances these wad materials must be lined with an inert product-
compatible facing.

When the resilient sealing compound forms a ring pressed against the
sealing edge of the container, the product comes into contact with the inside
of the closure that is encompassed by the ring. This applies particularly to
metal vacuum closures for wide-mouth jars. The inside lacquer or enamel
system with which the cap is protected must therefore also be inert to the
product packed, not only at room temperature, but also under the
processing treatment to which the pack is submitted.

Plastic closures are sometimes fitted with rubber sealing rings that expose
part of the inside of the closure to the product, and this again requires that both
the closure and product must be completely unaffected by each other. Plastic
closures are not usually suitable for processing treatments, although some of
the high-density polyethylenes, polypropylenes and phenolic thermosets
have at times been used for applications where some processing is involved.

Facing materials

Whenever a wad material does not provide the protection or inertness
required, then it must be faced with a material compatible with the product
packed. The facing material must also bed down well on the sealing edge of
the container to ensure a good seal.

Almost all facing materials are based on paper (bleached sulphite or kraft) either coated with white pigmented synthetic resins such as vinyl copolymer or polyvinylidene chloride (PVDC) resins; or laminated to plastic films such as PVDC, polyester (Melinex) and polyethylene. Metal foils (aluminium and tin) make excellent facing materials where maximum resistance to gases, water vapour or solvents is required. Lead foil is never used in the food, beverage, cosmetic, and pharmaceutical industries because of its toxicity. Metal foils are almost always backed with paper to facilitate their lamination to the wad material during the lining operation. Where there is risk of corrosion, the foil can be coated on one side with a lacquer or a plastic film.

The range of facing materials used to be more extensive than it is today, but an endeavour has been made to cut the range down in the interests of rationalization. Closure manufacturers have always sought to find a universal facing material or liner for sealing all products, but this ideal is not yet in sight.

Experience is usually the best guide in recommending a sealing system for any particular product, but extensive sealing tests could be required on a new product, or a significant alteration in the composition of an existing product before sound advice could be given on the best sealing material(s) to use.

Closure design

A closure is designed for its appearance and performance. Its appearance will depend upon its external shape, the material in which it is made, and the decoration required to achieve a particular effect or purpose. Its performance will depend upon ensuring that the specification of its bottle finish engaging features match the specification of the glass or plastic bottle finish, that the cap is fitted with the correct sealing system for the product packed, and that it will withstand any processing treatment involved.

Methods of application of closures

There are four principal methods (see Table 17.5).

Screw-on caps mainly have an external screw-thread cap which can be applied by hand; but this is a slow operation, and consistency of tightness falls off as the operatives become tired.

All standard screw closures can be applied automatically by a rotating head having a chuck that fits over and grips the loosely pre-applied cap and tightens it to the required torque which is regulated by means of an adjustable friction clutch fitted to the head. Hand-fed single-head semi-

Table 17.5   The four principal methods of application of closures.

Method of application	Types of closures
1. Screw on, screw in	Internal and external screw thread caps and lug caps
2. Push in, push on	Driven corks and flanged-top cork stoppers; plus; snap-on caps and other forms of press-on friction-fit caps
3. Crimp on	All closures which are secured to the rims of bottles by squeezing the skirt of the cap into a groove—under a bead on the side of the neck of the bottle; or on to screwing-off threads round the outside of the finish of the bottle
4. Roll on, spin on	All caps in which the screw threads are formed in the skirts by means of spinning rollers that shape the skirts to the contours of the threads of the bottle finishes; also all threadless closures in which the skirts are rolled under retaining rings on the necks of the bottles

automatic machines can apply caps at speeds of up to about 60 per minute whilst multi-head fully automatic rotary machines are capable of operating at speeds of 250 per minute or more. Caps can also be tightened by conveying the loosely capped bottles in a straight line between rollers or bands applied to their sides. Really fast speeds in excess of 400 caps per minute are possible with such a straight-line capper.

Screw-on lug caps are pressed into position on jars by a horizontal moving belt while the jars are moving along a conveyer; the necessary twisting action is obtained by dividing the belt in two lengthwise, and running one half, faster than the other. A normal capping speed is 250 per minute, but speeds up to 750 are possible.

Screw-in caps are now almost entirely superseded by either pre-threaded external screw or spin-on closures. The old internal screw stopper is the main example here, and it was either entirely applied by hand, or loosely fitted into the bottle and tightened by means of a rotating head. Its rubber washer made an excellent seal, and re-sealing was just as effective when replaced by hand. The main disadvantage was an unattractive appearance, and the lack of protection afforded around the rim over which the liquid contents of the bottle are poured.

Push-in corks were originally and still are largely represented by the driven cork, now mostly used for sealing vintage wines. A good many corks are still inserted by hand or mallet, but mechanical methods are also available. The modern example of a push-in cork is the flanged-top cork stopper extensively used for spirits and fortified wines, and which can be applied by mechanical means.

Push-in plugs and shives are pushed into the mouths of bottles prior to the application of the closures, and mechanical means are available to do so when the output warrants it.

Press-on plastic caps are represented by a wide variety of snap-on plastic closures. There are various means for mechanically applying them, the simplest being to pass loosely capped containers under a moving pressure belt, which presses the caps home.

Press-on metal caps are represented by the prise-off cap applied by a pressure belt using downward pressure only. The cap is retained on the jar by an in-pack vacuum developed by the use of steam flushing in the capping machine.

Crimp-on metal caps are represented by the crowns used for pressure-retaining seals and by the larger-diameter vacuum-tight closures commonly seen on jams, preserves and similar products. The crimping-on operation is effected by a shaped chuck that, when applied to the cap under top pressure, squeezes the skirt on to the cap retaining feature (groove, bead, or threads) on the neck of the container. Sealing equipment starts with a single-head semi-automatic machine and extends to multi-head fully automatic rotary machines capable of applying closures at speeds of up to about 750 per minute for crowns.

Roll-on or *spin-on metal caps* are represented by all those aluminium caps supplied to the bottlers in shell form, and in which the threads are developed in the skirts during the capping operation. During application, the liner in the closure is compressed by top pressure exerted by the capping machine head, while spinning rollers rotating round the skirt of the cap develop the threads in the cap using the thread on the bottle finish as a former. In the pilfer-proof versions, the pilfer-proof tell-tale ring is, at the same time, tucked under its corresponding bead on the bottle finish. Capping machines range from single-head to multi-head fully automatic rotary machines capable of applying caps at speeds of up to 400 per minute.

Sealing methods

There are a number of methods for sealing the mouths of containers in addition to the rim top seal. These other methods tend to be associated with the types of closures for which they are used. There are five main methods.

The *rim top seal* is the standard seal in which the sealing component in the closure, namely the liner, is pressed downward against the top of the sealing rim on the container finish.

In the *side top seal*, a rubber or synthetic rubber band, fitted on the inside of the skirt of a metal lug-type closure, mates with a slightly outward tapering seal edge on the outside wall of the top of the rim of the container finish. This type of seal is essentially a vacuum seal, since it relies almost as much on the external atmospheric pressure as on the downwards sealing pressure exerted by the lugs. It was at one time popular, particularly in the United States, for sealing baby food jars.

The most important examples of the *combined rim and side top seal* are the United Glass 'Eurospin' (Figure 17.23) and Metal Closures 'Flavorlok' roll-on aluminium closures. They consist of cap shells lined with an overall flowed-in plastisol gasket that rides up slightly on the inside of the skirt and which increases to a thicker annulus in the angle between the crown and the skirt of the cap. In the course of spinning the cap on the bottle finish, a specially designed pressure chuck reforms the top of the outside knurl section of the cap shell, so as to cause the plastisol liner to seal on both the top and side of the rim of the bottle finish before the threads are formed in the skirt of the cap by the standard spinning process. By this means, the pressure that a relatively weak aluminium closure can contain in a bottle is very much increased, and both the 'Eurospin' and 'Flavorlok' closures are extensively used for sealing carbonated beverages.

Normally it is not good practice to design closures to seat on two (or more) sealing surfaces at the same time since, almost invariably, mating occurs first on one sealing surface and, since any further sealing movement of the cap is then blocked, it prevents the other mating surface from bedding down properly. With the 'Eurospin' and 'Flavorlok' closures, however, the top of the cap is reformed while the whole cap is under top pressure, thus ensuring maximum sealing effectiveness on both the rim top (top pressure) and the side top (reforming action) of the bottle finish.

In the *bore* or *plug type seal*, an interference-fit cork or stopper is inserted into the mouth of the container. It is the most ancient method of sealing a bottle—natural corks have been used for over 300 years for this purpose. They are still used today for sealing vintage wines and champagnes in glass bottles. The modern counterparts to natural corks include hollow plastics plugs, with or without fins, or rings designed to conform more positively with

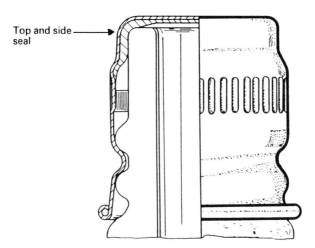

Top and side seal

Figure 17.23   The sealing principle of Flavorlok and Eurospin closures.

bore dimensional variations and thus ensure a more effective seal. Plastic shives are also used for making a plug-type seal in the larger-diameter mouths of (principally) glass containers.

The *snap-on seal* has resulted from the advent of resilient thermoplastics, such as low-density polyethylene and, to a lesser extent, polypropylene, increasingly popular closures have been designed to make a snap-on fit on the finishes of containers. A snap-on cap is forced over a cooperating bead on the mouth of the container, so that the cap hugs the bead tightly and uses the whole surface of the bead as a sealing area. The plastic cap is the resilient component; the tighter the fit it makes on the bead, the more effective is the seal.

Combination closures

In certain applications double closure systems are used for sealing (mostly) glass containers. Lead-tin, and pleated aluminium foil capsules, viscose bands and heat-shrinkable plastic bands are applied over the closures on to the necks of bottles, not only to dress them in an attractive manner, but to protect the primary closures from contamination and provide evidence of pilfering, since the closures cannot be removed without destroying the capsules or bands.

The plastic shive, mentioned above, which is not a pilfer-proof device in itself, is usually used with an additional screw closure which can be used on its own (without the shive) to re-seal the container after removal and disposal of the shive.

The membrane adhered to the mouth of a container (Figure 17.7) is also a combination closure in as much as, once the membrane has been torn off the rim of the container, the cap itself, be it a screw cap or a snap-on cap, provides a re-seal during use.

18
Decoration of packaging

Introduction

The power of packaging to influence judgement is great. In the developed world more than 2500 packages are purchased by every family every year, around 50 every week. Retail selling is a complex business and probably the greatest increases in the decoration of packages lie in the markets for food, household products, pharmaceuticals, cosmetics and toiletries.

Many of the packages selected will be obtained through self-service stores and supermarkets. Consequently the package has to provide the attraction and all the information needed by the purchaser to make that choice. Packaging design is frequently concerned with pictorial realism, visual clarity, bright colours and clearly recognizable symbols. Moreover the recognizability of a design, the colour or the image has to be attained on several different printed substrates; paper, board, plastics, metal foil, tin cans, etc. Any design adopted may be reproduced on every media. This calls for careful consideration. Since printing is a complex process with many steps before ink is finally placed on the packaging, developments cover a wide field from design and origination through production to final delivery. The demand is for high quality of both on-line work and illustration.

There are two aspects of quality in print: the level of quality and the repeatability. In reproducing an old master, like Constable's *Hay Wain*, the objective is to produce a good likeness of the original. Minor differences between any two reproductions are of little consequence as a rule since they are only very rarely examined side by side. In packaging, however, the emphasis may be the other way round. A reasonable likeness must not only be achieved but every reproduction must be very nearly indistinguishable from the others, particularly in respect of the depth of colour. Lighter shades may be equated with fading and hence the package rejected as old stock. Contrary to general belief the colour printing of packaging is more critical than the printing of publications. It must also be remembered that all packaging for foods etc must be odour free and compatible with the package contents.

Printing processes

Print production may be considered to include:
1. Artwork and copy preparation
2. Production of printing image carriers
3. Press operations
4. Finishing

Packages today are printed by all the major processes; Relief, Plano-graphic, Intaglio, Screen and Impactless (jet) printing. Each has its own peculiar characteristics, advantages and problems.

Relief printing

Letterpress printing has long been established (Figure 18.1a). Originally it used metal type but today much of the metal has been replaced by synthetic and/or photopolymer printing plates. The image areas which carry the ink are raised above the non-printing areas. Flexography is the other relief process in common use. Developed specifically for packaging printing, it differs from letterpress primarily in the nature of the ink used. While letterpress inks are based on thick pigmented drying oils, flexo inks are highly fluid, rapid drying and solvent-based. Because of this the multi-roller inking system used in letterpress to achieve a uniform film of ink is replaced by a mechanically or laser engraved metering roll (Anilox roll, Figure 18.1c). This receives ink from a rubber fountain roll and delivers it to the flexible elastomeric printing plates, the raised areas of which contact the anilox roll.

While letterpress printing is used less and less these days, flexography is still growing in applications because of its advantages. The process can be used on a wide variety of substrates, the inks are quick drying and can even be water-based; variable repeat lengths, quick changeover and pre-makeready are possible and costs are generally low.

Intaglio printing

Gravure or rotogravure printing uses an image carrier which holds the ink in recesses below the non-printing areas. It is the only process in which there are no plates to attach to a printing cylinder. The image areas consist of small cells etched into a copper cylinder. The non-printing area is the surface of the cylinder (Figure 18.2).

The cylinder rotates in a fountain of ink where the individual cells are filled and as it comes out of the fountain excess ink is removed from the cylinder by the action of a flexible doctor blade in intimate contact with its surface. This renders the non-printing area free from ink. The cylinder then contacts the substrate which is pressed against it by the impression cylinder and the ink in the cells is transferred by capillary action.

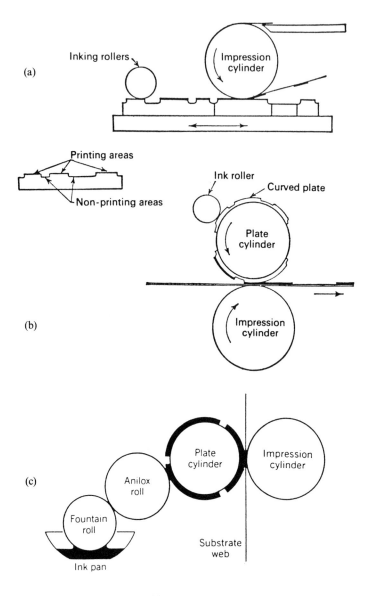

Figure 18.1 Relief printing. (a) Flat-bed letterpress. (b) Rotary letterpress. (c) Flexography.

Planographic printing

As the name implies planographic printing or lithography prints from a flat surface (Figure 18.3). Lithography works on the principle that oil and water do not mix. Offset litho plates (usually of aluminium, but paper and plastic

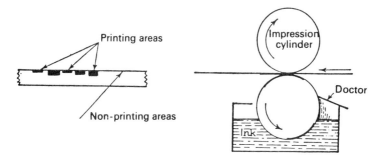

Figure 18.2   Intaglio method: gravure.

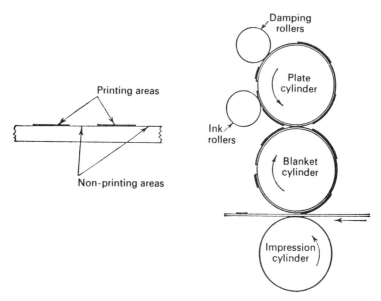

Figure 18.3   Planographic method: offset litho.

plates are available) are clamped onto the plate cylinder of the printing press and in operation are alternately wetted with water and then inked. The image area on the plate has been rendered ink receptive and the non-image areas are water receptive and repel ink. The ink which is of a similar viscosity to letterpress inks is distributed from the ink duct through a series of rollers which spread it evenly to the plate which has just previously been dampened with a water-based fountain solution from the damping system rollers. The image so laid down on the plate is then 'offset' on to a blanket cylinder covered with a rubber blanket. The substrate to be printed then passes between this blanket and an impression cylinder.

The letterset, dry offset or indirect letterpress process was developed to provide the letterpress process with some of the advantages of offset lithography without the problems of the damping system. It is generally carried out on an offset litho press bypassing the damping system and using a photopolymer relief plate offsetting the image via the rubber blanket. Typically it is used with water-based inks which have very low odour characteristics to print food cartons and wrappers.

Screen printing

Screen printing involves forcing ink through a screen-mesh-supported stencil of silk, synthetic fabric, or stainless steel onto a substrate (Figure 18.4). A stencil, prepared either by hand or photographically, becomes the image carrier, with the non-printing areas protected by the stencil. After attaching the stencil to the fabric and frame assembly, printing is accomplished by placing the substrate to be imaged under the screen, applying ink to the inside of the screen, then spreading and forcing ink through the openings in the stencil with a rubber squeegee.

Ink-jet printing

Unlike other printing processes that require contact between an inked image and the substrate, ink-jet is a non-contact printing process. There are several ink-jet technologies in current use, but they all operate on the same basic principles. Jets spray electrically charged drops onto a substrate while dot matrices, housed in the printer's memory, electrostatically direct these drops to form characters. More drops are generated than are needed and

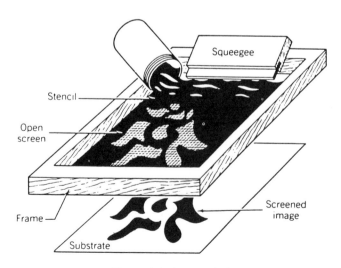

Figure 18.4 Screen printing.

those remaining fall into a gutter that returns them to an ink reservoir (see Figure 18.5). Available since the mid-1960s, ink-jet printing found an immediate application in coding and dating. Ink-jet's ability to project tiny drops of ink into various recesses, the availability of inks which adhere to non-absorbent surfaces, plus the low cost and high speed compared to labelling have firmly established this process in packaging.

Labels and labelling

Labelling is a means of performing the communication function of packaging, by which contents are identified, the customer is encouraged to buy, and the law with regard to consumer and carrier protection is satisfied.

A secondary function of a label is to close seal the package, for example by the application of a circular heat-seal label to the end of a biscuit wrap.

Labelling is used in preference to pre-printing a container under several circumstances:

1. When the required standard of graphics can be achieved more cheaply than by printing direct on to the container
2. When the same basic container is to be used for a number of different products or purposes

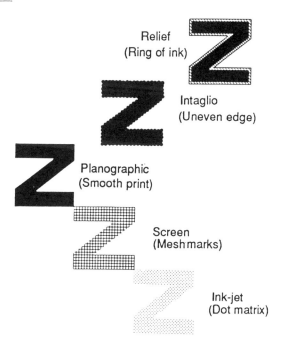

Figure 18.5 Distinguishing characteristics of printing processes.

3. When the exact nature of the information to be displayed is not known until after the containers have been filled

Moreover, the increasing use of plastic containers in the last decade or so has led to considerable changes in material surfaces, shapes and visual effects. This in turn has revolutionized the method of decorating containers. Metal, glass and plastic containers can be printed by several processes and where required can be decorated by labels. While labelling once referred solely to the application of a paper label it now covers all means of applying graphics to a package.

Types of labels

Glued-on labels are simply sheet material, often paper, printed and cut to size. Adhesive may be applied, either at the time of application to the container or at the time of manufacture of the label, the adhesive being activated (moistened) just before application to the container.

Thermosensitive labels (heat-activated) have a thermoplastic coating applied by the manufacturer of the label which is activated by heat at the time of application to the container. In general these are more expensive than glued-on labels but their application is simpler and higher speeds can be attained. They have a special application when performing the dual role of a label and a means of closure, as on biscuit packs and bread-wraps.

Self-adhesive (pressure-sensitive) labels are supplied coated with an adhesive on the unprinted side, and mounted on a release paper, which is removed to expose the adhesive immediately before application.

Self-adhesive labels may be mounted individually and regularly spaced on the release paper, which can then be cut out to facilitate their removal; or they may be mounted on a continuous web of release paper, perforated between each label so that the labels can be roll-fed to the application point.

Self-adhesive labels can be produced to adhere to a wide range of materials, either to give permanent adhesion or to provide the facility of easy removal. They are particularly useful when an addition such as 'Special Offer' or price labels have to be added at short notice to an existing pack. They avoid several operations necessary in the application of glued-on labels so that, although more costly than glued-on labels, in many instances the overall operation may be cheaper.

Tie-on labels are generally used for the utilitarian functions of addressing packages or marking items, the shape of which precludes other means of identification. The label itself will generally be made of strong paper, fabric or metal, and attached by twine or wire. There is, however, a use for tie-on labels on luxury consumer goods, for which an infinite variety of coloured threads and ribbons with seals, tags and labels of paper, plastics or metal attached have been devised.

Reversible labels required for returnable transit containers are generally affixed by insertion into a metal frame or a transparent plastic envelope, secured to the container.

Insert labels are inserted in transparent plastic packs and need no fixing, remaining visible to the customer through the pack.

Heat transfer labelling

A decoration process which has been used for many years is the transfer printing method. The design is printed on to a special substrate which is placed in contact with the surface to be decorated and the printed image is transferred by the application of heat. Excellent quality with dense colours is obtainable. Preprinting is accomplished by gravure, silk screen, or flexography on a paper or polyester substrate.

Two types of label are used, one employing a wax release and the other using a heat reactive adhesive. Wax release labels are applied through a combination of heat and slight pressure. This labelling system depends on the inks and/or lacquer to bond the image to the substrate. The systems using heat activated labels operate at lower application speeds than the wax release system and depend on the coating to form the bond. The bonding is usually stronger than that achieved with wax release labels.

Transfer in both instances takes place when heat and pressure are applied to the reverse (u' ¬rinted) side of the label, causing the inks and/or the coating to liquefy a₁₁₄ bond to the substrate. As the pad or roller carrying the label separates from the item being decorated, cooling occurs and the bond sets.

In-mould labelling

This is a process which is increasing in importance. Originally it was devised to eliminate a secondary process, improve the decorative quality and reduce cost. An example was the decoration of rounded square yoghurt pots. These are not easily printed by conventional means but in-mould labelling provided an answer. The pots are thermoformed and decorated at the same time and then filled in-line in the dairy.

The basic system uses a printed paper label that has a heat activated coating on the reverse. The label is placed in the mould before it closes around the hot extruded plastic parison and when air is injected to blow the parison to the mould contours, the heat activates the coating. The combination of heat, air pressure and the cold surface of the mould sets the adhesive and the label adheres to the container surface. The dimensional stability and the tolerances achieved on containers is good and since the label forms part of the moulding it contributes to the strength. This can result in a reduced polymer requirement and 10–15% savings have been claimed.

Shrink sleeve decoration

The shrink sleeve is becoming more widely used for the decoration of glass and plastic containers. One process uses mono-axially oriented PVC or PP film which is made larger than the container to be decorated and then heat shrunk to fit. A second method uses an LDPE sleeve which is smaller than the container and is stretched during application. The sleeve is held in place by the elasticity of the film. No heat or adhesive is used.

Both methods use gravure or flexo printed sleeves. Shrink labels have an extra advantage because if shrunk over the necks and closures of containers they provide a simple and reasonably effective tamper-evident closure. When the container is of glass there is an extra safeguard against fragmentation if the container is broken.

Materials for labels

Papers. Uncoated, clay-coated, cast-coated and metallized papers are used in combination with all the more common printing methods, and so one face of the label must have the characteristics needed by the printing method chosen to obtain the graphic effect required. After printing, good abrasion- and scuff-resistance are needed, which may involve over-varnishing. The characteristics of the other, inner face of a label must ensure correct bonding by the adhesive used.

The label needs to be opaque and, when it is to adhere to a curved surface, must be sufficiently limp to conform, and be sufficiently free from curl to adhere without lifting (Figure 18.6). Grammage is generally 80–100 g/m². The effects of curl can be limited to some extent by controlling the grain-direction of the paper, which is always parallel to the axis of curl. It is usual for the manufacturer of a labelling machine to specify the grain-direction of the labels to be used. Generally, but not always, it is parallel to the copy.

It is usual to specify that label papers should be flat within stated limits of relative humidity (e.g. between 35 and 80%).

Paper will probably continue to be the dominant material for labels for the foreseeable future but the use of plastic films is increasing particularly for cosmetics, toiletries, DIY, garden and car products where plastics have advantages on both aesthetic and durability grounds.

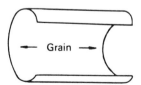

Grain

Figure 18.6   Finding the grain direction (curl). By moistening the back of the label paper, the direction of curl (and thus grain direction) can be determined.

Foils and laminates. As well as producing special decorative effects, such as metallized appearance and deep embossing, foils or laminates with foil may be used where there are special environmental conditions, as when labels need still to adhere when bottles have been immersed in ice.

Plastics. The most common use of plastic labels is to give the effect of direct printing on to the container by use of a printed transparent plastic film.

Label forms and shapes

Labels may be supplied on a roll, from which each label must be cut before application; or may be supplied ready-cut (either in a guillotine, or by die-cutting which is more usual when the shape is other than rectangular). Roll-cut labels are particularly used for pharmaceuticals, to avoid risk of wrong labelling.

Rectangular wrap-round labels are commonly used on cans. Labels on bottles may be wrap-round, or attached to the front, back, neck or shoulder of the bottle. Additional information indicating, for example, special offers are usually 'spot' labels. With 'combined' labels, a single label fulfils the purpose of two separate labels (see Figure 18.7).

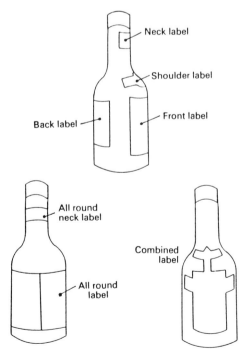

Figure 18.7   Types of labels for bottles.

Adhesion of labels

The initial action of the adhesive applied to the label is to bond sufficiently (initial tack) to the surface to prevent the label assuming its natural shape when pressed to the container and so lifting at the edges or blistering. This bond must be maintained for sufficient time (setting time) to enable the adhesive to set permanently (final set). Such a bond must then maintain adhesion in all conditions through the container's life, which may include standing in ice; but may also require to break down when a bottle is returned for re-use (wet-off).

Achieving a bond between a label (which may be relatively absorbent) and a tinplate or glass container (which is non-absorbent and may have been surface-treated) requires careful formulation by the adhesive manufacturer to 'match' the adhesive to label and container, within any constraints imposed by the labelling machine.

It follows that, having once achieved satisfactory performance, further supplies of label, adhesive and container should vary as little as possible from the original. This not only requires attention from the suppliers concerned, but from the user, to ensure that labels and adhesives are properly stored while awaiting use, and are used before they deteriorate.

Selection of adhesive. Usually a machine manufacturer, given samples of the label to be used and the container to be labelled, will make recommendations as to suitable adhesives. Such recommendations are sometimes too specific to permit buying in the cheapest market, or in times of short supply. Established adhesive manufacturers, provided with samples of labels and containers, and information on the machine to be used, can also advise on suitable formulations. Subsequent changes in label, adhesive or container, should be made only after a trial to check performance.

Further information on adhesives may be found in Chapter 8.

Labelling machinery and equipment

Labelling machinery can be categorized generally by the label/adhesive/container combination for which it is designed. As with other machinery, the higher the output required, the less versatility must be expected. If the production rate is low, most combinations will be satisfied with no more than a glue-pot and brush. A through-put of 60 000 bottles an hour will only be achieved, however, when label, adhesive and container are all carefully specified.

In addition to the specific operation of applying the label, the machine may be required to code, count, weigh, price or overprint the label.

It follows that there is a great variation in the degree of automation offered to meet the requirements of different industries. The bottling industries are generally concerned with very high-speed lines, and every sub-operation needs to be fully automated. The pharmaceutical industry has a main concern for permanence of label, and absolute reliability that the correct label for the content has been used. In the self-service store there is a need for equipment that can be brought to the stock-in-hand, but first over-printed with the price, whereas a labelling machine for meat packs that are to be sold in this market will probably need to weigh, cost and over-print on a production line. The simplest equipment may do no more than dispense a label, ready for the operator to place it.

The complexity of the machine is influenced by the design of both label and container. It must be recognized that the proper balance must be maintained between the marketing advantages of unusual and original designs, and the difficulties of fast reliable application.

Reliability is an important feature of any labelling machine, since a hold-up or breakdown at the point of labelling can cause stoppage of the whole line.

The principal operations performed by a labelling machine are:

1. Feed label from magazine or roll.
2. Pick up label generally by suction cups, compessed air or by secondary adhesive.
3. Apply adhesive with either full coverage, vertical or horizonal stripes, generally from rollers on to either label or container.
4. Press label to container by pressure pads, compressed air, belt or brushes, during which process the container will be moved into position, held firmly while the label is applied, and then removed. This may be achieved by a rotary movement, the containers being held in a rotating turret: or by a 'straight line' movement, by conveyor star-wheel or screw mechanism. 'Straight-line' operation is generally more versatile, particularly where containers of unusual shapes are concerned. Most bottle-labelling machines hold the bottles vertically, but some machines, particularly for cans, hold the containers horizontally.

Purchasing, installation and operation of labelling machinery

The first step to successful labelling is to ensure that the machine supplier is aware of the exact requirement, and therefore should be supplied with samples of the containers and labels to be used, throughput required and other details. An acceptance trial should be conducted, and a careful record made of the details of the label, adhesive and containers that were successfully used. The machine must be operated and maintained strictly in accordance with its Operating Manual, which should be readily available. Control should be kept of labels, adhesives and containers, to ensure they are as close as possible to those successfully used in the acceptance trial. Finally, when the same machine is to be used for a new design of label and/or container, sufficient time must be allowed for trials to find a new, probably different, adhesive.

A check list of points that may need consideration is given in Table 18.1.

Table 18.1   Check list of details for consideration when selecting and ordering labelling machinery.

Containers to be labelled
  Shape
  Sizes
  Dimensions
  Materials

Labels—User's requirements
  Number and type
  Dimensions
  Materials including treatments, e.g. whether embossed

Adhesive and application system
  Cold gluing   Full coverage
  Hot-gluing   Stripe application
  Heat-activation   Machine maker's recommendation on adhesive supply
  Self-adhesion

Payment and delivery
  Cost
  Method of payment
  Weight of machine
  Point of delivery (machine site? delivery bay?)

Installation
  Services needed:
    electricity (voltage, frequency, power, phase)
    compressed air (pressure and flow rate)
  Weight and dimensions of machine
  Heights of IN and OUT feeds for containers

Responsibilities: supplier/user for
  Preparation of floor
  Provision of electricity, air, etc.
  Move machine to site
  Install machine
  Conveyors to and from machine
  Provision of safety devices (switches, guards, etc.)

Labelling machine capacity
  Maximum/minimum label widths
  Maximum/minimum label heights from base of bottle
  Maximum/minimum width and depth of rectangular bottle to be labelled
  Maximum/minimum diameter of cylindrical bottle

Attachments needed
  Coding
  Counting
  Wrong label detector
  Over-printing
  Special fittings for odd-shape bottles
  Logical sequence control to keep output speed at optimum, influenced by number of bottles
    on line before and after labelling machine

Table 18.1 *cont'd.*

Operation
  Roll or magazine (hopper) feed
  Normal, maximum and minimum speeds required
  Left-to-right or right-to-left operation
  Labour skills and training arrangement for:
    operation
    maintenance
    changeover
  Changeover times
  Changeover parts:
    provision
    identification
  No bottle/no label, or no label/no bottle devices
  Re-load capacity during operation

Servicing
  Labour skill and training required
  Recommended spares holding by user
  Terms of machine manufacturers' repair service

Records
  Layout plan
  Electrical wiring diagram
  Operational manual
  Maintenance instructions
  Parts list

Security
  Label machinery manufacturer to maintain security on any information concerning the
    product to be bottled

Acceptance
  Bottle/label combinations to be used on acceptance trials
  Length of trial run for each bottle/label combination

Minimization of labelling faults and problems

Each foreseen combination of label, adhesive and container should be run successfully during the acceptance trials of the machine. Samples of labels and containers used on the acceptance trial, and subsequent successful operation, should be retained for comparison in event of future breakdown.

The labelling machine must be run, maintained and adjusted strictly in accordance with the Machine Handbook, of which there should be at least one copy kept close to the machine (and a second copy in the Works Engineer's records). If either is lost, it should be replaced.

Suppliers of adhesives and labels should be asked to advise at once if, owing to shortages or other reasons, supplies cannot match previous consignments. This information must be passed by the Buyer to the shop floor.

All supplies must be kept in good storage conditions, with labels packed in moisture-proof wraps when necessary. Rapid changes in temperature before use should be avoided. Containers to be labelled should be clean and dry. When a new label or container is introduced, the labelling-machine manufacturer should be asked whether the particular machine is capable of the new requirement, and for recommendation as to adhesive, adjustment or modification to the machine. It follows that sufficient time must be allowed (and the machine made available) for trials prior to production.

A system of Quality Assurance on labels, adhesives and containers should be operated and amended as experience is gained of the particular characteristics required.

Correction of labelling faults

The more common labelling faults are:

1. Tearing or creasing of the label
2. Inadequate adhesion, lifting of edges, blisters
3. Staining of labels
4. Scuffing of labels
5. Fading of labels
6. Incorrect positioning of labels
7. Failure to 'wet-off'

Provided that the initial trial has been successful, failure will take place only when some change has occurred. Such changes may occur in the area of the machine (e.g. a new operator), of the label (e.g. new paper), of the adhesive (e.g. new supplier), of the container (e.g. new surface finish). The first step is therefore to narrow the likely source of trouble by inquiries as to what changes have taken place. The nature of the fault may also indicate the likely area of change.

Where the machine is the possible source of trouble, the most likely cause is that its operation or adjustment is not in accordance with the Manual. Where the label is the possible source, the approach should be to compare 'good' and 'bad' labels for such properties as grain-direction, curl, stiffness and absorbency. Most adhesive troubles come from using an adhesive different from that which has run successfully, e.g. by unauthorized adulteration with water or thinning agent.

In all investigations concerned with adhesion, it is worth attempting to make the label/container bond by hand since, if it cannot be done by hand, it is certain it will not be done by machine.

PART D
Distribution Packages

19

Cylindrical shipping packages

Wooden casks and plywood kegs

Wooden casks and plywood kegs include three main types of construction:

1. Wet cooperage: intended for the transport of liquids; the wooden beer cask is probably the best known example
2. Dry cooperage: made from wooden staves and suitable only for dry or near dry products
3. Plywood barrels and veneer casks: straight-sided cylindrical drums manufactured from sheet plywood and from a double row of veneer staves respectively; veneer casks are no longer manufactured in the United Kingdom

The trade terms are often quite distinctive and the following definitions will assist in understanding the subject.

Wet and dry cooperage

A *wet cask* is a round bilged coopered wooden container made of staves, heads and hoops, and constructed to hold liquids without leaking (Figure 19.1).

A *dry cask* is a round bilged coopered wooden container made of staves, heads and hoops, but constructed to hold powders and commodities other than liquids.

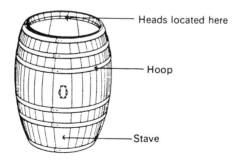

Figure 19.1 A wet cask.

A *stave* is the name given to the arched vertical components forming the walls or sides of a cask, which are held together with *hoops*. The hoops are made of steel or wooden strips passing round the circumference of the cask.

The wooden disc forming the end of a cask is called the *head*. The *bilge* is defined as the largest circumference of the cask, located at the horizontal plane through its centre.

Plywood barrels

Plywood barrels are cylindrical containers, each made from a plywood sheet butt-jointed and held in position either by means of a plywood strip on the inside, or a metal strip on the outside. The head and bottom are discs of plywood.

The reinforcing plywood bands riveted to the outside of the barrel, normally at each end, are known as *body bands*. Intermediate bands may be used if necessary.

The plywood band riveted inside the body wall and set down from the end to form a seating for the head is called a *lining hoop*. They are generally positioned one at each end of the body.

After inserting the bottom (a disc of plywood which is seated on the lining hoop) another plywood band is riveted inside the body wall against the outside of the bottom head. This plywood band is called the *closing hoop*. The top closing hoop is supplied loose, for fixing by the customer after the barrel has been filled.

A lifting ring of metal is fixed to the outside of the lid to enable it to be removed. The ring is normally secured by a clenched staple.

Production methods

Wet and dry cooperage. The first operation is to shape the staves using the axe, knife, hollow knife and jointer in that order. Having shaped the requisite number of staves, they are placed side by side in a raising hoop, then further truss hoops are placed on the staves to hold them firmly, these then being placed over a cresset of wood chips and timber until the staves are warm enough for bending. When the staves are bent and while the cask is still warm, the chimes are made by using the adze to slope the tops of the staves, the inside then being levelled with a chive so that a perfect groove may be cut with a croze.

Some explanation of these terms is necessary: a *raising hoop* is usually a thick ash hoop of the required circumference for the finished cask, i.e. it acts both as a tool and a template. A *cresset* is an open fireplace rather like a watchman's fire. A *chive* is a tool with a curved blade which is swung around the inside of the chime and cuts the staves level. A *croze* is a tool which looks similar to the chive but which holds three small blades, two of which cut the wood while the third clears the groove.

The next operation is to make the heads, which is done by flattening and jointing the heading timber on the heading jointer, after which the pieces are dowelled together forming a rough square. Next the grooves are compassed, and the rough heads are marked for sawing into a circle. When sawn to the round shape, a bevel is cut with the heading knife. The heads are then placed in position, the hoops are made and driven into place, and the cask is completed. This method is used with minor variations for the making of all bilge casks in the United Kingdom, whether for wines, sprits, beer, cider or dry goods.

Plywood barrels. The components are cut to size by power-driven saws. It is therefore possible to produce the containers in an almost unlimited variety of sizes, without special tool charges. Broadly speaking, the upper limit of the size range is determined by handling and weight considerations, and the lower limit by the minimum diameter to which the plywood can be bent.

The components are first dimensioned by power-driven saws, and bent to shape after steaming. They are then ready for assembly. The vertical body joint (or joints if more than one panel is used) is secured by a metal or plywood jointing strip, to form the *body cylinder*.

Assembly after bending is by riveting. Bifurcated rivets (which have two legs which penetrate the plywood and clench on the inside) are normally used. This makes mass production quick, easy and simple, and enables construction to be altered as necessary to meet specific consumer needs. The strips of plywood which are to act as lining hoops and body bands are then placed into position. The bottom is inserted, followed by the bottom closing hoop. The barrel is then closed by the lid, after the loose top closing hoop has been placed inside. Apart from the top closing hoop which is left loose for use by the customer after packing, all components are riveted.

Types and styles of container

Wet and dry cooperage. Wine casks are all made in the country of origin from native or imported timber, and it would be fair to say that almost all the better quality wines are shipped in oak casks, and the cheaper wines in chestnut casks.

Ninety per cent of port and sherry is shipped in oak casks and 90% of vermouth in chestnut. Bordeaux, Burgundy, Rhone wine, etc., vary, the better wines being shipped in casks of oak and the cheaper in chestnut casks. The casks in general use for sherry are the butt (110/112 gal capacity = 500 litres) and the hogshead (54/55 gal capacity = 250 litres). For port the corresponding sizes are the pipe (116/118 gal capacity = 530 litres) and the hogshead (58/59 gal capacity = 265 litres). Vermouth, Bordeaux, Burgundy and other wines travel in hogsheads, that for Vermouth holding 54/56 gal (250 litres) and the others 46/48 gal (215 litres).

All *spirit casks* are made of oak and with the exception of brandy, for which French oak is used, gin, rum and whisky are all filled in American oak. Whisky and gin casks have the same measurements as sherry casks. Rum is usually imported from the West Indies, principally from Jamaica in puncheons (110/115 gal capacity = 520 litres) and in hogsheads (54/58 gal capacity = 255 litres) and more recently in 40-gal barrels (180 litres). Brandy is almost always imported in hogsheads of 62/63 gal capacity (280 litres) but occasionally in quarter casks of 31/32 gal capacity (140 litres).

Beer casks are made in these capacities, butts (108 gal capacity = 500 litres), hogsheads (54 gal = 245 litres), barrels (36 gal = 165 litres), kilderkins (18 gal = 82 litres), firkins (9 gal = 41 litres) and pins (4.5 gal = 20 litres). Beer casks are mainly replaced by metal containers.

Dry casks are made to any measurement required, some to hold a specific weight, some a particular volume. It is worth noting that very few dry casks are made nowadays.

Plywood barrels. There are two types:

1. Those which employ an unsealed vertical joint. These have either a metal body jointing strip on the outside of the body wall, or a plywood jointing strip on the inside. There is a tendency toward the former if bag linings are to be used, and the latter if direct packing is intended.
2. Those which have a sealed vertical joint and have a sealing gasket between the body and the ends. These are mainly employed for direct packing of powders and similar products which would 'sift' through unsealed joints.

In each instance, the number and thickness of the components may be varied according to the size of the container, its content weight, and whether it is for home or export use.

Types of closure and methods of closing

Wet and dry cooperage. A wet cask is always closed by a wooden shive driven home with an adze, and smoothed off flush with the bung stave. (A shive is a wooden stopper for a bung hole; a cork stopper is called a bung.) For solid products the two top hoops are removed so that the head is loose enough to be lifted. After packing, the head and hoops are replaced.

Plywood barrels. The type of closure generally employed is the closing hoop. This may be secured by nailing, wing nuts (for returnable use) or staples (driven by a hand-stapling machine).

Other methods of closing which are employed according to circumstances are:

1. *Griffiths fasteners:* a special fitting, sometimes demanded by Government Departments.
2. *Toggle fasteners:* these are used in conjunction with double lids, the under lid being to the internal diameter of the drum, and the top lid to the external diameter.
3. *Swivel bars:* metal bars are fixed to the lid, and located to engage corresponding slots in the body wall. If required, they can be secured in the locked position by a screw.
4. *L-shaped lugs:* secured by a bolt through holes in the body band.

In all of these instances, the lid must be firmly and evenly seated when the drum is closed.

5. *Body slots:* slots cut in the body wall to enable tension strapping to fit flush over the heads.

Drums

Classical definitions of the word 'drum' usually refer to straight sided cylindrical containers: with the development of modern manufacturing techniques capable of producing different shapes to meet particular needs the term 'drum' can be applied to all essentially straight sided rigid containers intended for the transport of industrial solid or liquid products. The usual size range is 5–45 gal, but both larger and smaller sizes are found in service.

The words 'pail', 'keg' and 'jerrican' are also used at times to describe specific types of drum. A pail is a straight sided metal or plastic container with a reduced diameter bottom, i.e. tapered sides to facilitate stacking; it is usually fitted with a bail handle and a full aperture head. The term 'keg' is often applied to small open-end drums made of metal or of wood. The word 'jerrican' is often applied to rectangular plastic containers which resemble the famous German Army container of World War II.

Whatever their intended contents—liquid, powder, paste, articles or inner packages—drums are most frequently referred to in terms of their nominal cubic capacity, e.g. 5 gal or 25 litres. Although the nominal capacity is used as an indication of the size of packages it must be remembered that the brimfull capacity of the drum, with closure fitted may be different. This difference between brimfull and nominal, which is termed 'ullage', is very important in order to allow for the thermal expansion of liquid contents; it is also useful where foaming liquids have to be packed. Ullage may account for 5% of the total capacity, and therefore to safely pack say 10 gal of a liquid, one would require a drum with an actual capacity of 10.5 gal.

In the packaging sense, the term 'drums' covers all drums, jerricans, pails and kegs made from metal, plastics, fibreboard and plywood or reconstituted wood. Coopered wooden barrels are excluded. It is convenient to categorize them by material and type of construction, i.e. non-removable head with relatively small apertures, and removable head with larger apertures, and refer to them by codes similar to those used by the United Nations for packagings for dangerous goods (Table 19.1).

Table 19.1 United Nations codes for packaging for dangerous goods.

Material	Construction	Code[a]
Steel (incl. tinplate)	Non-removable head	1A1
Steel (incl. tinplate)	Removable head	1A2
Aluminium	Non-removable head	1B1
Aluminium	Removable head	1B2
Plywood	Removable head	1D2
Fibreboard	Removable head	1G2
Plastics	Non-removable head	1H1
Plastics	Removable head	1H2
Steel/Plastic composite	Non-removable head	6HA1
Fibreboard/Plastic composite	Non-removable head	6HG1

[a]Where appropriate a symbol may be added to indicate the capacity range:
S = small, ranging from 20 to 30 litres;
M = intermediate, typically 45–60 and 100–120 litres;
L = large, ranging from 180 to 250 litres, but principally 210 litres/ 55 US gallons.

Metal drums

The steel drum took its name from the musical instrument when in the 1840s a parallel-sided iron cylinder of riveted construction was first made as a substitute for the traditional bilged (or bulged) wooden cask. Over the next 60 years various developments were made, such as welding and double seaming instead of riveting; and the terminology and the definitions of the trade became a combination of those used by blacksmiths and those used by coopers. The First World War and the expanding petroleum industry no doubt created the demand which turned the trade into a mass-production industry.

After the war many innovations appeared, including pouring pails, agitator drums for paints, and new, colourful decorating techniques. Manufacturers began to use steel containers for products other than petroleum and chemicals. Toward the end of the 1930s, the steel container industry started gearing up for a second wartime effort. However, the demands now were far more stringent. Production considerations became top priority to supply war time machines on the ground and in the air, consuming vast amounts of fuel and chemicals. The 210 litre, 18- and 16-gauge drums were indispensable to the supply of fuel in the Pacific areas, the frontline mechanized operation in Europe as well as air and ground operations in Asia. Apart from its strength, the fact that a cylindrical drum can be rolled by one man was an important feature.

Although a downturn in steel drum and pail production occurred at the end of World War II, it was of short duration. Resumption of business created a demand from industry, agriculture and consumers. An increase in chemical and pharmaceutical product development and output provided new markets for drums and pails. Paints, lacquers and varnishes, adhesives, inks, foodstuffs, and other products made the steel drum industry the second largest user of sheet steel. Despite competition from other materials, U.S. production increased from 2.4 million drums of all types per annum in 1922 to 40 million 210-litre drums alone by 1984.

In the United Kingdom, it is estimated that approximately 45 500 000 metal drums were produced in 1985. Of these, 35 000 000 were small drums below 350 mm in diameter and 7 500 000 were large drums. The use pattern of these drums is similar to that in the United States (Table 19.2).

In addition to the production of new metal drums in the United Kingdom, there is a highly efficient drum reconditioning industry which is estimated to recondition about 4 000 000 units every year, primarily large drums. Many of these are used for lubricating oil and only need cleaning, inspecting and repainting, but other drums are completely refurbished by chemical and mechanical treatments and can be almost as good as new drums.

Raw materials. Steel drums are normally made of low-carbon mild-steel sheet, traditionally called *black iron* because of the blue-black scale that covered it in the days when it was produced by hand-rolling. Today drum steel is rolled either hot or cold on continuous strip mills. In the United Kingdom and the United States most is produced by cold-rolling steel strip

Table 19.2   Estimated end use of new steel drums in the United States in 1985.

Use	Large drums %	Small drums %
Chemicals	50	21
Petroleum products	27	19
Paints and printing inks	7	29
Antiseptics, detergents and disinfectants	3	6
Foodstuffs	4	–
Adhesives, cement, ceramic and roofing materials	–	13
Miscellaneous	9	12

which has been cleaned by pickling before rolling; it is subsequently annealed in an inert atmosphere to prevent oxidation, and consequently has a bright silvery appearance. For some purposes the steel is coated in various ways. The most common of these are tinplate, terneplate and galvanized steel. Terneplate employs a coating of a mixture of tin and lead, while galvanized steel is produced by coating the steel with zinc, either by hot dipping or electroplating. It should also be remembered that some of the drums with tin, lead or zinc coatings may have had the coatings applied not to the sheet steel but to the partly finished drum.

When drums are made from stainless steel, the alloy selected is usually an 18/8 austenitic variety, probably with a stabilizer against weld decay. Aluminium drums are made from commercially pure aluminium or a magnesium alloy.

By tradition the thickness of metal was described in terms of *gauge*, mild steel being given in Birmingham Gauge (B.G.) and stainless steel or aluminium in Standard Wire Gauge (S.W.G.). These gauges were defined in terms of weight per unit area, which after metrication, meant grams per square metre. For a given specific gravity this can be converted to thickness. So gradually the term gauges was dropped and replaced by millimetres (Table 19.3).

BS 1449 covers the kind of steel used in drum making and mentions thickness tolerance. The relationship between the gauge of the sheet metal employed in its construction and the strength of the finished drum is not a simple one, because the drum can become damaged in different ways. We can relate the difficulty of piercing a hole in the metal to its tensile strength, but resistance to deformation is a different matter—not merely stiffness of

Table 19.3 Thicknesses of steel sheet for drums, etc.[a]

mm	B.G.[b]	in	Use
0.40	28	0.015 6	Used only for small, light drums
0.45	27	0.017 45	
0.50	26	0.019 6	
0.625	24	0.024 8	
0.80	22	0.031 25	
1.00	20	0.039 2	
1.125	19	0.044 0	
1.25	18	0.029 5	
1.575	16	0.062 5	
2.00	14	0.078 5	Used only for heavy duty, fully welded drums
2.50	12	0.099 1	
3.20	10	0.125 0	

[a]These values are approximate equivalents only. The nominal thickness of any gauge of steel sheet will be subject to the rolling tolerances appropriate to that gauge as set out in BS 1449, *Steel plate, sheet and strip*.
[b]B.G. = Birmingham Wire Gauge.

metal but also design and mode of construction must be taken into account. In fact, selection of suitable gauge is usually done on a trial-and-error basis. A trend over the years has been to use lighter gauges. Thus up to the mid-1960s, 18 gauge was normal for 210 litre drums. Since then both in United Kingdom and the United States there has been a trend to use 18/20 gauge, 20 for the side walls, 18 for the ends, and for some products 20 gauge all round.

Where other metals are used (e.g. aluminium alloys of a given chemical constitution; aluminium as commercially pure sheet; stainless steel, usually an austenitic 18/8 nickel chrome stabilized steel), these also should be specified by thickness in millimetres, but note that traditionally the thickness of aluminium and its alloys was quoted in S.W.G. which does not correspond with the B.G. given for steel sheet in Table 19.3.

Manufacturing processes. The exact method of manufacture varies from manufacturer to manufacturer, but in principle centres round two machines —one to produce a tube by joining the two end edges of a sheet and forming a side seam, and the second to fix the ends to the tubular bodies (Figure 19.2). Drums with a folded side seam have been produced for molten solids such as rosin or bitumen, but usually side seams are welded by one of two mass-production methods. Both are electrical, one being *resistance lap welding* and the other *flash butt welding*.

The principle of flash butt welding is to strike an electric arc between the two edges and then butt them together hot. In the case of small tinplate drums, soldering can be used. The cylinder for large drums may or may not have corrugations but it will have rolling hoops, typically I-section, either by pressing them out of the cylinder or by fitting separate bands . If the drum is to have a square cross-section, the circular cylinder is squared, and then the reinforcing corrugations are pressed in.

The ends of the drum are usually fixed to the bodies by a process known as *seaming*. The double-seaming process consists of rolling the edge of the flange of the ends round the flange of the body, and then folding this seam flat against the body (Figure 19.3a). It is usual to put a layer of seaming compound between the two flanges to ensure a liquid-tight seam. This seam may be reinforced by inclusion of another band of metal, or it may be made more liquid-tight by resistance welding the two single thicknesses on the drum side of the seam (Figure 19.3b). More recently, new designs of end seam have been developed, whereby the metal of the end flange and that of the body flange are rolled together to form a spiral (Figure 19.3c). Drums with round seams have greater resistance to damage when dropped. Heavy gauge drums (say 1.9 mm or thicker) do not have rolled seams but have the ends welded to bodies, usually by the electric arc process.

Heavy duty aluminium alloy drums are also used when packaging detergents. They are made in a slightly different manner to steel drums. The ends, which are shallow pressings, are butt welded to the body giving a

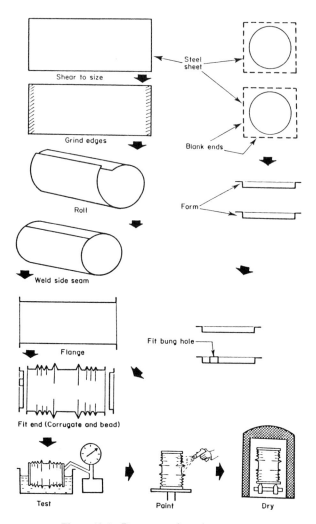

Figure 19.2 Drum manufacturing process.

completely crevice-less interior. The chimbs are reinforced with mild steel
skirts, and the drums fitted with inverted 'U' rolling bands made from heavy
gauge mild steel. A typical construction is shown in Figure 19.4.

Drum types. Drum manufacturing grew up with certain arbitrary dividing
lines, e.g. small drums and large drums, divided at about 15 gallon (70 litre)
size; large heavy-gauge drums and large light-gauge drums divided at about
the 16 B.G. (1.5 mm) thickness. The heavy-gauge drums were once further
divided into bilged and straight-sided drums, but the bilged drum is now no
longer made. These classifications are no longer so clear and, for example,

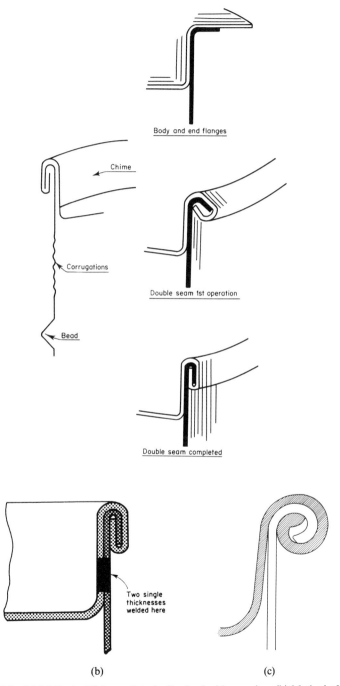

Body and end flanges

Chime

Corrugations

Bead

Double seam 1st operation

Double seam completed

Two single
thicknesses
welded here

(b)   (c)

Figure 19.3   (a) Method of fixing ends to bodies by double seaming. (b) Method of making double seam liquid-tight. (c) Van Leer Spiralon®.

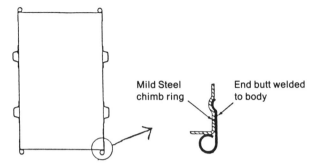

Figure 19.4 Crevice-less aluminium alloy drum, with mild steel reinforcement.

large and small now tend to be judged by diameter. Both large and small drums are available, with either a tight head or fully removable head and in the small drum range the latter may be subdivided into kegs (Figure 19.5) and pails (Figure 19.6). Large tight-head drums may have pressed out rolling hoops, with or without reinforcement (Figure 19.2), but if they are of heavy gauge, it is usual for them to be fitted with I-section rolling hoops (Figure 19.7).

In the range of small drums, those with conical top sections are handy for pouring, but uneconomic of space, so they have almost disappeared from the market. There are other small drums which have specialized uses and are available in a very restricted size range, examples in the rectangular shape being the Robbican and the Tandrum. Another development applicable to small pails and kegs is a slight taper to the body which permits empty drums to be nested one inside another, and thus to reduce the space occupied in the storage and transport of empty drums. 'Nesting' should not be confused with 'stacking'. Whether the sides are tapered or parallel, if the outside diameter of the base fits within the inside diameter of the top when closed, more stability in stacking is achieved.

As the ends for drums are produced by presswork involving the use of expensive dies, so each manufacturer would operate a limited range of diameters, and achieve a range of capacities by varying the height of the drum. This resulted in a very wide variety of drum sizes. Since the 1930s rationalization has been going on and, with the assistance of the British Standards Institution, three sets of specifications have been issued. Classification is by *duty*, so in the specification for light-duty drums with fixed ends (BS814) both large and small drums are referred to, but in a restricted range of diameters. Similarly BS 2003 covers the range from small kegs to full-aperture 210-litre drums. BS 1702 covers heavy-duty drums with fixed ends. With the adoption of metrication in the United Kingdom, many drum capacities have been changed to litres. Initially this led to extra variety in the range, but now the 'gallon' capacities are disappearing.

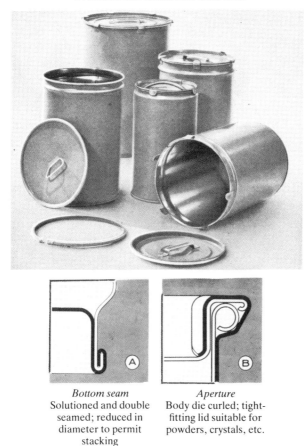

Bottom seam	*Aperture*
Solutioned and double seamed; reduced in diameter to permit stacking	Body die curled; tight-fitting lid suitable for powders, crystals, etc.

Figure 19.5 General-purpose steel kegs. *Side seam*, electrically welded; *Heads*, formed to fit curl and retained by (i) bend-over clips attached to body (B), (ii) bend-over clips attached to head, (iii) latch and eyelet closing ring (left front foreground); *Finish*, interior plain or lacquered steel, exterior painted, body decorated by litho or screen printing; *Strengthening*, one pressed-out bead near top of body.

Closures. Large tight-head drums for liquids are normally fitted with two small screw-threaded openings, one with a 2-in and the other with a ¾-in pipe thread (called G2 and G¾ in BS 2779). Into these are screwed a steel, zinc or plastic plug with external thread to match. The 2 in is typically for filling and the ¾ in for emptying by screwing in a tap. Interestingly, even when the drums are metric the closures are still in Imperial units even in French specifications.

A few all welded drums may still have a welded-in flange, but the majority of drums have a pressed-in flange with plug and cap to match, the most widely known being the 'Tri-sure' closure (Figure 19.8).

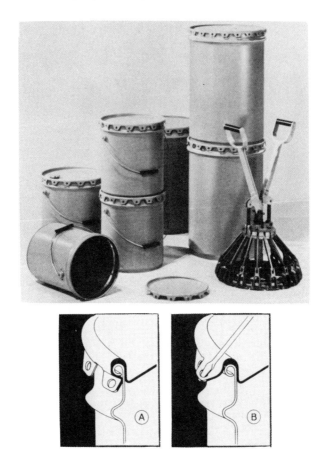

Figure 19.6 'American' pails. *Body seam*, soudronic or conventional, electrically welded; *Bottom seam*, solutioned and double seamed, reduced in diameter to permit stacking; *Aperture*, body, die curled; *Handle*, pail handle with grip attached to body if required; *Head*, a gasket is incorporated in the multi-lug cover (A) which is easily opened (B); *Closures*, various neck fittings possible in the lid; *Strengthening*, one pressed-out bead near the top of the body. This package for powders, semi-solids, greases, etc. becomes a liquid pack by the addition of a conventional fitting in the cover.

Small-size liquid-tight drums often have tinplate closures such as a threaded screw cap or a press cap, or a friction plug with spun-on cap. The most popular friction plug is the 3-in diameter (76 mm) size (Figure 19.9). Other typical closures are shown in Figure 19.10a, b and c. There are also closures which combine with pouring devices such as the 'Flexspout'. Drums for powders can be quite satisfactorily closed by large diameter (7- or 9-in push-in friction lids. If the flange on the head is downwards into the drums, these friction lids can be expanded in position, but, of course, once removed cannot be resealed (Figure 19.11).

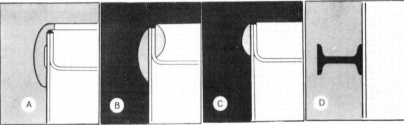

Figure 19.7 Steel drums for a closed returnable system. *Body seam*, electrically welded; *Top and bottom seam*, submerged arc welded and reinforced with lip section (A) or inner and outer convex bands (B) or a convex outer band (C); *Rolling hoops*, two I-section rolling hoops shrunk on to the body (D).

There is also a range of drums with fully removable heads. With the larger drums these tops are held on by a closing ring, consisting of a channel-section steel band held together by a bolt, latch or lever. Usually a gasket made of expanded or foam rubber acts as a seal between the lid and the curled-over top of the drum. For small drums, an alternative to the lid held on by a closing ring is the lug type lid, sometimes known as the 'American pail' (Figure 19.6). In this type the washer or gasket is usually cast and cured in situ.

Protective and decorative finishes. At one time a thin coat of cheap bitumen paint was considered adequate for a drum, but today both interior and exterior surfaces are the subject of much specialized attention. The subject is best considered under three heads; painting, decorating and internal coating.

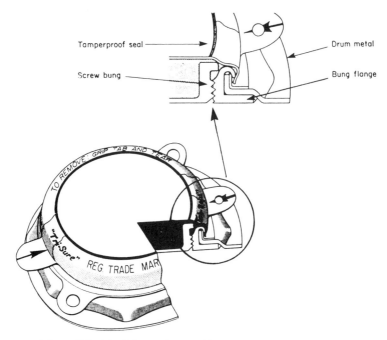

Figure 19.8   The Tri-sure closure with lacquer coated steel or zinc plug.

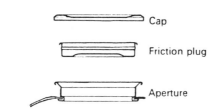

Figure 19.9   Friction plug for small liquid-tight drum.

Wherever practicable, drums are sprayed with stoving paints of a quality somewhere between that of a toy enamel and that of a refrigerator finish, according to circumstances. It is not usual to phosphate the outside of drums in the United Kingdom (although it is in the United States because of an inside treatment frequently given there). Air-drying paint is still used on some smaller production lines. The advent of roller coating to provide a base on which to print has led to many drums being painted entirely by roller coating the sheets before the drums are formed.

For many years it has been possible to print, by lithography, body sheets for small drums on tinplate printing machines, but it is only more recently that litho machines robust enough to deal with sheets for larger drums have become available. Even so, there are disadvantages in the lithographic

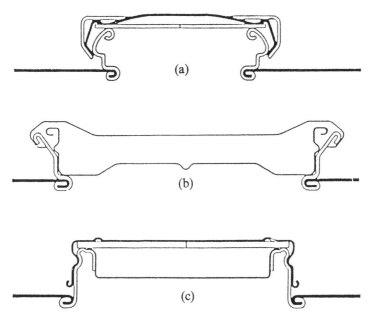

Figure 19.10 Typical tinplate drum closures. (a) Press cap with inner wad and tamper-proof overseal. (b) Inner friction plug and spun-on capsule. (c) Inner plug with screw cap and liner.

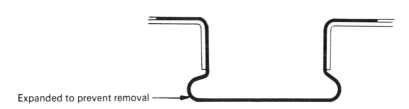

Expanded to prevent removal

Figure 19.11 Standard expanded lid.

process for small orders, and the silk screen process on the finished drum is used by many drum makers.

Probably the oldest and most expensive interior coating for mild-steel drums is pure tin (tin-lined drums should not be confused with tin-plate drums). The coating thickness of tinplate is often a mere 0.25 μm, whereas in tin-lined drums the coating is nearer 25 μm. Unfortunately, tin is cathodic to iron, so that any pinholes in the coating will have a tendency to turn brown when the iron base rusts. Lead-lined drums and drums made from terneplate (95% lead, 5% tin) were used in the past, but are now less common. Another popular metallic lining was zinc; galvanized drums are still used where long life is demanded and zinc itself does not create new problems. More recently, however, zinc has been replaced by organic lacquer linings. First of all the drum trade used the oleoresinous-type lacquer used by tinbox makers to line food cans. These were known as *sanitary lacquers*. Then in the 1930s these were followed by phenolic resin coatings with good acid resistance but poor flexibility, vinyl resins with good flexibility but poor solvent resistance and, after the war, by epoxy-phenolic resins aimed at the best all-round properties of flexibility and chemical resistance. These were followed by lacquers based on polyurethane.

There is, of course, no universal coating and, even within each group, there can be a wide divergence of properties. Part of the success of a lining lies in the correct surface preparation of the underlying metal, which may consist of zinc or iron phosphating or some form of roughening of the surface such as grit blasting.

Some small drums are, like tin cans, made from pre-lacquered sheet, but the majority of drums are pretreated and lacquered in the prefabricated state, i.e. body and two ends separately. All conventional lacquer linings are heat-cured, and so the components are stoved before final assembly. This restricts the possibility of having all-welded lacquered drums, as the lacquer will get burned in the region of the weld. The Chemical Industries Association once discussed lacquer lining pros and cons, and have published the opinions expressed. Conventional lacquers consist of synthetic resin in solution or dispersion, and one that had a limited use was a PVC plastisol. This led to the idea of making a lined drum from PVC-coated steel. In the United Kingdom the idea was abandoned, again because of the high cost and the limited chemical resistance of plasticized PVC. Recently the idea has been developed again, this time using un-plasticized PVC and applying it to the components, as is done with spray lacquer instead of to sheet steel.

Another process involves the application of a dry powder to steel, then sintering or fusing it. This is especially suited to a thermoplastic resin like polyethylene, though epoxy resin powder coatings are now available. Loose plastic liners can also be fitted to drums to give a composite having the strength of steel with the chemical resistance of plastics.

Fibre drums

Fibre drums are essentially non-returnable transit packages. Their size can vary very considerably; the diameter ranges from 180 to 660 mm, while the height can be anything from 75 mm to 1.80 m, or more in special instances. Although they are designated one-trip containers, many manufacturers using their own transport regularly have found they can be employed safely on a returnable basis for several journeys.

Manufacture. As the name implies, they are made from fibrous material (paper or paperboard) and since this is of lower tensile strength and rigidity than the materials used for their predecessors (conventional wooden barrels or steel drums) it is essential that the maximum strength be derived from the shape of their final form. Fibre drums are, therefore, generally cylindrical with parallel side walls and a cross-section taken across the drum at 90° to a vertical axis is usually, and preferably, a circle. Fibre drums always have the side walls made from fibrous material (paper or paperboard); the ends at the top and the bottom may also be constructed of fibrous material, but can equally well be produced in other materials. Steel, plastics, wood, plywood and hardboard are the common alternatives, and almost any combination of these may be used; e.g. we can have a fibre base with a steel lid, or a wooden base with a fibre lid.

Body construction. The side wall construction is achieved by winding onto a mandrel one or more plies of fibreboard, the plies being firmly adhered to give a solid side wall. All fibre drums, irrespective of their final form, start in this way from a cylinder or tube of laminated fibreboard. The side wall material used can be wound from material in reel form, from several individual sheets, or from a single sheet. When a reel is used, the side wall of the drum may be either wound straight (often referred to as convolute winding) or spirally wound.

Spiral winding is a process of producing a cylindrical tube by winding several plies of paper around a stationary mandrel, each ply being wound in a helix or spiral overlapping, superimposed upon, and bonded to the ply below.

Straight winding is a process of producing a cylindrical tube by winding several individual rectangular sheets or a web of paper around a rotatable mandrel, so that the several sheets or the several turns of the web are directly superimposed upon and bonded to the sheet or ply below.

Whichever method of winding is used, it is essential that all plies are firmly glued together overall to form a solid side wall. The adhesive used will depend upon the use for which the drum is being made. The normal adhesive used is sodium silicate, but starches, dextrines or special waterproof adhesives are also employed.

Spiral winding is a more rapid operation than straight winding, and therefore is usually cheaper, but the method does not give such a strong cylinder as straight winding. Usually spiral cylinders require more layers or plies of paper to give the same strength as straight-wound cylinders. It is also important to note the regulations within overseas countries when exporting goods in fibre drums. In some countries drums with spiral-wound side walls are not acceptable.

Drum ends. The materials used for the making of the end are often determined by the type of product to be packed, and also the relevant handling conditions for filling, storing and transporting. It is sufficient to say that, if fibreboard ends are used, and they are made from two or more discs, they must be secured together with a suitable adhesive, and fixed into the ends of the cylinder sufficiently firmly to give the finished drum a satisfactory performance. The method of fixing the ends varies between manufacturers, and basically the method used gives two distinct types of drum (Figure 19.12). The first is a flush-ended drum in which the flat part of the top and/or base is level with the edge of the drum side wall (Figure 19.13a, b and c). In the other construction the flat part of the top or bottom is recessed below the edge of the drum side wall (Figure 19.13d).

If steel ends are used, they must be suitably ribbed to give additional rigidity. They can be attached to the side wall either with adhesive (Figure 19.14a) or by crimping the metal (Figure 19.14b), thus trapping the ends of the side wall firmly by purely mechanical means. Laminated fibre bases can be secured in a similar manner by crimping a metal hoop onto the drum base (Figure 19.14c).

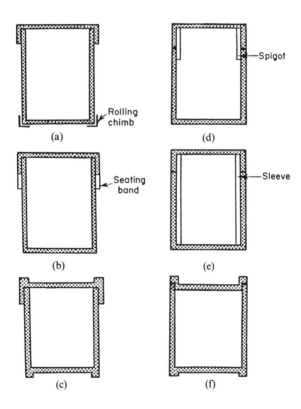

Figure 19.12 All-fibre drum construction. (a) Flush-ended; (b) flush-ended with lid seating band; (c) recessed end, pill box lid; (d) flush-sided with spigot; (e) flush-sided with inner sleeve; (f) recessed end, plug lid.

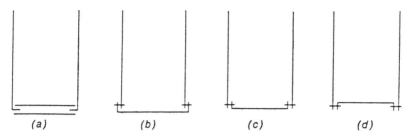

Figure 19.13 All-fibre drums, types of end construction. (a) Flush end, inner and outer discs glued to the flange with adhesive. (b) Flush end, pill box style with end secured with staples or rivets. (c) Flush end, reverse plug style end secured with staples or rivets. (d) Recessed end, plug style end secured with staples or rivets.

Metal lids are secured by several means, e.g. if the lid edge is straight, it can be secured to the side wall with self-adhesive tape or gummed tape, as in securing a fibre lid of the pillbox type. It is also possible to use lugs—strips of metal fitted at equal intervals near to the top edge of the side wall, around the periphery, which are turned over to hold the lid in position. This is similar to a traditional closure used on some types of steel drum. The third method of securing a steel lid is by means of a *locking band*, sometimes known as a *closure ring* or *hoop*. In this method of closure it is necessary for the top edge of the drum side wall to be reinforced with a metal ring, to give a lip under which the locking ring is clamped when closed (Figure 19.15a). The closing of this ring can be effected either by a clip type of closure, or by a lever-action closure (Figure 19.15b, c and d). Both of these methods are also used on steel drums with full-open tops.

With all-fibre lids, various forms of tape can be used, the method selected depending upon the weight of the contents in the drum. For extra-heavy weights it may be necessary to use a gummed cambric tape rather than self-adhesive plastics tape used for lighter loads.

When sealing drums with any form of tape, it is preferable to have a 'flush' side wall to avoid wrinkling the tape at the juncture between the lid and the drum side wall. This flush side can be achieved by several methods. The first is by the use of a spigot (Figure 19.12d) which consists of a short collar attached to the top end of the drum inside the side wall, so that it protrudes a few centimetres. A pillbox lid then fits snugly over this spigot so that the side of the lid and the side wall of the drum give a flush joint. The second method is to insert a snug-fitting sleeve into the full length of the drum (Figure 19.12e), which again protrudes a few centimetres. This method is more expensive, since more material is used, but at the same time it gives additional strength to the complete side wall. The third method, which is the least common, is to use a plug-fitting lid (Figure 19.12f). This, as the name implies, refers to a lid which pushes inside the drum rather than fits over it. This method does not afford such a strong closure, and is usually limited to drums of small capacity.

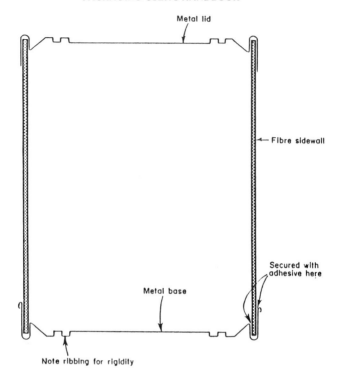

Figure 19.14   (a) Drum with metal base and lid.

Use of fibre drums. Originally fibre drums were used only for the carriage of dry powdered and granular solids, including chemicals, pharmaceuticals and foodstuffs. It was soon found that they are ideal for solid cylindrical objects, e.g heavy rolls of such material as plastic sheeting, linoleum, expensive papers, aluminium foil, etc. Fibre drums with recessed ends are usually used for these materials, since this decreases the risk of damage to the edges of the rolls.

With improved methods of manufacture and more styles of drum available, it was found that various refinements can be incorporated to extend the range of products for which drums would be suitable. The first improvement was the introduction of a moisture vapour barrier into the side wall to permit the packing of hygroscopic, deliquescent or other moisture sensitive products. These barriers may be of polyethylene-laminated kraft paper or, for very sensitive products, aluminium foil laminated to a kraft paper. The barrier is 'buried' in the side wall of the drum, i.e. it is wound in the side wall so that it is not immediately adjacent to the inside surface or the outside of the drum. Its position within the side wall can be varied, and will depend upon the purpose for which the barrier is used.

The second improvement consisted of extending the use of the drum to the carriage of semi-solids and even liquids which are unsatisfactory in plain fibre drums.

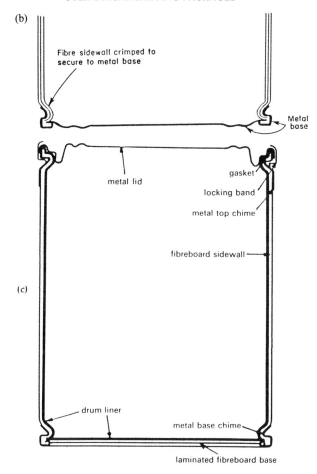

Figure 19.14 cont'd. (b) Fixing a metal base by crimping. (c) Fibre drum with laminated base secured with metal hoop (chimb or chime).

Fibre drums can be lined with various materials, e.g. polyethylene, aluminium foil and vegetable parchment. This permits drums to be used without the separate loose polyethylene liner for the carriage of semi-liquid materials. Excellent examples of materials which fall within this definition are lard and shortening, paste adhesives, certain printing inks and thick syrups.

The inside surfaces can also be treated by spraying with waxes or special polyethylene wax blends, or with a group of materials loosely known as *release coatings*, which include silicones, alginates and polyvinyl acetate emulsions. These are used if drums are to carry materials having a melting point of 50°C or higher, e.g. bitumen, waxes, resins and lanolin. They are all filled after liquefying by heat, and the special interior treatment prevents them from bonding to the fibre side walls when the liquid sets solid on cooling. The solidified contents are easily removed by cutting away the container with a sharp knife.

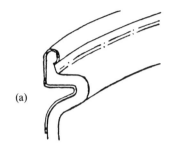

(a)

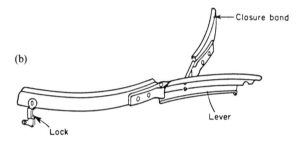

(b)

Closure band

Lock

Lever

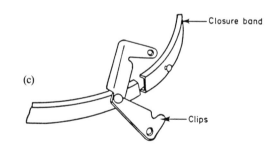

(c)

Closure band

Clips

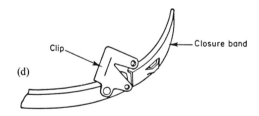

(d)

Clip

Closure band

Figure 19.15   (a) Metal ring for gripping closure band. A top reinforcing band, after shaping in a press tool, provides a top curl and facility for a closing hoop to grip the drum body. (b), (c), (d) Types of catch for closure bands.

As well as semi-liquids certain liquids can also be carried if a separate container is used inside the fibre drum, or if the side wall and base are lined with special duplex linings (laminates of polyethylene-fibre-polyethylene-fibre). The inner container can be of thin-walled plastics, e.g. blow-moulded polyethylene or closed-bag liners made from welded flexible PVC sheet. If duplex linings are used, the lining joint must be secured with a suitable heat-sealed tape, and the base sealed with an inert caulking material. Liquids packed in this way include sodium hypochlorite, liquid detergents, liquid fertilizer and insecticides and concentrated fruit juices.

A further use of fibre drums is for irregularly shaped solid objects by the use of internal fitments and cushioning materials. Users are finding that a cylindrical package is easier to handle than a square one. The cylindrical drums can be rolled 'milk churn' fashion by one person, whereas a square box or wooden case requires at least two persons to lift it, or some form of mechanical handling.

The external surfaces of fibre drums are decorated when required by spray painting or varnishing, and by printing in various colours. The only printing method is usually silk screen. The painted surface not only decorates but also provides a certain resistance to moisture.

Plastic drums and composites

This section covers two forms of transport packagings, plastic drums and jerricans and composite packagings.

Plastic drums range up to a nominal 210-litre capacity though occasionally may be up to 450 litres; typically they have a circular or *square round* cross-section, smaller sizes may be fitted with a handle. Plastic jerricans typically are up to 60-litre capacity, have a rectangular cross-section and are fitted with a handle; they are not described separately here.

Composite packagings (plastics material) comprise an outer packaging and an inner plastic receptacle so constructed that together they form an integral unit which, once assembled, remains integrated; it is filled, stored, transported and emptied as such. The varieties described here are in the shape of a drum with others made of steel, plastic or fibre.

Plastic drums. Plastic drums may be considered in three size ranges:

1. *Large:* 200 litres and above with a limit to 450 litres; 210 litres nominal is a common size
2. *Small:* 60 litres and below. 20 and 25 litres nominal are common sizes; while there is no bottom limit, this section does not cover sizes below 10 litres
3. *Intermediate:* between small and large

Methods of manufacture include:

1. Extrusion blow moulding which has several variants
2. Continuous extrusion of a body tube which is then cut and has ends welded on
3. Rotational-moulding
4. Injection-moulding

The first three methods are used for all sizes of drum while injection-moulding is confined to small drums plus of course components such as lids for larger full-aperture drums.

Normal *extrusion blow-moulding* consists of five essential stages and can result in a complete drum including a threaded closure opening.

1. Plasicization or conversion of plastics granules or plastic powder to a molten state in a screw extruder
2. Extrusion of the molten plastics as a parison into a mould; extrusion may be continuous—typically for small drums—or may first be into an accumulator.
3. Closure of the mould around the parison and its separation from the extruder
4. Blowing air or other gas into the molten parison via a blowpin so that the plastic is expanded to the shape of the mould cavity
5. Cooling, mould opening, removal of the moulded drum, trimming off the flashing, fitting closures, etc.

Figure 19.16 shows finished 25-litre *square-round* drums for liquids. Figure 19.17 shows a 210-litre, full aperture drum typically used for solids

Figure 19.16.   Square-round drums (25 litres) manufactured by Van Leer (UK) Ltd.

Figure 19.17 Full aperture drum (210 litres) manufactured by Harcostar Limited.

and also manufactured by normal extrusion blow-moulded techniques; the lid however is injection-moulded. Variations on normal extrusion blow-moulded techniques include:

1. In one design of a 210-litre tight-head drum, the Mauser L-ring, there is an additional stage carried out in conjunction with (4) above; the mould has moving ends with a compressive capability for forming the rings top and bottom (Figure 19.18). The L-ring design was developed to have storage and handling characteristics similar to those of the equivalent steel drum. The L-shaped rings permit handling with parrot-beak and other mechanical devices.
2. Co-extrusion to provide an inner layer of polymer having specific properties related to the characteristics of the substances to be contained.
3. A third variation is extrusion blow-moulding an integral base and body with the top welded on at a later stage. This method could be referred to as two-piece construction and is not illustrated here.

The *extrusion and welded ends* method involves:

1. Plasticization or conversion of plastic granules or plastic powder to a molten state in a screw extruder, the continous extrusion of a tube and its cutting to the required length. Each length of tube forms the drum body.

Figure 19.18   Mauser L-ring drum (210 litres) manufactured by Harcostar Limited.

2. Separate injection-moulding of the two ends.
3. Heat welding the ends to the body tube. This is a critical process and manufacture may include an automatic check on the weld quality.

This method could be referred to as three-piece construction and is typified by the ranges of drums illustrated in Figure 19.19.

Rotational moulding, also known as rotational casting or roto-moulding (Figure 19.20), basically consists of three stages:

1. Loading a metal mould with thermoplastic powders such as polyethylenes, polyamides, ECTFE.
2. Heating the metal mould containing the powder in an oven with hot air, or by gas flames, while rotating the mould in one or two axes depending on the required shape. During this stage the powder melts and fuses as a reasonably uniformly thick coating on the inside of the mould.
3. Cooling the mould at the end of the heating cycle, when the powder has been completely distributed and fused, and finally extracting the moulding.

Figure 19.19 Extruded and welded end drums manufactured by Van Leer (UK) Ltd. (a) 25 litre Valerex drum. (b) 210 litre Valerex drum.

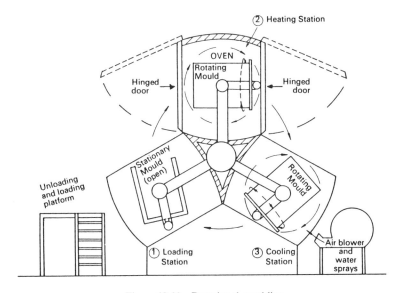

Figure 19.20 Rotational moulding.

Figure 19.21   Warboy (30 litres) rotationally moulded drums manufactured by Francis Ward
Limited.

The appearance of a rotationally moulded drum is shown in Figure 19.21.
Injection moulding consists of three stages

1. Plasticization or conversion of plastic granules or plastic powder to a molten
   state in an extruder which may be of the screw or ram type
2. Injection of the melt or plasticized material into a space defined by the male and
   female parts of the mould
3. Cooling of the plastic at the end of the injection cycle and extracting the
   moulding

Blow-moulded drums and drums made by extrusion and with welded ends
are typically made in high density, high molecular weight polyethylene.
High density provides relatively high tensile strength, stiffness, hardness,
resistance to weakening by chemicals and resistance to gas/vapour
transmission. High molecular weight (i.e. the size of the molecule or the
number of ethylene groups in it) provides relatively high toughness,
resistance to fatigue and resistance to environmental stress cracking.

Polypropylene is also used and has many valuable properties; it does not
for example exhibit environmental stress cracking. Low temperature impact
performance, even of the copolymers, tends however, to restrict its use.

Since the advent of LLDPEs, rotational moulding employs high
performance grades of these which have a low melt flow index (MFI).
Medium to high density grades of polyethylene used in the process have
excellent low temperature impact strength and good environmental stress

cracking resistance and conform at least to the requirements of ASTM and BSI specifications.

In the early 1970s plastic drums were regarded as relatively new packages for which there was little service experience. The performance tests in the transport of dangerous goods rules were used as a measure of their robustness during storage and distribution—often whether or not they were to contain dangerous goods. For the transport of liquids, the design type tests comprise:

1. Drop tests after conditioning cold at $-18°C$; drop heights are typically between 1.2 and 1.8 m
2. Leakproofness and hydraulic pressure tests
3. Stack tests for 28 days at 40°C under a load equivalent to a stack 3 m high

As new packages, there was concern that they should not be unduly weakened by the liquids they contained—the principal processes being stress cracking, swelling from absorption of solvents, oxidation and molecular degradation. A system was developed of storing sample drums with the liquid to be transported for a period of 6 months before carrying out the tests. This was cumbersome and a modified system is now often applicable in Europe; this uses, to cover many but not all situations, standard liquids and an accelerated storage period of 21 days at 40°C. Apart from assessing the drum itself, such testing has drawn attention to the importance of the closure and its gasket also not being weakened.

Progressive weight reduction—light-weighting—has been an on-going feature of much packaging development. With drums it may be noted that, while a 210-litre steel drum for liquids has a tare weight of around 20 kg, the equivalent plastic drum is now less than half that weight. Light-weighting plastic drums not only reduces the quantity of polymer used but also speeds up the moulding time.

Composite packagings. Composite packaging in the shape of a drum, used to contain liquids, typically consists of:

1. A relatively rigid outer which may be made of steel, plastic or fibre
2. A semi-rigid plastic inner with a neck protruding through the top of the outer; a non-pigmented, extrusion blow-moulded, polyethylene inner is often used

Where the outer is steel, it is often made by the conventional methods used for steel drums. While the top is seamed on, the bottom is left off at this stage. The inner is made by conventional moulding techniques and fitted into the outer; finally the bottom is seamed on. Composite packagings of these types are available in similar size ranges to those described for plastics drums.

Some composite packagings are shown in Figures 19.22–19.24.

Figure 19.22 Steel/plastics composites (25 litres) manufactured by Van Leer (UK) Limited.

Figure 19.23 Steel/plastics composites (210 litres) manufactured by Van Leer (UK) Limited. The drum in the foreground shows the plastics inner receptacle, with the steel top fitted, being lowered into the steel body before final seaming.

Figure 19.24   The Acitainer manufactured by Harcostar Limited, a plastics/plastics composite.

20
Rectangular shipping containers

TIMBER AND PLYWOOD CASES AND CRATES

The very earliest types of shipping containers were bales and bilged casks, made to facilitate man-handling. The Industrial Revolution, and the resulting improvements in transport, the building of railways and better roads led to the development of wooden boxes and crates as the first modern shipping containers. Timber was then plentiful, and so inexpensive that wooden box makers, like most other users of wood, demanded high grades and ignored inferior material. Little consideration was given to wastage, either in forestry or in box making, and even less to questions of design to give maximum strength for minimum use of material.

Today the cheaper grades of timber are used for box making, and questions of design and cost are of major importance. Wooden shipping containers are used only when their strength characteristics are needed to carry their contents safely to their destination, or when the value of the contents warrants the extra protection afforded.

Basic types

Wooden shipping cases came into general use after the introduction of lifting appliances at the major ports in the United Kingdom and overseas. They can be divided into three general types: nailed wooden boxes, open or sheathed crates and wire bound boxes.

Just as lifting appliances brought about the change from casks to cases, so other factors have intervened to influence export packing-case design. Thus the shortage of imported softwood in World War II and the increasing cost in post-war years have compelled economy in its use, and paved the way for the wider use of plywood or other sheet materials forming the whole or part of the case. In addition to this, the increased use of Container Traffic has reduced the need for handling and therefore the strength needed in the pack, when these are being delivered on a door-to-door basis. It should, however, be remembered that like airfreight, these packs often face severe journey hazards across country before they reach their final destination, and goods should be protected accordingly.

Most export manufacturers seek to neutralize rising production and freight costs by comparable economies in packaging, and this has helped to accelerate the use of lighter and cheaper packaging materials. It is important to note that shipping charges are based on the actual gross weight of case and contents, or on weight which used to be calculated at 40 cubic feet per ton, whichever is greater. The metric equivalent is 1000 kg (= 1 tonne) per cubic metre. Volume is reckoned on 'outside dimensions', so that a battened wood case normally adds four thicknesses of timber to each dimension—length, width and depth.

Because of their strong joints and rigid construction, together with great flexibility in size and load-bearing capacity, outside-battened cases are used for a wide range of export goods. The girth battens ensure that the body of the case is above ground-level, thus encouraging safer handling and stowage.

Extra girth battens and/or diagonal braces are introduced to stiffen the assembly where necessary and, when the nature of the contents permit, current practice frequently uses thicker battens and reduces the thickness of the sheathing timbers, thus reducing the overall timber content and the weight of the case. With the coming of 'stress-graded' timber, this may well become more common, for, although the timber may cost more initially, it may be possible with good design and accurate knowledge of its structural strength to reduce the amount required. Large cases require the addition of lid bearers (crusher bars) to resist the compressive action of slings. With the exception of fork-lift battened cases the other types developed from the battened wood case are modification designed to secure economy of cost, tare-weight and shipping volume (Figure 20.1).

Nailed wooden boxes

These are very common and satisfactory containers for moderate to heavy weight commodities, particularly for overseas shipments. They have high compressive strength, high retention characteristics, ability to contain difficult loads and protect from punctures and other external damage.

To prevent warping, they should be made from seasoned wood (12–18% moisture content) and the components should be knot free to guarantee uniform strength. To guarantee proper stacking strength, components should be properly nailed and corners properly constructed.

The types of load may be classified as follows:

1. *Easy load:* low density commodities that fill the box (clothes, cans, etc.).
2. *Average load:* moderate density commodities (hardware, paint in cans, etc.)

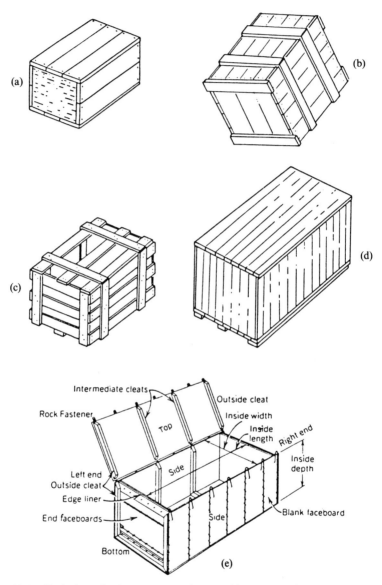

Figure 20.1 Typical wooden boxes, cases and crates. (a) Flush-side (merchant case): close-boarded box construction without battens, combining physical and climatic protection (with Kraft-union liner) for 'easy-type' distributed loads up to 100–150 kg. (b) Battened case (belted case): close-boarded assembly reinforced with end and girth battens (belts or cleats) for 'average' and 'difficult' loads beyond the safe capacity of the flush-side case in terms of volume and/or weight. (c) Skeleton crate: open slatted construction affording purely physical protection to one or more self-supporting items not liable to climatic deterioration in the course of transit. (d) Skid-base case: closed case with crosslaid base-boards or planks and longitudinal skids or 'runners' fixed underneath to support heavy loads and (originally) to enable movement of the case on rollers or snow if necessary. (e) wirebound box.

3. *Difficult loads:* high density products and those that require a high degree of protection (castings, stampings, etc.) and impose high stresses on the box at the joints, e.g. where a large proportion of the weight is placed on a small area of the case side or base. The term is also applied to lighter loads of a fragile nature, of an awkward shape or difficult to protect.

Materials. As in any structure, knowing the characteristics of the material used is very important. The factors that affect the strength are as follows.

1. *Moisture content:* Normal fibre saturation point is considered to be 30% moisture content for most species. Reduction of moisture through drying increases strength, reduces weight, but increases tendency to split. A change from green condition to 12% moisture content results in 30–100% increase in strength.
2. *Duration of load:* Duration of load application affects load carrying capacity of wood. A wood member can support continuously for 1 year only about 66% of the load required to cause failure in a standard strength test of only a few minutes.
3. *Knots:* Knots are a cause of discontinuity of fibres and therefore cause stress concentration. Size of knots and their location must therefore be considered.
4. *Slope of grain:* When the grain is not parallel to the longitudinal axis, it is said to be cross grained. When cross grain is steep, it reduces the strength of the members. A grain slope of 1 in 8 will reduce bending strength by nearly 50% (Table 20.1).
5. *Other factors:* Checks, splits and other defects also affect the strength of the structural members and should be avoided in critical spots.

To perform properly, the nails must be correctly used to achieve maximum holding power. Effectiveness of nails depends on their resistance to withdrawal and to their shear strength, which is proportional to the

Table 20.1 Reduction factors used to compute working stresses from basic stresses.

Strength-reducing characteristic	Strength in (%)			
			Bending	
	Tension	Compression	Flat	On edge
Knots				
One-quarter of the width	25	13	25	43
One-third of the width	33	17	33	53
One-half of the width	50	25	50	74
Cross grain:				
1 in 20	0	0	0	0
1 in 15	24	0	24	24
1 in 10	39	26	39	39
1 in 8	47	34	47	47
1 in 6	60	44	60	60

diameter of the nail. For maximum holding power, the nail must be driven perpendicular to the grain of the wood and its withdrawal resistance is then dependent on the specific gravity of the wood. Softwoods have specific gravity varying from about 0.33 to 0.38 but hardwoods may have specific gravities as high as 0.55 or more. The latter are of course far more brittle than the softwoods. Precautions must be taken not to split the wood during nailing and if the wood is dense then pre-drilling is recommended.

Nailing rules. To achieve the best results with nails the following practices are recommended:

1. Unless nails are clinched, use cement coated nails.
2. When possible, the nails should be driven through the thinner member into the thick one.
3. When the combined thickness of the two pieces is less than 3 inches, the nails should be clinched.
4. Nails should be driven no closer to the edge than one half of the members' thickness.
5. Nail two members in two rows or stagger the nails to reduce the chances of splitting.
6. To increase the resistance to racking, the rows of nails should be spaced at least 25 mm apart.

The withdrawal strength of a nailed joint depends on the type of nail and the angle of penetration. The lowest strength is given by staples and the highest by annular ringed-shank nails; the latter are the only type where withdrawal strength can equal or even exceed the shear strength (see Table 20.2).

Crates

The difference between a box (or case) and a crate is that a box is a rigid container with closed faces that completely enclose the contents. A crate is a rigid container of framed construction. The framework may or may not be enclosed (sheathed) (Figure 20.2). Purely physical in function, it does not offer much scope for improvement, and little for economy except by widening the spaces between the crate members. Lining with waterproof paper does not convert a crate into a closed case, but large crates may be lined with thin plywood, hardboard or fibreboard as a cheaper alternative to a closed wood case when the load is suitable (e.g. a large open tank).

For larger and heavier equipment and depending on product characteristics, open or sheathed crates are used. The crate differs from nailed wood boxes in that it has a structural framework that is designed to resist and withstand outside forces, and provides support and attachment for the lading. This framework can be compared to truss structure used in buildings and bridges.

Table 20.2 Design working loads for fastenings (SWL)
from BS 433, section 8).

	Maximum safe working load per 30 mm fastening penetration (kg)	
Fastenings[a]	Withdrawal	Shear
Round plain: 90°	8	23
80°	12	23
70°	12	23
Round plain: gun driven 90°	8	20
Cement-coated 90°	8	23
Round plain galvanized 90°[b]	20	25
Round plain tee head 90°	6	20
Square twisted 90°	18	23
Annular ringed shank: 90°	30	31
80°	28	29
Annular ringed shank: gun driven 90°	30	29
Driven woodscrew	29	34
Hammer driven woodscrew[c]	26	31
Staple: chisel-point	4	14
Coated staple: chisel-point	7	14

[a]All nails are 50 mm×2.65 mm diameter; all screws are 50 mm×3.4 mm
(No. 6) diameter countersunk head; all staples are 50 mm×9 mm×1.63 mm
diameter.
[b]Clout nails give a similar performance to bright or galvanized nails.
[c]Hammered in, driven for the last two turns.

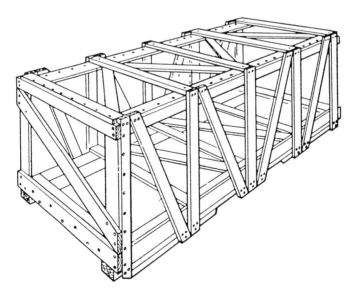

Figure 20.2 Simple wooden crate (unsheathed).

As in any structure, when the crate is designed, one must consider the following:

1. Establish the hazards (forces) from which the product must be protected.
2. Make sure that the material (wood) used in the structure is adequate to take these forces.
3. Design the lightest adequate structure, as in shipment, the cost often depends on the weight.
4. Design the crate for minimum volume. In overseas shipment, charges are based on cube or weight, whichever is the higher.

A properly designed crate gives many economic advantages as well as excellent protection.

Designing a crate. In designing a crate, one must appreciate several basic concepts:

1. A rectangular structure, while it provides compressive strength, is poor under racking conditions.
2. A triangular structure is good in resisting racking.
3. Combination of rectangular and triangular structure by use of a diagonal member combines the two properties.
4. A diagonal member is much stronger in tension than compression as when in compression the member can be flexible.
5. A rectangular structure with crossed diagonals gives maximum strength and resistance to racking.
6. All sides of the crate structure should be cross braced to give uniform resistance to twisting and diagonal distortion.

Applied loads. In developing the specifications for the crate, one must consider the following factors:

1. *Characteristics of the product:* does it need weather protection in addition to protection from rough handling? How to provide adequate anchorage?
2. *Length of journey and mode of transport:* for journeys of long duration requiring transport by rail or ship, size and resistance to handling during loading, transfer and unloading, are very important. Ask such questions as: Is the load to be handled manually or by rolling, by lift truck or overhead crane? Are adequate means provided for lifting the load? If overhead crane and slings are used, can the crate take the crushing forces of sling and grabhooks?
3. *What will the storage conditions be like during shipment?* Is there enough compressive strength in the crate to withstand stacking at the destination or in the ship's hold.
4. *Can the crate withstand accidental impacts?* These may happen in

transport handling or mishandling. What safety factors should be designed into the crate?

Wire bound boxes and crates

Nailed wooden boxes and crates can be designed and put together as individual and custom built containers as well as mass-produced in large numbers. The wire bound container is usually factory manufactured, shipped in a flat form and is designed to combine the strength of wood and steel wire.

The first wire bound container was designed in 1891 in Chicago using a continuous sheet of veneer with wire stapled to it. An improved version in 1904, attached the wires to pre-cut wood sections and in 1906 machinery was designed to produce wire bound containers.

They are now made in various sizes and shapes and because they are not rigid, they absorb shock and thus provide some protection from impact in addition to containment of load and stacking strength.

Wirebound containers have several cost advantages:

1. Delivery in knocked down condition
2. Minimum storage space
3. Simple assembly
4. Simple means of closure

Materials

1. *Faceboards* are boards that form the six sides or faces. These boards provide containment and stacking strength. Faceboards vary in thickness depending on application and are made from wood veneers and re-sawn wood.
2. *Binding wire* provides the load carrying capacity and must be selected to avoid being stressed beyond the elastic limit. Adequate strength can be provided either by a few large diameter wires or a larger number of small diameter wires. Which is used depends on the manufacturing method and on particular applications.
3. *Staples* are used to secure the binding wire to either the faceboards or the cleats. They are made from bright and galvanized wire. Both types of wire are free of scale and can operate in the close clearances of the stitching operation

Styles. The basic material used in wirebound construction is wood integrally combined with steel binding wires fastened to the wood elements by staples.

The many design options available by combining the various components of the composite container present an opportunity to custom design wire bound containers for specific products, weight, sizes and distribution hazards. The high speed automated machinery used to fabricate wire bound containers is custom built for specific size ranges and styles.

The most frequently used style is the All-Bound (Figure 20.3) used in crate form for shipment and storage of fresh fruits and vegetables, and in box form for industrial and military applications. The openings between faceboards in the crate style provide ventilation for produce such as sweet corn, celery, beans, cabbage, citrus, etc. Outside the United States this style is often called the Bruce box. Both styles are characterized by binding wires on all six faces of the made-up container, and a cleat framework in the vertical plane. Both assembly and closing is accomplished by engaging the wire loops at the ends of each binding wire. These loops are called Rock Fasteners. Two simple hand tools are required to engage the Rock Fasteners.

A completely different style, the wire bound pallet box (Figure 20.4) is

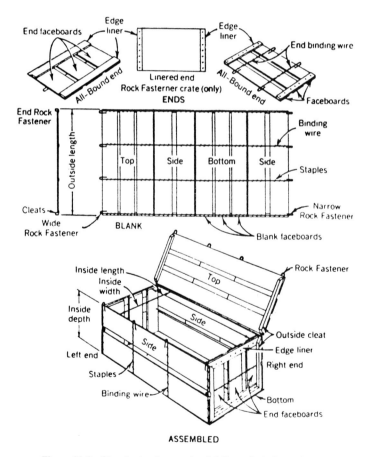

Figure 20.3 Standard reference for All-Bound wirebound crates.

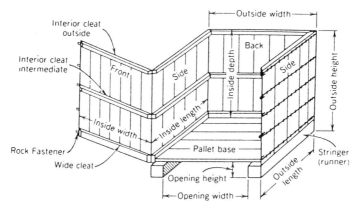

Figure 20.4 Wire bound pallet box.

used for shipment of industrial products including auto parts, chemicals, castings, and forgings, etc. The bottom horizontal cleats are either fastened to or lock onto a two- or four-way pallet base for fork truck handling. Cleats may be located inside the pallet container as shown or outside if the nature of the load requires a smooth interior.

Box specifications. Since they are factory produced, the manufacturer must be given sufficient information to decide on a satisfactory design. He needs to know the type of load, as previously defined for nailed boxes. In addition, requirements as to protection needed by the contents such as cushioning and surface protection must be specified. Hazards likely to be encountered in distribution must be specified, e.g. expected stacking heights, methods of transport, type of handling, etc.

Sill and skid-base cases and crates

There is a sharp distinction between battened wood cases for loads up to 250 kg and 2 m^3 and skid-base cases for heavier loads. In practice, however, skid-base battened cases are often built for loads of 5 tonnes and more, and are seldom used for loads under 1 tonne, but the following types are constructed on engineering principles, and are therefore used when the cost and the need is justified.

1. *Interior-framed (sheathed) wood case.* A large wood case consisting of substantial frame members designed to withstand the load to which the case is subjected. Vertical wood sheathing is applied to give strength and complete covering on the outside of the frame.
2. *Interior framed plywood sheathed crate.* This is constructed with internal framing members in the same way as the sheathed wood case, but plywood is substituted for the sheathing. The use of large panels, as opposed

to narrow boards laid vertically, enables quicker assembly, gives better moisture resistance because of the absence of joints and, when 9-mm plywood is used in lieu of 19 mm boards, there is some reduction in volume.

Both these may be constructed with either a skid-base or a sill-type base, depending on the nature and disposition of the load. Skids are fixed underneath the base in conjunction with a flat deck or floor: sills are raised above floor level to support the load and permit the use of light rubbing-strips beneath instead of skids. The factors governing the need for each type may be defined thus:

1. *Sill-type:* Whenever a supporting point or points of the object being packed are above the floor line, and proper bearing can be made so that the cubic displacement is conserved, a sill-base should be used providing that no weight is borne by the floorboards.
2. *Skid-type:* The skid-type base should be used for heavy loads when the supporting area bears on a large flat surface or is concentrated at a few points on the floor line.

Waterproof linings

Few cases can claim to be waterproof in their own right. Shrinkage occurs across the boards of butt-jointed wood cases, and the joints of battened plywood cases may be loosened by various forms of handling during transit. It is therefore necessary to introduce a physical barrier strong enough to give 'raincoat' protection to the contents, more especially if they are liable to corrosion or climatic deterioration.

Many special barriers are employed for packing goods, and for lining wooden cases and other types of shipping containers, particularly where packing for the Defence Services is involved. These are produced to Defence Specifications and are referred to by the appropriate number. The most usual ones are a waxed wrapping paper and a complex laminate of foil, cotton scrim, cellulose film and paper. The laminating material is waterproof (usually a hot melt) and the foil is coated with a heat sealing agent. A crêped kraft paper coated with a special wax compound on both sides can be moulded around irregularly shaped articles, as can a mouldable waxed grease-resistant wrapping made by coating a laminate of cellulose film and cotton scrim with wax. Kraft paper coated on one side with grease-resistant lacquer is also used. In addition, various types of coating are applied to kraft paper and used instead of union kraft. The most common coatings are based on polyethylene, polypropylene and PVC. However, for commercial use the lining materials most commonly used are:

1. *Waterproof paper (kraft union or pitchpaper):* two thicknesses of heavy kraft paper with a centre core of bitumen, stapled or adhered to all interior faces of packing-case (laid between framing and sheathing of interior-framed cases).
2. *VCI-coated waterproof paper:* adds chemical action of inhibitor to the physical protection from water afforded by the case liner.
3. *LDPE cover* over the top of the cargo and down the first few centimetres of each side for rain protection.
4. *Bag liners:* bags with sealed closure, which conform to the inside dimensions of the packing-case; made in various materials including crêped and plain waterproof papers, special laminates and polyethylene.
5. *Tinplate linings (tin shells):* enclosure shells with hermetically sealed lid and soldered joints, usually made with tinplate, terneplate or zinc and fitting snugly inside the packing-case. Mainly used for tropical cargo, moisture-sensitive goods, scientific instruments and chemicals liable to seepage. The use of desiccant material within the shell needs to be considered.

Closure of cases

Returnable wood and plywood cases and crates are often fitted with closure devices to avoid damaging the lid in conditions of re-use. Methods include the fitting of captive-screws, palm bolts, strap bolts and various types of fastener, many of which are patented.

The lids of export cases are normally nailed (using nails two and a half times the thickness of the lid), and care should be taken that battens are 'locked' where they butt. Lids should be screwed if the case is to be subjected to Customs inspection during shipment.

CORRUGATED AND SOLID FIBREBOARD CASES

The first use of corrugated material for packaging purposes was the subject of a patent (No. 122023) granted to an American named Albert Jones in December 1871. It claimed 'A new and improved corrugated packing paper' that could be used for wrapping fragile objects such as glass vials and bottles to 'present an elastic surface . . . more effective to prevent breaking than many thicknesses of the same material in a smooth state'.

In August 1874, Oliver Long patented an improvement on this, adding a lining to prevent stretching, etc. The Jones patent was acquired by Henry D. Norris and the Long pate it by Robert Gair. At about the same time, Robert Thompson was also making a cork or shredded-paper lined packing material and, in 1875, Thompson and Norris combined forces to develop these materials.

Between 1878 and 1888, a legal battle developed between Gair and the Thompson and Norris combine, the latter charging the former with infringements. In 1888, a compromise settlement was reached, in which Gair recognized the validity of the Thompson and Norris patents in return for manufacturing rights on a royalty basis. The compromise may have been reached because the litigants had discovered an English patent of 1856 in the names of Healey and Allen concerned with the corrugating of paper and other materials for lining or cushioning the sweat bands of hats.

The best material found in these early days for making corrugated paper was butchers' strawpaper, and the development of machinery to add one or two liners to the corrugated sheet was the principal concern of the manufacturers up to 1890. About this time, the idea of using the material for making cases rather than limiting its use to inner packagings was conceived and, in 1894, the Thompson and Norris Company introduced its 'cellular board boxes'.

By 20th May, 1895, these boxes were becoming accepted by the carriers, as is evidenced by a letter from a Wells Fargo agent that states: 'The new cellular board shipping box which you sent to our office has been tested and I am pleased to say has borne, without damage, such handling as it would probably be called upon to stand in ordinary transportation. I consider it superior to the wooden crates commonly used around pasteboard boxes. Our drivers will be authorized to receive goods packed in these boxes for transportation.'

It is interesting to speculate on the tests that were made and to note that they would seem to be performance tests. The prominence gained by the successful use of these boxes for express shipment, together with the expiry of the original patents, left the way clear for rapid development.

There was considerable secrecy in the early days about the methods of making corrugated paper, but it seems likely that the idea was copied from the fluted rolls used in laundries for ruffled lace collars. The sheet of paper was first dipped in water and then passed through a pair of meshed rolls driven by a hand crank and heated, first by gas jets and later by steam. One of the first machines made produced 6 m of fluted paper per minute about 0.6 m wide.

Considerable advances in technique have been made, and today a modern corrugating board machine can produce double-lined board 2.5 m wide at speeds up to 280 m/min.

Solid fibreboard cases were first produced in the United States about 1902 and were introduced to the United Kingdom some seven years later.

The use of both types of case was influenced by what is now known as the Pridham decision in 1914. The Interstate Commerce Commission of the United States at this time made a ruling that the railways would no longer discriminate against fibreboard cases, and that they they could be carried for the same charge as timber cases. From this time onwards the development of fibreboard was rapid and in the United Kingdom fibreboard cases account for 15% of the total spent on packaging. A ratio of about 2:1 existed for some time between the production of corrugated as against solid fibreboard for packaging purposes, but this has changed in favour of corrugated board and in the United States solid board cases account for under 1% of the total fibreboard case products.

Manufacture of corrugated containers

Solid fibreboard is made by pasting plies of board (usually chipboard) together, and lining one or both faces with either a kraft or a test sheet. Corrugated board is made by running the fluted sheet through corrugated rollers and applying a facing sheet of either kraft or 'test' liner to both sides of the fluted sheet. Once the board has been produced, the processes of conversion into cases are very similar for both materials. Many variations of style, quality and properties are possible in both materials.

Modern box plants use a variety of equipment for corrugating, printing, scoring, slotting, die-cutting, slitting, folding, gluing, taping, stitching, coating, impregnating, and performing special operations on corrugated board. Some plants are supplied with corrugated sheets made in another location. These are known as sheet plants. Large plants with corrugators usually have the same finishing equipment as sheet plants, so almost all produce finished boxes. There are a growing number of corrugating operations that simply specialize in producing sheet for sheet plants.

Corrugators

A modern corrugator is a large and expensive piece of equipment (Figure 20.5). It corrugates the fluting medium, glues it to a liner, adds another liner, makes double- or triple-wall board if needed by adding extra flutings and liners and then scores and cuts the corrugated sheet to produce a blank for the corrugated container. Starch-based adhesives are used to glue the flute tips to the liners.

Corrugators vary considerably in width and speed, but typical widths lie between 1.4 and 2.7 m and line speeds from 50–280 m/min. Normal practice is to run several orders or multiple widths of the same order to make maximum use of the width of the machine.

The basic operations carried out in 1890 have not changed in principle. The fluting medium is plasticized by steam showers (Figure 20.6) and a pre-heating roll, passes into the corrugating rolls, after which the tips of the flutes are coated with adhesive. The first lining sheet, which may be passed

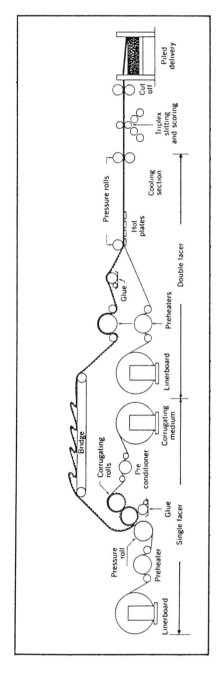

Figure 20.5 Schematic of a corrugator (simplified and condensed).

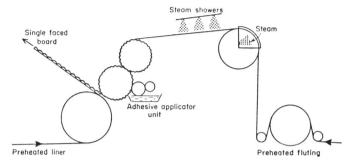

Figure 20.6 The single facer.

over one or more pre-heating rolls, is stuck to the flute tips. The single-faced board so produced passes up into and over a bridge between the single facer and double-backing unit. This bridge holds a store of material festooned on a belt conveyer to allow time for adjusting the machine at points between the various units of the corrugator and for reloading fresh reels of the components.

The single-faced material is drawn from this 'store' and passes, corrugations downwards, over an adhesive applicator that coats the flute tips with adhesive (Figure 20.7). The second web of liner passes over pre-heated rolls, is then stuck to the flute tips between a belt conveyer and the top of a heated plate or 'table'. The sheet is now no longer flexible, and must be kept flat as it travels over the heated table beneath the 'blanket' belt conveyer which holds the board under a slight pressure. After passing through this drying section, the board enters the second half of the maturing section, where it cools as it passes between 'blanket' belt conveyers towards the slitting and scoring units and the cut-off.

Thus, the corrugator produces 'blanks' cut to the outside dimensions

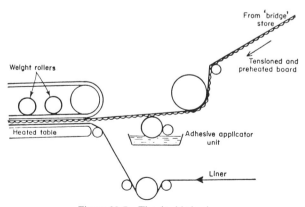

Figure 20.7 The double backer.

required and, if the blanks are for cases, with the machine-direction scores already made.

Printer-slotters

The blank produced by the corrugator is cut to the correct size, and the scores parallel to the machine direction which define the flaps and depth of the box are usually made on the corrugator. Scoring for the length and width panels, the cutting of slots to separate the flaps, the cutting of the flap for the manufacturer's joint and the printing of necessary information or decoration must now be added. The printer-slotter performs all these functions in a single pass. Although they are being replaced by flexo folder-gluers, these machines are still widely used. A simple version is the two-colour unit, made in a variety of sizes.

The board enters the machine as a rectangular blank with the two scores produced on the corrugator, and emerges as a printed 'cut-out' that merely requires folding and joining to turn it into a case. Figure 20.8 shows a section through a two-colour printer-slotter, but three- or four-colour prints are possible. The blanks are fed one at a time into the machine by a kick or suction feed from the bottom of the pile, and pass between a series of rollers which control their progress. They are first passed beneath the printing cylinders of the rotary letterpress type, using rubber or synthetic dies, and then between two pairs of shafts, where they are scored and slotted.

In addition to scoring, slotting, and printing, these machines usually have an automatic feeder, an ink washup device and an automatic stacker. Resilient letterpress-type printing plates mounted on rotary cylinders are used for both oil-based inks and flexographic water-based inks. The flexo inks have largely replaced oil-based inks because they dry much faster.

Flexo folder-gluers

The introduction of flexographic printing, using fast-drying easy-to-clean inks, made it possible to bring the printing operation in line with the folder-

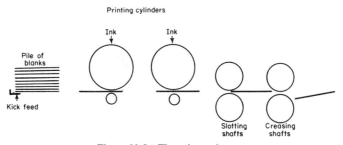

Figure 20.8   The printer-slotter.

gluer. The flexo folder-gluer takes scored blanks from a corrugator and can perform all of the operations to produce finished boxes. Only two machines are needed to produce long runs of slotted boxes with one glued joint, the corrugator and the flexo folder-gluer. The increasing efficiency of these basic machines is making much of the older equipment obsolete.

Die-cut cases

So far we have only discussed the conversion of board blanks from the corrugator into regular slotted cases (RSCs). Other styles of case are possible, and these are produced by platen presses that cut and score the board in a manner similar to that used in making folding boxboard cartons.

Die cutting of corrugated blanks has traditionally been done on flatbed presses, which can give great accuracy even for the most complicated design. More recently rotary die cutters using cylindrical dies have become popular because of greater productivity. Modern flatbed machines can produce several thousand die-cut sheets per hour. Rotary die cutters can process large sheets with outputs exceeding those of flatbed machines. They are used in many installations in conjunction with flexo folder-gluers. Separate rotary die cutters with flexographic printing units and complete automation are now quite common.

Manufacturer's joint

Most boxes, such as slotted containers, are folded at two score lines so that the ends of the blank are brought together to form the manufacturer's joint. In this way a single joint is formed where one end panel meets one side panel Three types of joint are possible: glued, stitched or taped (Figure 20.9).

Glued joints on slotted boxes that use a cold aqueous resin adhesive and run on high speed folder-gluers are considered to be the least expensive manufacturer's joint and are the most common.

In making a stitched joint, stitches from a continuous coil of metal wire join the flap to the adjacent panel of the box at about 6–7 cm intervals. Stitched or stapled joints are normally used for large boxes with dense loads because there is greater resistance to breaking of the joint. Stitched or stapled joints cost more than glued joints.

The glued or stitched flap is usually placed inside the box so that the outside panels are smooth for printing and ease of handling. If an inside flap causes problems with the contents, it can be placed on the outside of the box.

Taped joints are usually made with laminated, reinforced tapes. Taped joints use less board because no overlapping flap is needed. However, they are more expensive than glued joints, the main advantage being a smooth surface inside and outside.

Fitments are also produced from board for strengthening the case for

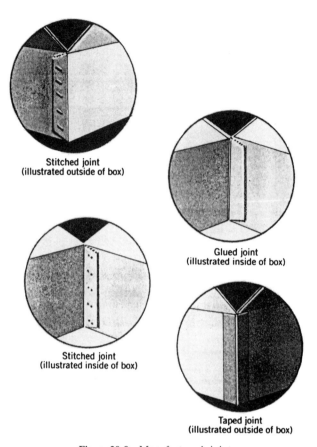

Figure 20.9   Manufacturer's joints.

supporting various articles packed inside. Figure 20.10 shows a typical flow chart from raw materials to finished products.

Partition makers

Partition slotting and assembly machines are used extensively in box plants that supply the glass bottle industry. The special slotting machines for partitions are readily adjustable so that slots of correct width are made to match flute size. For high volume runs, automatic partition assemblers are used to slot the pieces together, collapse them, and bundle a number together for despatch.

Inserts and interior packing pieces. Inserts and interior packing pieces are used to keep items in place inside a container. They may also protect goods from shock by separating them one from another.

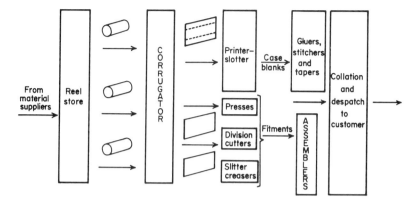

Figure 20.10   Flow chart for container production.

Trays and other folded sheets can be formed inside a box to give clearance for the product or to suspend it away from the box panels. Corner protectors used for furniture and large appliances are typical.

Die-cut fitments allow adaptation of irregular shapes to regular containers. These are especially useful for facilitating assembly-line packing methods for odd-shaped items. The ingenuity of the box plant designer in protecting a difficult-to-pack product with the least amount of material and number of pieces is often the key to success for corrugated packaging.

Manufacture of solid fibreboard cases

The manufacture of solid fibreboard containers is very similar to that of corrugated containers. The board itself is made on a machine called a *paster*. Most pasters are equipped to run as many as five plies. The board combinations are usually a kraft liner at the top (and sometimes the bottom) and one, two or three plies of chip paper for the centre fillers. The reels of board are placed on reel stands, one reel being pulled from the top, the next from the bottom, thus causing the edges of the web to push against each other to prevent curling. Modern machines have two adhesive stations. In making five-ply board, no adhesive is applied to the top liner; generally, the adhesive is applied to the top and bottom of the first ply of chip, and to the bottom of each of the second and third plies. The top and bottom liners are joined to the fillers just prior to a series of five or more squeeze rolls. Each set of rolls has a higher degree of pressure applied consecutively, in order that the adhesive is not squeezed out along the edges of the board. To keep the board even between each set of rollers, the rolls differ in diameter by approximately 0.15 mm with the larger roll first, making each set of rolls pull just a little faster than those preceding. The trimmed sheets come off the cut-off on a conveyer and are piled in loads ready for the next operation.

The caliper of the finished board is the result of the number and thickness of individual plies. It varies according to the needs of the customer. For shipping containers, board thickness varies from 1.2–2 mm.

Fibre presses perform three operations: creasing, slotting and printing. Some machines are hand-fed, others have a slitting attachment so that there can be two or more blanks cut out at the same time. The original sized sheets are fed on to a conveyor that has spacer bars to guide the blank on to the rolls of the first section of the press. Thus the horizontal scores, the tapered slots and the stitch flaps are accomplished in one operation. Rolls then pull or carry the sheet under the printing cylinders, which have either metal or rubber plates, commonly known as *dies*, which have been previously mounted on a cylinder.

The printing process is exactly the same as that of a corrugated container. So far as this type of board goes, the manufacturer's join is usually stitched.

Closures

To provide a useful transit container, a fibreboard case must have a satisfactory and effective closure. There are three main ways of providing this. The flaps of the case can be secured in place by the application of adhesive tape, by stitching or stapling, or by gluing.

The adhesive tape can be of the self-adhesive type, with a plastic film base, or based on kraft paper requiring water to activate the adhesive prior to sticking. For heavy case weights (25 kg or more) heavy-duty adhesive tapes are available based on polypropylene and polyester films or kraft paper, with reinforcing filaments. Adhesive tapes are usually applied along the line where the case flaps meet. For greater strength H-taping along the flap join and the case edge can be used. In most operations, case-sealing tapes are applied by machines which can seal cases with tape at a rate of tens of cases per minute, applying tape to top and bottom of the case in one pass.

Some closures are made to fibreboard cases by applying stitches or staples to the case flaps, thus securing the inner flaps to the outer ones. The stitches can be applied either by hand-operated guns (working on spring pressure or pneumatically) or by machinery similar to that used for stitching the manufacturer's joint. Before deciding on a case closure by stitches, it is necessary to consider the nature of the contained product, since the application of the stitches or vibration of the product against the stitches during transit can cause damage to certain products.

In some instances the closure is made by applying adhesive to the surface of the inner flaps of a fibreboard case, and pressing the outer flaps against them. This can be done either manually or by machine, and is usually the cheapest closure system. Generally, if more than about 2000 cases of a given type are dispatched weekly, it is economic to install a gluing unit.

Suppliers of fibreboard containers are always able and willing to provide information about the best methods of closure, according to the quality of case used and the contents to be packed.

Paper, plastics and fabric sealing tapes

Adhesive sealing tapes have long been accepted in packaging as a speedy, clean and efficient means of closing and sealing fibreboard cases, fibre drums, paperboard boxes and cartons, and paper-wrapped merchandise of all types. The correct selection of tape is of paramount importance, and careful consideration must be given to the specific pack on which it is to be used. Such matters as the width of tape, its strength, and its adhesive grip must be assessed with reference to the weight, bulk and value of the pack, and the transit hazards involved.

In addition, the means of application warrant careful thought on the grounds of both efficiency in sealing and economy in the use of labour. Efficient sealing necessitates a careful study of all these aspects, and consultation with the tape manufacturers in the preliminary stages is advisable.

Three main types of tape are used:

1. *Water activated:* glued or gummed tapes, paper or fabric-based, plain or reinforced.
2. *Already activated:* self-adhesive or pressure-sensitive tapes, paper, plastics or fabric-based.
3. *Heat activated:* heatfix or thermo-sensitive tapes, paper or fabric-based.

Gummed tapes. Gummed tapes will adhere satisfactorily to all packages or containers made from paper or paperboard, but are not satisfactory on metals or plastics. By virtue of their adhesive bond, they become, when properly applied on paper board, an integral part of the pack, providing considerable reinforcement to weak points. The added protection provided by mitred corners, as recommended for the traditional H taping of a fibreboard case, and the 'gaitering' of large kraft-wrapped bales are examples of this. Tapes manufactured from hard-sized kraft paper and efficiently applied, provide reasonable moisture- and water-resisting properties, adequate for normal transport hazards.

Base papers should consist of a strong paper hard-sized to resist adhesive penetration, and flexible enough to conform with the shape of the package. Where exceptional strength is required, heavy-weight duplex glass-fibre-reinforced base papers may be used. The alignment of the glass-fibre strands may be varied according to requirements to the degree that they are virtually impossible to tear or rupture. Such tapes are used for single-strip case sealing and to produce the manufacturer's join on many fibreboard containers. Fabric and cloth glued tapes are available for specialized uses, mostly where reinforcement is required.

Adhesive coatings are mainly non-protein, based on modified starches, on which much progress has been made and which have now largely superseded animal glue.

Application and uses. Most gummed tapes are obtainable in different widths ranging from 24 mm up to 288 mm. They can be printed in several colours with a trade mark, slogan or other information relative to the contents of the package. Colour codes which can be used immediately to identify a batch or quality are also available.

The efficient application of gummed tape depends on correct moistening, either too much or too little moisture being detrimental to ultimate adhesion. Modern dispensing machines for gummed tape employ automatic moistening devices and will cut pre-set lengths of tape, correctly moistened for immediate application. Electrically operated and remote controls enable both hands of the packer to be free for the application of tape. Pre-determined lengths of tape are used when rectangular-shaped packages are being sealed by the H method and, for large-scale production, fully and semi-automatic case taping machines are available, both for single-strip or H-method applications (Figure 20.11).

Storage. Gummed tapes should be kept in the original moisture resistant wrappers until required for use, and should be stored in a dry place. Storage in cold damp conditions, or on concrete floors in unheated store rooms, should be avoided.

Pressure-sensitive tapes. Pressure-sensitive tapes will adhere satisfactorily to packages or containers made from paper, paperboard, metal, glass or plastics. Such tapes need only the application of pressure to cause them to

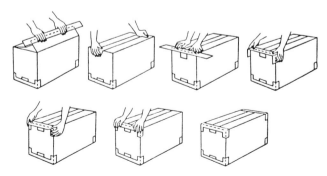

Figure 20.11   Method of applying gum strip to fibreboard cases.

adhere to almost any surface. Apart from seeing that the surface is free from dust, oil, grease or moisture, no special preparation before application is necessary.

Application and uses. The major use of pressure sensitive tape in packaging is the closure of regular slotted containers. Figure 20.12 depicts a typical construction of a box-sealing tape; a plastic film is coated on one side with pressure-sensitive adhesive. The film may have a release treatment on the backing to allow easy removal of the tape from the roll during dispensing.

The choice of tape is very important and affects the performance of the entire package during storage and distribution. The most common backing films are biaxially oriented polypropylene or polyester although some unplasticized PVC film is also used.

A second broad category of tape is pressure-sensitive filament tape, sometimes known as 'strapping tape'. It consists of a film backing (usually

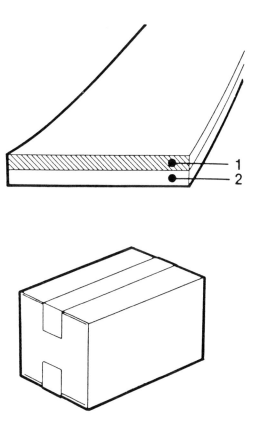

Figure 20.12 A typical box-sealing tape consists of a backing film (1) and a layer of pressure-sensitive adhesive (2). It is used most often as a closure for regular slotted containers.

polyester) with reinforcing filaments embedded in the pressure-sensitive adhesive (Figure 20.13). The most common reinforcement is fibreglass, which provides a high tensile strength with very little stretch. Some tapes have polyester or rayon filaments.

Filament tape is used in many packaging applications including general purpose, bundling and pallet unitization. If two packages are being sent to the same destination, the shipping costs can often be reduced by bundling them together with filament tape for the shipment.

Availability. The greater proportion of all self-adhesive tapes can be printed and are obtainable in several colours. Most tapes are supplied in several widths up to about 100 mm and, as with gummed tape, there are many dispensing machines available, ranging from small hand applicators to fully automatic case-sealing machines.

Storage. Bales of self-adhesive tape should be stored in their original wrappings until required for use, and should be protected from dust, solvent fumes and direct sunlight. High temperatures and humid conditions will also accelerate tape deterioration.

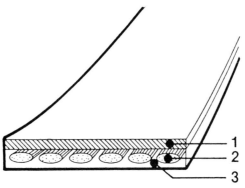

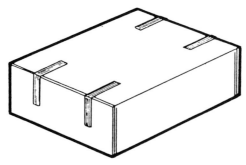

Figure 20.13   A filament tape has a backing film (1) and filaments (2) embedded in a pressure-sensitive adhesive (3). One use is closure of full overlap boxes.

Heatfix tapes. Heatfix or thermo-sensitive tapes will adhere satisfactorily to all packages or containers made from paper or paperboard, glass and a wide range of plastics, provided the surface is clean and dry. Such tapes, once activated by heat, need only the application of pressure to cause them to ahdere.

Materials. Heatfix tapes are in the main coated on to a kraft-base paper. The adhesive coating is normally formulated from a thermoplastic emulsion which, once activated by heat, becomes immediately self-adhesive and retains this property for several hours, before finally 'setting off' to give a dry permanent bond. This particular type of adhesive is referred to as a *delayed-action heatfix*. Other types of adhesive formulations are described as *immediate action* and require the tape to be activated by heat and applied to the packaging surface in situ.

Application and uses. Thermo-sensitive tapes are obtainable in several different widths and can be printed in colour. The equipment to apply heatfix tape correctly is available both in the form of a bench dispenser, where an operative can dispense a required length of tape by hand, as well as both semi- and fully-automatic case taping machines. The latter are designed to apply automatically a single strip of tape centrally to both the top and bottom of a carton or case.

Storage. Thermo-sensitive tapes should be kept in their original wrappers until required for use. They should be stored in a dry place and be protected from solvent fumes and direct sunlight. High temperatures must be avoided, as they could result in tape 'blocking' if present for long periods.

Costs. The correct selection of tape can be determined only by consideration of the specific packaging application. Pressure-sensitive tapes are more expensive than gummed tapes but, since in packaging the overall cost of sealing rather than the cost of the material alone is important, it is necessary to take into account such factors as labour, productivity, and the relative lengths and widths of the possible tapes employed.

Heat-activated tapes offer many of the advantages of self-adhesive tapes and are priced midway between gummed and pressure-sensitive tapes.

Generally a detailed study of the problem will highlight an obvious choice. The manufacturers of sealing tapes are always ready to give advice, demonstrations and to cooperate in field tests in order that the efficiency of any tape to fulfil its function may be confirmed.

Stapling

In recent years the King Size staple has gained favour as a closing method for corrugated fibreboard cases. These large preformed staples are clinched on a rectractable anvil, thus enabling the case to be closed after it has been packed.

The main advantages are:

1. *Economy and speed of closure:* the average case requires only four staples in the top, and the same number in the bottom, and can be closed in as little as five seconds.
2. *Security:* it is almost impossible to pilfer a case closed in this way without detection, and the staples give an extremely strong closure which is not affected by atmospheric conditions.
3. *Efficiency:* closure can be made in the production area, and there is no problem with moving or storing pre-assembled cases. No drying time is involved as with some glues and paper tapes, and the case is thus ready for dispatch as soon as it is sealed. The strength of the closure readily lends itself to stacking, with little fear of the case flaps opening.

A wide range of equipment is available, including portable machines which can be either hand or air-operated. There are also production-line machines which will staple top and bottom, or either, or both ends of a case simultaneously, and there is also a completely automatic model available. The machines are adjustable, so that it is possible to control the depth to which the anvil hooks penetrate the board, and also the tightness with which the staple is clinched. There are several leg lengths of staple available, the length used being governed by the depth of fluting in the board, and also the case design. The most common is a 15 mm leg which would, for example, be used on a centre seam of A-flute corrugated cases.

This type of closure does, of course, have limitations. In the first place it is not possible to drive the staples unless the case flaps are supported, i.e. unless the goods are packed very close to the top of the case, particularly with the B flute board. It is sometimes necessary to use a top pad of corrugated board to prevent damage to the contents.

King Size staples are most commonly used on corrugated fibreboard cases but, in ideal conditions of internal packing and with the use of an air-powered machine to drive and clinch the staple satisfactorily, they may be used with solid fibreboard cases.

The hazards of transport and the assessment of fibreboard packages

The four main hazards of transport (drops and impacts, vibration, vibration under load and compressive forces in stacks) apply almost universally in all forms of transport. The hazards that cause most damage to corrugated cases are drops and compressive forces. It is not intended to deal with the subject of package performance, but it should be borne in mind that, for the best use of fibreboard cases, a careful study of the hazards occurring in the particular distribution system concerned is necessary, so that the laboratory transport test correlates fairly accurately with the actual conditions of transport.

Essentially, four correlations need to be known before a complete understanding of what is required can be obtained. The first correlation is between the actual distribution system and package evaluation tests made in the laboratory on a filled container. A second correlation between the laboratory transport test on the filled container and simpler tests on cases without the merchandise is then required for quality control purposes. The test on the emply containers (or tests, if more than one is required) must then be correlated with tests on the materials used for making the container.

Having ascertained those properties of board that contribute towards the strength of the case, a fourth and final correlation between these properties and the variables of board-making that control them would provide a complete understanding of what is required between the board-making operation and the performance of the package in the distribution system. These factors are not always appreciated.

The first correlation between actual journey hazards and laboratory transport trials is reasonably well understood, and the performance of a package which has been tested in the laboratory can be predicted fairly accurately. The second correlation between the laboratory journey and some simpler test or tests on an empty container is by no means so well understood. For fibreboard cases it would probably be generally agreed that some form of drop test with a compression test is necessary. There is no simple relation here, however, since the conditions that cases meet in transport vary considerably. Nevertheless, the importance of the compression test in evaluating corrugated fibreboard cases is universally recognized.

Workers in this field have largely concerned themselves with relating the compression strength of a sleeve of rectangular cross-section:

1. The compression strength of a case
2. The 'stiffness' of the board

The difference between the sleeve and a case is provided by the flap-forming operation at the top and bottom, for the bending scores lower the stiffness of the board in this area, allowing it to roll and bend under compression. Thus, the load that the case can sustain is less that that of the corresponding sleeve by an amount dependent on the quality of the materials used in making it and on the efficiency of the score-forming operations.

Uses of corrugated and solid board

Corrugated fibreboard packaging is widely used for the carriage of all types of merchandise (Table 20.3). Foodstuffs and household goods generally are probably the largest field of use, but considerable quantities are absorbed by such diverse goods as furniture, domestic appliances, chemicals, nails in bulk, light engineering components, etc. It is probably true to say that, apart

Table 20.3 Typical uses of fibreboard packaging

Industry	% of corrugated usage
Non-durable goods	
Food products	37
Textile mill products	2
Apparel and fabrics	2
Paper products	13
Printing and publishing	2
Chemicals	7
Petroleum products	1
Rubber, plastic and leather products	7
Subtotal	71
Durable goods	
Furniture	3
Stone, clay and glass products	8
Metal products	4
Non-electrical machinery	2
Electrical machinery	5
Instruments, clocks, etc.	1
Miscellaneous manufacturing	6
Subtotal	29
Total	100

from heavy machinery and the like, corrugated fibreboard, is a suitable packaging medium for almost any product. The most common packages, however, lie within the weight range 2–50 kg. Special types of fitting and reinforcement extend this range to about 250 kg for normal goods; and for the bulk packaging of powdered commodities, special packages in corrugated board will carry up to 1 tonne of material. Limitations on the use of fibreboard may be imposed by its resistance to climatic conditions. Weatherproof boards of various types are made, but the degree of weather resistance varies from type to type, and also increases the cost.

Solid fibreboard containers are used almost exclusively for applications in which container return and re-use are desirable and where return can be controlled by the distributor. Without such control, the impetus to use such multi-trip shipping containers, which are more costly than corrugated boxes, would not exist. Solid fibreboard boxes cost two to three times as much as a similar size corrugated case. However, the solid fibre board container can be used 10–15 times before it is discarded. The economics are obvious, but only in a 'closed' distribution system.

Generally speaking, fitments to secure the goods within the outer case are constructed from corrugated board. In a number of instance other materials, such as timber, hardboard, insulating boards and various cushioning materials, are employed. As usual, the final cost of the package frequently determines whether or not the particular material used will be successful.

Ever since the early part of the century, when paperboard packing cases began to be used instead of the more traditional timber, wickerwork or steel containers, manufacturers have been comparing the advantages and disadvantages of solid board as against corrugated board. It will generally be agreed that, from a technical point of view, either solid or corrugated fibreboard can be used to give a satisfactory package for almost any type of commodity, and normally the deciding factor as to which is preferable is the relative cost. In certain instances, however, there may be an advantage in using one material rather than another, and the following considerations which can apply in specific circumstances are worth noting (see Table 20.4).

Table 20.4   Comparison of solid and corrugated packagings.

1. Solid fibreboard is more resistant to puncturing, both from external hazards encountered in transport, and also when the article packed possesses sharp projections. Very irregularly shaped objects, such as engineering components, steel bars, nuts and bolts, are more likely to puncture a corrugated case than a solid one.

2. Solid fibreboard is more resistant to wet or damp conditions than corrugated fibreboard, although the water resistance of both materials can be improved by various technical means. In particular, wax impregnation and coating with polyethylene are used to improve the weather resistance of board.

3. Although both types of case are principally used as single-journey containers, there are a number of instances where several journeys are required from one container. This type of usage occurs principally where the manufacture of the product has a closed distribution system in which the chances of loss of an empty container are relatively small. In this type of distribution system the case is delivered with its contents to the retailer, and empty cases from the previous delivery collected at the same time. Solid fibreboard generally gives better results for two main reasons. Firstly, it last longer, since the board cannot become crushed and, secondly, slight damage to the container is more easily repaired.

4. When the density of the product is high, solid fibreboard frequently gives better results, because under these conditions corrugated board is liable to become crushed and to lose its strength.

5. Where the contents of the package do not fill the case completely, corrugated fibreboard may have some advantage because of its greater rigidity. This greater rigidity is also of advantage in respect of the resistance of the container to the crushing forces produced by stacking, and therefore such contents as cartoned goods are preferably packed in corrugated board.

6. When the product to be packed is relatively light in weight, e.g. such commodities as breakfast cereals or soap flakes in cartons, the corrugated fibreboard case is almost invariably cheaper than the lightest-weight solid fibreboard case producible. Efficient protection of the contents is achieved with a very light corrugated board because of the great rigidity of this material.

7. In certain instances, a combination of solid and corrugated is a better solution than either alone. Corrugated cases with solid board divisions, and solid cases with corrugated surrounds, are examples.

   In modern industrial packaging applications, corrugated cases are often used with plastics liners for liquids. There are also systems using fibreboard trays and shrink wraps, e.g. in the packaging of glass bottles and jars it is possible to eliminate dividers by holding them tightly together in a tray with a shrink wrap.

Case styles

An international fibreboard case code has been agreed between the various fibreboard case-making organizations of many countries. The code is known as the FEFCO/ASSCO International Fibreboard Case Code and is widely used as part of a case specification, and for description and ordering purposes.

The regular slotted container (RSC) is the most common corrugated container. It is economical to produce and makes efficient use of board. The design has four flaps of equal width, at the top and bottom, and thus is made from a rectangular blank. The outer flaps meet; the inner flaps do not. As a result, parts of the top and bottom of the box have only a single thickness of corrugated board (see Figure 20.14, 0201, 0205).

The RSC is used for products where economy is important. Where more protection or special features are needed, there are several variations on this basic design, all of which require more board for the same size box (Figure 20.14, 0203, 0206, 0208).

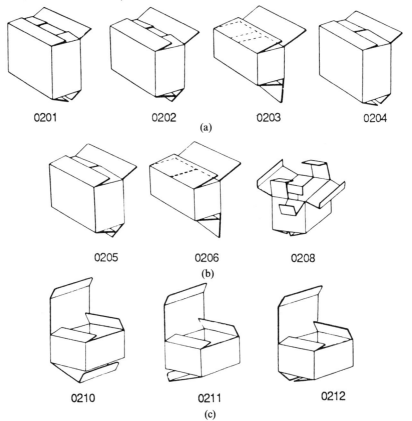

Figure 20.14 The international case code for slotted boxes, 0200 code.

Slotted and scored blanks can be formed into trays or boxes by the packer with gluing or stitching machines. They differ from half-slotted boxes and covers in that the tops and bottoms have only a single thickness of board and the side walls are at least partially reinforced with the overlapping flanges (Figure 20.15). This is a more efficient use of board for good stacking strength.

The half-slotted box (HSC) is an RSC with only one set of flaps plus a cover. The cover may be a telescoping half-slotted style or a tray as illustrated in Figure 20.16.

A half-slotted inverted cover may come down only part way to allow for settling of the contents, such as textile or paper products. It may also be a full telescopic cover to give extra wall strength. Fresh fruits and vegetables are often shipped in this type of box.

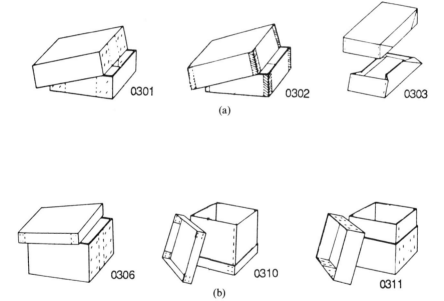

Figure 20.15   Trays and telescopic two-piece boxes.

Figure 20.16   Half-slotted boxes.

One-, two-, and three-piece folders are used to ship books, apparel, and other products via express or parcel post. All are scored sheets with tucks of specified lengths. All are easy to store and set up by taping or with staples (Figure 20.17).

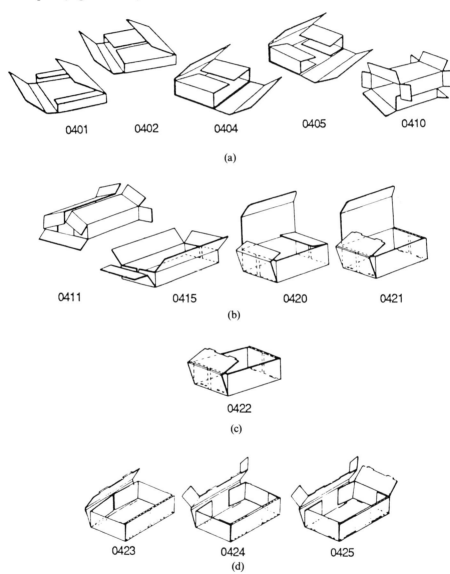

Figure 20.17   Folders, styles 0400 code.

Wraparound blanks are pre-scored and slotted for use in automatic gluing equipment. They provide a tight pack for canned goods preventing product damage while using a minimum amount of board. The blank is automatically wrapped around the collated product which has been placed on the bottom panel. The machine glues both the joint flap and the top and bottom flaps to close the case (see Figure 20.18).

Two designs for special applications are the Bliss box and the recessed-end box. The Bliss box (Figure 20.19, 0601, etc. and Figure 20.20) has good stacking strength because all four corners are reinforced. It also uses less board than most slotted boxes because the bottom is of single thickness and there is less flap area. However, it is a three-piece die-cut box which requires special set-up equipment at the point of use. Machines using hot melt adhesive are available to set up Bliss boxes. These boxes are used for bulk packs of meat, explosives, fresh fruits and vegetables, and articles of concentrated weight.

A recessed-end box is made by stitching a scored body sheet to two flanged endpieces (code 0615). It is used to ship long, fragile items (fluorescent bulbs) or items of the same width or girth but varying lengths. The body sheets can easily be cut to fit.

There are also fibreboard codes for pads, surrounds and dividers (Figure 20.21) and platforms, pads and corner pieces (Figure 20.22).

Industry trends

Automation

For many years most users of corrugated boxes set up, filled and closed boxes by hand, but automatic case openers, palletizers, and machinery associated with filling boxes are comparatively recent. Today, many box users have automated packaging lines and most users have at least a case sealer.

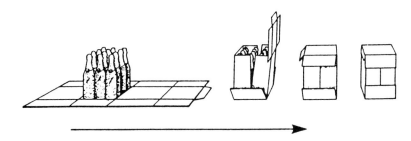

Figure 20.18 Wraparound blank.

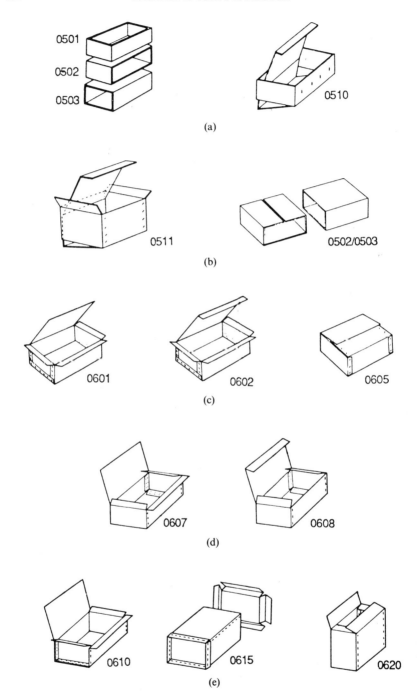

Figure 20.19 Fibreboard case code, styles 0500 and 0600 codes.

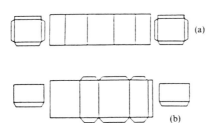

Figure 20.20 (a) Blank for No. 2 Bliss box (code 0605). (b) Blank for No. 4 Bliss box (code 0601).

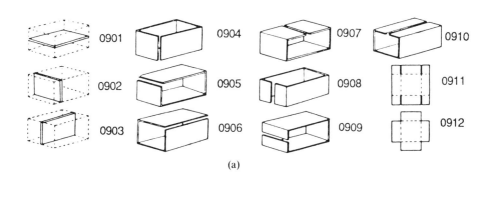

(a)

(b)   (c)

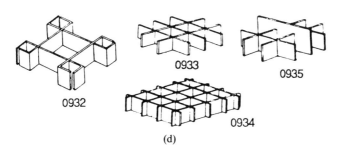

(d)

Figure 20.21 Fibreboard case code—fitments (pads, surrounds and dividers).

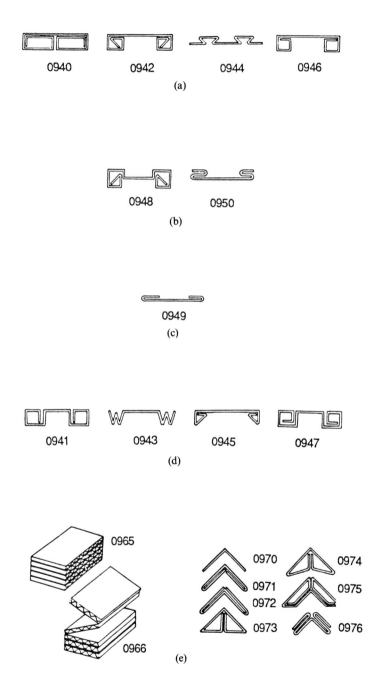

Figure 20.22   Fibreboard case code—fitments (platforms, pads and corner pieces).

Automatic machinery for handling corrugated boxes is not as forgiving of box defects as manual handling. Warping, dimensional variations, bad joints and weak bonds are not tolerated by box-handling machines. Significant improvements in quality control and product uniformity have resulted.

Retail packages

The use of corrugated boxes as retail packages for some consumer products is growing. The graphics become more important to attract customers. Pre-printed litho labels and preprinted linerboard provide higher quality printing. Litho labels are glued to one or more box panels on special machines. Pre-printed linerboard produced on large flexographic or gravure presses requires accurate registration with the box dimensions.

Competition

Plastics pose the most significant competition to corrugated market growth. Plastic film wrapped tray packs for canned goods are a good example. A corrugated tray containing cans overwrapped with plastic film has replaced a full corrugated case for many canned foods and beverages. The loss to corrugated has been estimated at less than 1%, but it is a proven alternative for certain applications.

Case quality and quality control

The performance of the fibreboard case pack is dependent on the quality of the materials and the efficiency of processing at three stages in production.

1. Before processing—raw materials: liners, corrugating medium, and adhesives
2. After combining into board
3. After case-making is finished

The sheet materials used for making both solid and corrugated board are usually bought on a description, basis weight and caliper (thickness) specification. The tests normally carried out are (in addition to the basis weight and caliper) for bursting strength and ring stiffness. Some method of assessing behaviour in respect of absorption of water is also used. The Cobb test, the water drop test, and other sizing tests are common for this last property.

Fluting medium is also assessed by the Concora fluting medium test (CMT). It consists essentially of taking a piece of the fluting medium and passing it through heated corrugating rolls at a rate of about 3 m/min to produce a fluted sheet. This sheet is then affixed to a backing medium, usually adhesive tape, and is then subjected to a flat crush test.

The combined board is usually assessed in terms of its basis weight, caliper and Mullen burst strength, its materials of construction and the adhesive used to bond its components.

Corrugated cases are often specified in terms of the style, internal dimensions, and board grade and quality. However, any experienced fibreboard case user will inspect closely a sample of the cases coming in to his plant to ensure that the print quality is good (class, correct colour, in register), that the case is squarely and cleanly cut and assembled, that there are no splits in the inner liners, and that the manufacturer's joint has been properly made.

If the right materials have been specified and good-quality processing performed, both corrugated and solid fibreboard cases offer economic and effective transit containers for the storage and transport of a wide range of products.

Packaging applications of hardboard and insulating board

Since an appreciation of the characteristics of fibre building boards as packaging materials is comparatively recent, it is clear that their application in this field will still be liable to development. Currently, packaging applications fall into a number of well-defined areas.

Case and pack reinforcement. The strength and rigidity of hardboard can be easily employed to provide reinforcement to conventional styles of case. Unless the contents are themselves sufficiently rigid (canned goods, for example) the case may require considerable strength to withstand stacking in the warehouse and during transit. Hardboard or softboard are cost-effective materials for reinforcing corrugated cases, both to protect the products and improve stack stability. Reinforcement can be in the form of side panels or dividers. Tests and service experience show that these can provide considerable improvements in compression failing load (CFL) and enable relatively lightweight cases to be used.

The details given in Table 20.5 show the degree of improvement and the relative costs. Table 20.6 demonstrates that the high CFL can be maintained under wet conditions.

It is essential that the reinforcement is designed to suit the pack, the stacking pattern used, the type of pallet employed, and the journey conditions. The panels must be positively located so that they cannot be easily displaced. In some instances, the contents of the pack will locate and hold the panel, and this method is particularly suitable in boxes containing a small number of bulky articles. An example is a case holding four 5-litre polyethylene bottles (Figure 20.23) where rigid hardboard panels located between the bottles provide the reinforcement. A single dividing piece of 3.2 mm hardboard can raise the Compression Failing Load of a 350 mm cube-

Table 20.5 Effect on compression failing load (CFL) of corrugated fibreboard case by addition of reinforcing pieces made of hardboard (internal case dimensions 460×280×380 mm).

	CFL, at 23°C and 50% relative humidity		Estimated cost 50 kg per CFL
Case make-up	kN	lb	(pence)
Corrugated fibreboard only			
1. 'B' flute 125 gsm kraft facings	1.8	400	1.5
2. 'B' flute 400 gsm kraft facings	4.5	1000	1.2
3. 'BC' flute	5.8	1300	0.9
4. 'AA' flute 400 gsm kraft facings	17.8	4000	0.9
Corrugated fibreboard reinforced with hardboard			
5. Case of Type 1 above with two end panels of 3.2 mm standard hardboard	9.3	2100	0.4
6. Case of Type 1 above with end and side panels in 3.2 mm standard hardboard	15.6	3500	0.4

Table 20.6 Effect of temperature and humidity on compression failing load of corrugated fibreboard case reinforced with 4.8 mm thick hardboard pieces in 'H' configuration (internal case dimensions 520×250×275 mm).

Case style	CFL	
1. Corrugated fibreboard case with corrugated board dividers	9.7 kN (2200 lb)	—at 23°C, 50% relative humidity
2. Corrugated fibreboard case with 4.8 mm thick hardboard dividers	20.3 kN (4600 lb)	—at 23°C, 50% relative humidity
3. Corrugated fibreboard case with 4.8 mm thick hardboard dividers after conditioning for 7 days	20.3 kN (4500 lb)	—at 38°C, 90% relative humidity

shaped corrugated box from 500 to 1000 lb. Where such location by the contents is not possible, spot gluing, taping or stapling the hardboard to the case may be required. Alternatively, it may be located by sandwiching it between the case flaps. A further development envisages the production of a case made from both corrugated board and hardboard. This would have four faces of corrugated and two faces of hardboard (in a style similar to the 'Ratio-Pak' system shown in Figure 20.24).

Softboard (insulating board) may also be used for reinforcement, and in some situations may be preferable. It provides a cushioning effect. Moreover, should failure of the pack occur, the edges of the board deform rather than the panel fracturing. Consequently, even at 'failure', a panel will still substantially support the load, reducing the risk of complete collapse of the stack. On the other hand, softboard has a lower resistance to moisture than hardboard: a factor which may be important in high humidity situations.

Figure 20.23 Hardboard and insulating board used as compression pieces in corrugated cardboard packing cases. Such compression pieces improve the stacking performance of the outer cases, and are being used for this purpose in the packaging of chemicals in plastic bottles.

Figure 20.24 'Ratio-Pak' style of box with hardboard end panels to provide good stacking performance.

Reinforcement of this type is not restricted to corrugated cases alone. For instance, a shrink-wrap pack containing four plastic bottles can be given added strength by the incorporation of a cruciform hardboard separator. In these circumstances a load-spreading device or collating tray is usually incorporated, and this may also be of hardboard.

As well as their use to provide vertical support, hardboard and softboard have a number of uses in horizontal reinforcement. A typical example is in cases containing heavy equipment, such as addressing machines or typewriters, where point loadings could damage the pack. A simple base plate of hardboard or softboard effectively distributes such point loads. With hardboard, locating and fastening holes enable the base plate to be positively fixed to the equipment packed, whilst softboard also provides cushioning.

Fruit and vegetable trays. The fruit and vegetable trade has for many years used wooden trays. Essentially, the trays require a high degree of moisture resistance, should allow free air circulation to the contents and have an ability to absorb any moisture released by the fruit and vegetables. In recent years, the thin timber laths have been replaced by hardboard pieces. The availability of larger size pieces of hardboard improves the rate of assembly of the trays, and at the same time, the hardboard pieces are free from splits, knots and splinters. Hardboard trays are made either by nailing hardboard to solid timber end pieces or, if hardboard is used for all side panels and the base, by using a corner stapling technique. In the latter instance, timber corner posts are usually included to provide the required stacking pattern (Figures 20.25–20.27).

When changing from one packaging material to another, it is often advantageous to start with the pack requirements and examine how they can best be met with the new materials. Commonly, hardboard has been used to replace timber in trays fabricated in the traditional manner. Whilst this has been successful, it is possible that developments in forming and jointing techniques could show improvements in convenience and cost.

Hardboard cases. A number of styles of case have been developed based on hardboard or medium density board panels jointed in various ways. Wood-framed hardboard panels are commonly used in the construction of large cases. The hardboard is nailed to a timber framework and six composite panels are joined together to form a case, often incorporating a pallet style base. The fastening together of the panels can be effected by nailing, banding, the use of spring tensioned corner clips or profiling/machining the panel edges (Figure 20.28). Smooth faces can be internal (as in Figure 20.39) or external.

As mentioned in a previous section, mechanical 'through fastenings' may be replaced by surface fastenings, i.e. solvent-based or pressure sensitive

Figure 20.25   Apple box made from hardboard and triangular timber corner posts.

Figure 20.26   Dutch trays with hardboard sides and bases wire-stiched to timber corner posts to give the required stacking pattern.

Figure 20.27 Wire stitcher fixing hardboard end and side pieces of produce trays to the timber corner post. Finished frames (stack at left) are then similarly stitched to a hardboard base.

Figure 20.28 Flush sided cases showing profiled or castellated edges which provide accurate location. Courtesy of Lamibox.

adhesives. For large cases, this technique facilitates the laminating of several pieces of hardboard at strategic points to enhance the rigidity of the panel.

If a surface fastening is used in conjunction with paper, textile or plastic tapes, or with metal angle strips, for the jointing system, much reliance is place on the tensile strength perpendicular to the board surface (Z-strength) of the hardboard.

Certainly, for smaller cases, this system is a possible alternative to the common technique of riveting or stapling to a metal strip 'tea chest style' where lines of weakness can be produced by perforating the hardboard (Figure 20.29).

Satisfactory transit performance depends on both the strength of the jointing material and on suitable methods of enabling shock to be distributed between adjacent panels. A style of pack which particularly meets the latter requirement has been evolved as the result of detailed study and practical testing. This 'PIRA-style' fold-flat case (Figure 20.30) consists of six hardboard panels joined with flexible tapes. The edge of each panel is bevelled at 45°, enabling the production of mitred joints which both provide a self-locating feature during erection and also perform the function of distributing shocks received by one panel to the adjacent sides. This style has been found to be especially suitable for long thin packs (Figure 20.31) and for other applications where pack rigidity is particularly required.

Figure 20.29 Export pack for machine tool parts (50 kg load) consists of hardboard sides and back joined with riveted tinplate strips 'tea-chest' fashion.

Figure 20.30 This fold-flat case, developed by FIDOR and PIRA, consists of six pieces of hardboard with bevelled edges joined by adhesive tapes.

Hardboards in 3.2 and 4.8 mm thicknesses are suitable for many cases of this type. As the dimensions increase, however, the use of medium board (panelboard) in 9–12 mm thickness becomes more appropriate.

Taped cases with medium board panels and the mitred corner techniques are extremely robust; the ultimate strength is usually related to the transit stresses that the tape can withstand. Furthermore, this type of box provides very good water resistance, the homogeneous sheet material and the completely sealed taped edges giving an unusually high degree of protection against water penetration to the goods being transported.

Rigid boxes. Small rigid boxes made from hardboard panels nailed or stapled to timber end pieces are used in the fish (Figure 20.32) and other trades. The V-grooving technique is used to produce a complete box from a single piece of hardboard; plain hardboard may be used, or flexible PVC laminate, or tape may be added at the fold lines to provide more robust hinges during assembly (Figure 20.33).

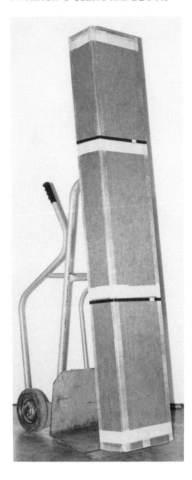

Figure 20.31 PIRA-style case with long panels, each cut with bevelled edges and joined by tapes to produce a long thin package for rotary printing screens.

Drums. Hardboard is also used for the construction of drums. A simple bend without the use of heat or moisture is adequate for the larger drums, where a small radius of curvature is not required. In the most basic situation, the fastening mechanisms tend to be very simple, with overlapped and metal-stitched side seams. The bottom is held in place by a simple metal chimb, also stitched into place. Such a drum is used for the transport of natural asphalt, where the hot liquid is poured into the drum and then allowed to set solid. It is not uncommon for the contents of this type of drum to weigh more than 200 kg, but the thixotropic bitumen and other materials require leak-free containers, consequently sealed joints are desirable.

One very large drum style container with a captive pallet, the Pallbin, is used for the packaging of plastic granules and other free flow powders

Figure 20.32 Fish cake boxes—timber sides with hardboard top and bottom panels.

Figure 20.33 Multi-trip transit and storage pack. The sides and base of the box are hardboard, V-grooved to form corners and channelled to accommodate hardboard side pieces.

Figure 20.34   Pallbin liquid container. 1000-litre drum style container with integral pallet and bottom discharge valve.

(Figure 20.34). The rigidity and compression resistance of the hardboard, even at high humidity levels, facilitates 'two high' stacking of containers each containing one tonne loads. A version of the container with a double wall of hardboard is used for the transportation of low viscosity liquids.

Cable reels. Cable reels are regularly constructed from hardboard, both the flange and the core being produced in this material (Figure 20.35). To obtain

Figure 20.35   Cable drum with flanges made from hardboard.

the shape of the flange it is usual to die-stamp the hardboard, this process also facilitating the inclusion of the necessary perforations to accept the locating lugs of the core. The thickness of the flanges can be as much as 8 mm. Because of the small radius of curvature of the core pieces, soaking the board in water is necessary before bending is carried out.

Pallets. Sheet materials such as hardboard and medium board have a number of applications in pallet construction. These include the maintenance of true 'squareness' where this is required, e.g. in automatic palletizing and depalletizing machines, and certain types of pallet racking where out-of-square pallets will jam. Current commercial styles of pallet include:

1. Conventional timber pallets incorporating a sheet of hardboard either above or beneath the top deck boards. The function of the hardboard may be to provide squareness, a continuous surface, or to repair a damaged pallet.
2. Top and bottom decks of hardboard fastened to timber blocks (4-way entry) or bearers (2-way entry) giving a fully reversible pallet as in Figure 20.36.
3. Top deck of hardboard with plastic feet. 3.2 mm hardboard is used when continuous or secured loads are involved, and the pallet is often expendable. Thicker medium boards are used where the loading is more severe, or the pallets are returnable.

Development work is in progress on a number of other styles. These include 'sandwich' top decks (similar in construction to flush doors with a

Figure 20.36 Four-way entry, fully reversible pallet.

core between two sheets of hardboard). These are extremely stiff and, when tempered hardboard is used, can be highly weather-resistant. Moulding and forming techniques may also have application in pallet construction.

Palletized boxes. Fold-flat palletized boxes using the techniques discussed earlier can be produced. An example of such a box capable of carrying approximately 0.5 tonne is shown in Figure 20.37. The pallet itself can be made from timber or hardboard as appropriate. The side panels and top of the box are hardboard with bevelled edges and flexible joints. However, for rigidity, the bottom edges of the side panels can have square edge boards and rigid through fastenings. Rigid containers on palletized bases are illustrated in Figures 20.38 and 20.39.

Processed and composite products. Being rigid products, softboard and hardboard materials are usually used in situations requiring flat panels or large radii of curvature. However, it is possible to process the boards to make them flexible and give them properties not usually associated with hardboard or softboard.

In the case of softboard, impregnation with bitumen (giving the board water resistance durability) following the machining of paralleled grooves

Figure 20.37   Palletized box made from hardboard panels with bevelled edges and a heavy-duty tape undergoing testing in the laboratory.

Figure 20.38   Rigid box made from 9-mm thick medium-density hardboard fastened mechanically to a timber base.

Figure 20.39   Timber frame and hardboard panels form a 1-ton leakproof case into which material in liquid form can be poured. The container is used for liquids which solidify before distribution.

at frequent and regular intervals, produces a flexible protective wrap with some cushioning properties.

One commercially available product, Lamiflex, is produced by cutting hardboard (either 3.2 mm or 4.5 mm thickness) into 35 mm wide strips which are then sandwiched between speciality papers. Small gaps between the rigid hardboard strips allow the product to be sufficiently flexible to bend through 180°. The surface papers, in addition to acting as the hinge, can incorporate rust inhibitors, moisture barriers and tear resistant fabrics.

Such products are used extensively for the protection of power and telecommunication cables; tubes, rods and bars; and coils of aluminium and steel. The low compressibility of the Lamiflex material makes it particularly suitable as wrap-around protection for heavy industrial items (Figures 20.40 and 20.41). On heavy items it has been usual to use the hardboard-based wrap in conjunction with steel strapping. However, a recent technique of combining the Lamiflex with stretch film has produced a pack demonstrating high moisture resistance, surface and edge protection and high degree of presentation.

Miscellaneous uses. Packaging uses of fibre building boards are varied. Discs of insulating board are used as end protecting pieces in paper reels. Reeded hardboard is used as an overwrap for coils of stainless steel. Standard hardboard is used for protective side pieces in loads of sacks and bales. Strips

Figure 20.40   Power cable drums wrapped (lagged) with Lamiflex. The outer cover of the Lamiflex is UV and moisture resistant. Courtesy of Scanpac Ltd.

Figure 20.41 Aluminium coil wrapped with Lamiflex and stretch film using an automatic roll wrapper. Note the way the Lamiflex is folded over the edge of the coil by the tension of the stretch film during wrapping. The Lamiflex provides both surface and edge protection. Courtesy of Scanpac Ltd.

of softboard have been used for many years in the packaging of plate glass. Laboratory studies of the cushioning performance of softboard indicate that its effect would be similar to that of bonded cork, and that it would be most suitable for the cushioning of dense and heavy items, as in the postal pack of a house name plate made in slate.

PLASTIC CRATES

Plastic crates can be divided into three categories, and these three categories depend upon the type of product to be packed rather than the method of manufacture of the plastic crate. The type of product to be packed, and therefore the usage which the crate will have to withstand, determines both the style of the crate and the material from which it is manufactured. The main advantage of plastic crates over wooden and metal crates is that plastic crates are easily washable and therefore hygienic. There are no maintenance costs with plastic crates.

The three categories of product for plastic crates are milk, beer and soft drinks, and agricultural products and foodstuffs. Historically, crates for milk were originally of metal, while crates for beer, soft drinks, agricultural products and foodstuffs were originally made of wood.

Various methods of production have been developed for plastic crates, including rotational-moulding, blow-moulding, injection-moulding and thermoforming. However, although some half-depth cases have been produced by extrusion blow-moulding, virtually all crates are now produced by injection-moulding.

The thermoforming (or vacuum forming) process consists of four stages:

1. Production of a sheet of plastic material
2. Heating of the plastic sheet so that it is soft and mouldable
3. Forming of the softened sheet in and/or around moulds by vacuum and/or air pressure, sometimes together with movement of the moulds
4. Cooling and removal of the moulding

Materials

The material used in the production of plastic crates is determined by the specific requirements of each market. We will consider each of the three markets in turn.

Plastics crates for milk

Plastics crates for the milk market must be able to withstand caustic and detergent washes at 80°C, and must also have good impact strength, particularly at low temperatures (this is necessary as milk is frequently delivered at sub-zero temperatures). Hence high-density polyethylene is used for the production of milk crates, as this is the only material which meets the requirements.

Plastics crates for beer and soft drinks

Beer crates are stacked higher and for longer periods than milk crates, and hence a more rigid crate is required. In addition the sub-zero impact strength of beer crates is not as critical as that of milk crates. Polypropylene copolymers have been found to have the necessary balance of properties, and are used for this type of crate.

Plastic crates for agricultural products and foodstuffs

The crate market for agricultural produce and foodstuffs is extremely diverse and the usage very variable. For example, there are delicate lightweight high-cost products, such as flowers and hot-house salad produce, and weighty products such as meat, sausages and fish. Plastic crates for products such as flowers are handled extremely carefully, but plastic crates for fish receive considerable abuse and are also subjected to sub-zero temperatures.

The materials used and the design of crate employed for this sector are therefore equally diverse. For example for light-framework crates for vegetables, polypropylene or toughened polystyrene is used, and the crates are single-trip. For bread, meat, sausages, etc., crates are produced in high density polyethylene or polypropylene, and are of robust construction so that they can withstand many trips.

Plastic crates are now firmly established for a wide range of products and usages, since these injection-moulded crates can easily be coloured and embossed to provide instant identification of ownership and product (Figure 20.42).

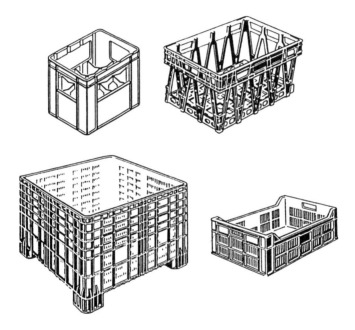

Figure 20.42   Typical plastic crates.

INTERMEDIATE BULK CONTAINERS

In the 1950s and 1960s there was a rapid growth in trade for the bulk carriage of liquids and solids by road, rail and sea with a requirement in many instances to effect transfer between transport modes with the minimum of manual handling. This led to the development of a larger size of package of a capacity between that of the full bulk vehicle or tanker and the 220-litre drum. These became known as intermediate or semi-bulk containers (IBCs). They are now defined as rigid, semi-rigid or flexible portable packagings that:

1. Have a capacity of not more than 3.0 m^3 (3000 litres)
2. Are designed for mechanical handling
3. Are resistant to the stresses produced in handling and transport as determined by prescribed tests

Most IBCs are used for granular and powdered solids, viscous liquids and free flowing liquids with a vapour pressure not exceeding 110 kPa (about 1 atmosphere) 50°C (i.e. the boiling point is below this temperature) and which can be filled and emptied without using pressure.

They are available, as the definition suggests, in many materials. Those carrying solids are usually constructed in metal, plastic, corrugated fibreboard, wood, heavy duty coated fabric or woven plastic tape fabric. Those for liquid are more often made from rigid metal or plastics although some flexible plastic coated fabrics are also used. They may be returnable, rigid or collapsible or expendable after a single trip and typical examples are shown in Figure 20.43.

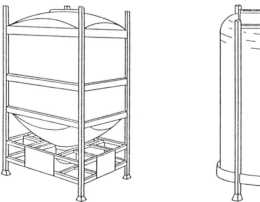 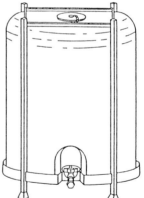

Figure 20.43 Intermediate bulk containers.

21

Flexible shipping packages

TEXTILE SACKS, BAGS, BALES AND BALING

Introduction

Jute, flax and hemp, when used for packaging purposes, are normally spun into yarns by dry-spinning processes, as distinct from the wet-spinning method used for finer flax yarns. Jute-substitute fibres are also used for sack production in certain areas of the world, although the quantities consumed are small in comparison with jute. Flax and hemp fabrics are normally too expensive to be used for other than special purposes, e.g. flax waterbags for use in hot countries, and tarred hemp sacks for the retail coal-carrying trade. Cotton is extensively used for the packaging of many commodities.

The larger part of the jute weaving industry employs plain looms of simple construction, although these are nowadays fitted with automatic weft replenishment and warp stop motions. In addition, the industry employs circular looms for formation of tubular fabric and for production of fully woven, i.e. seamless, sacks.

Jute sacks and bags are used for packing a wide range of materials, including powdered and crystalline substances, agricultural produce, textiles and small engineering parts. Apart from their intrinsic strength, they have a particular value in that they can be re-used in many instances. Even where re-use for the same purpose is not possible, the sack usually has a value for other purposes.

Sack manufacture

Flat fabric with a selvedge at each side may be made up into sacks in many ways. The choice of the method will usually be determined by the seams specified by the user, the width of the fabric which is available in relation to the size of the sack required, and the need to turn and hem all raw edges unless they are incorporated into sewn seams. The basic methods are:

1. *Where the fabric width is twice the length of the sack:* a piece equal to

the desired width of sack is folded in the middle, selvedge to selvedge forming the mouth; both pairs of raw edges are sewn, forming the side seams.

2. *Where the fabric width is equivalent to the length of the sack:* a piece twice the width of the sack is folded raw edge to raw edge, these raw edges are sewn together forming a side seam. One end comprised of two selvedges is also sewn forming the bottom seam, whilst the other end, which is also comprised of two selvedges, forms the mouth.

These are illustrated in Figures 21.1a and b, and it will be seen that as the selvedge is at the mouth of the sacks it needs no further securing, although it may be specified that the mouth be hemmed for additional strength. It will also be seen that in both instances the weft yarns are running along the length of the sack, not across the width of the sack as in the case in method (3). It is important to know how warp and weft threads are orientated, particularly if it is desired to build in extra strength in parts of the sack.

3. *Where the fabric width is equivalent to the width of the sack:* a piece twice the desired length is cut, and folded in the middle, raw edge to raw edge; both sides each comprised of two selvedges are sewn forming two side seams, and the raw edges are turned and hemmed forming the mouth of the sack.

Circular woven fabric requires seaming at one end only, and hemming at the mouth. There are no side seams but the flat width of the woven tube must be the required width of the sack. The warp yarns run along the length of the sack.

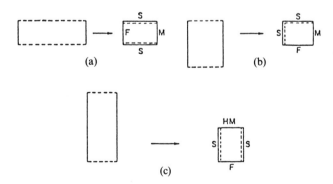

Figure 21.1 Sack manufacture from flat fabric. (a) Fabric width = twice sack length. (b) Fabric width = sack length. (c) Fabric width = sack width. F = Fold; M = mouth; S = sewn seam; HM = hemmed mouth.

Nomenclature

Jute fabric is specified by:

1. The width and length of the piece required
2. The weight of a given area
3. The place of manufacture (e.g. Calcutta, Dundee)
4. The type of weave (e.g. hessian, tarpaulin, bagging or sacking)
5. The *porter*, or threads per unit length in the warp
6. The number of *shots* (threads per unit length in the weft)
7. The finish

Jute fabrics* used for sack making are essentially simple in construction, and are classified as follows:

1. *Hessians:* plain-weave single yarns in both warp and weft
2. *Tarpaulins and baggings:* plain weave, warp threads laid double, i.e. double warp (DW)
3. *Twilled sackings:* twilled 2:1 weave, usually with double warps, but sometimes single warps (SW)

Threads per unit length in the warp are commonly specified by the porter system, which specifies the number of splits in the loom reed. Thus a one-porter reed contains 20 splits in 37 in of reed width; ten porter contains 200 splits and so on. Hessians are usually drawn 2 warp threads per split, tarpaulins 4 threads, and twilled sackings 6 threads (DW) or three threads (SW) per split. If P = porter and N = threads per inch then for

Hessians $\qquad N = (2 \times 20/37)P = 1.08P$

DW tarpaulins $\quad N = (4 \times 20/37)P = 2.16P$

DW twills $\qquad N = (6 \times 20/37)P = 3.24P$

SW twills $\qquad N = (3 \times 20/37)P = 1.62P$

Since fabric normally shrinks in width when removed from the loom, however, the actual warp threads per inch in Dundee jute fabrics are slightly higher than when calculated by these formulae. Calcutta fabric, on the other hand, uses *porter* in reference to finished cloth. *Shots* or *threads per inch* is the common designation for weft.

Hessians are generally specified by width, weight in ounces per yard linear at some fixed width and also by place of origin. For example: '60 in, 6oz/40 in Dundee' means that the cloth required will measure 60 in wide and weigh approximately 6 oz per linear yard of 40 in material (i.e. 9 oz per linear yard of 60 in material) and originated in Dundee.

---

* The North American sack industry uses the term 'Burlap' to denote jute and in particular hessian sacks.

Tarpaulins are of double warp construction (DW), as also are baggings, but the latter are usually coarser fabric than tarpaulins. All are specified by the width required, the porter and the average weight per linear yard of whatever width is normal for that construction. For example: '45 in 6/14/27 (or 45 in 6/27/14 oz) twilled sacking Dundee' means that the cloth required will measure 45 in wide, be of 6 porter construction and weigh approximately 14 oz per linear yard of material 27 in wide (i.e. 23.3 oz per linear yard of 45 in wide material.

Finishing treatments

Mechanical

For the most part jute fabrics receive only simple mechanical finishing treatments prior to being converted into sacks or bags. These may be classified as follows:

Cropping and singeing: the loose hairs on both faces are sheared off by rapidly rotating helical knives working against fixed blades, as in a lawnmower. Singeing is less common.

Damping: the fabric is passed through machines which add water in the form of a fine spray produced either by jets or by long cylindrical soft-bristle brushes rapidly revolving in contact with rotating metal rollers semi-immersed in water. The amount of water added is controlled, and varies according to the fabric and the finish required.

Calendering: calendering is a process for flattening the threads of a cloth, and is normally achieved by running the cloth, which may be previously damped, between two or more alternate steel and compressed-paper cylinders, to which pressure is applied. The machine used is known as a *calender* and the cylinders, usually five in number, are referred to in the trade as *bowls*. The central bowl is of steam- or gas-heated steel with paper and steel bowls alternating above and below. Heat, water, and pressure combine to flatten the threads and thereby reduce the size of the interstices, improving the appearance and ease of handling in later processes. The fabric face last in contact with steel is more polished than the other face. Speeds of up to 30 metres per minute are common.

Chesting: the degree of flattening may be increased by allowing the fabric to wind up on the top or second bowl after passage through the lower nips. In contrast to calendering, this method is discontinuous and uses shorter lengths of fabric. The finish obtained lies between that given by calendering and that associated with mangling.

Mangling is a process somewhat akin to chesting, but in this case the previously calendered cloth is rolled on a steel pin and rotated backwards and forwards between two steel bowls of the 'mangle' under heavy pressure. The method closes the interstices more effectively than chesting.

According to the degree of damping and tension employed, calendering and chesting may tend to increase piece length and slightly reduce piece width; mangling on the other hand may tend to reduce piece length and to increase width. The effect is sufficiently well known to be taken into account when weaving, in order that the final width may be correct. Sackings and baggings are commonly calendered only, whereas hessian may be chested or mangled.

Chemical

The bulk of jute sacks and bags are used without chemical treatment, but a definite demand exists for specific treatments, more especially for imparting a more attractive appearance or resistance to certain adverse influences. In most instances these treatments are applied in the piece prior to mechanical finishing, but may also be applied at some earlier stage of processing or alternatively to the finished sacks. Such treatments include proofing against water, rotting, insects, rodents and fire.

Sacks and bags

The words *sack* and *bag* are often used synonymously, but the trade usually applies the first to those made from twilled sacking, and the latter to those made from hessian, bagging or tarpaulin. Sacks and bags, which are described by size, weight and the material from which they are made, are normally supplied with plain open mouths, but where necessary 'valved' sacks can be specified. The two types of valves commonly in use are the Bates valve and the Sleeve valve. Although a little more costly, the latter allows a greater volume content and gives a better shape to the sack when filled.

Sacks and bags may be hemmed and seamed in about a dozen different ways, and a variety of stitching is employed for these. It is not within the compass of this chapter to describe all of these, but definitions of a few are as follows:

Plain seam: the seam is made through two thicknesses of cloth and the bag is turned after sewing. One or two lines of stitches may be used; plain seams are used only on selvedge edges (Figure 21.2).

Counterlaid seam: here the edges are turned outwards; the seam is formed by sewing through four thicknesses of cloth, and the bag is turned after sewing (Figure 21.3). Again one or two lines of stitches may be used.

'M' seam: this seam is formed by first turning in the edges to be sewn and then sewing through four thicknesses of cloth. The bag, which is not turned, is that commonly used for paper-lined bags.

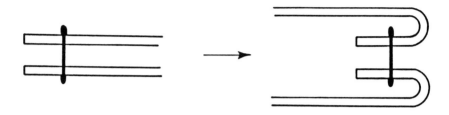

Figure 21.2   Plain seam.

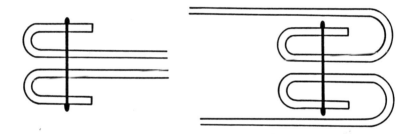

Figure 21.3   Counterlaid seam.

Splay seam: the selvedges of the cloth are laid together, one selvedge overlapping by about half an inch, and then sewn through two thicknesses of cloth, leaving the overlap protruding. The latter is then folded over, and again stitched through two thicknesses of cloth, the bag being finally turned.

Hemming is normally carried out in one of two ways, (a) using raw edges of the cloth, in which the fabric is turned in twice, and (b) on selvedges where the cloth is turned in once.

Sizes of sacks and bags

Sacks and bags can be made in any required size. However, a large variety of 'standard goods' are normally available from stock. A short selection of these is given in Tables 21.1, 21.2 and 21.3.

Table 21.1 Typical Calcutta standard sacks and bags (hessian).

Name	Size	Weight	Porter and shot	Stripe	Sewing	Packing (yards)
Hessian bags	54 in × 27 in hd. w.i.p.[a]	7½ oz/40 in	9×9	Plain	Dry	1000
Hessian bags	56 in × 28 in hd. w.i.p.	7½ oz/40 in	9×9	Plain	Dry	1000
Hessian bags	56 in × 30 in hd. w.i.p.	7½ oz/40 in	9×9	Plain	Dry	1000
Hessian bags	56 in × 36 in hd. w.i.p.	7½ oz/40 in	9×9	Plain	Dry	1000
Hessian bags	47 in × 30 in hd. w.i.p.	10 oz/40 in	11×12	Plain	Dry	500
Hessian bags	47 in × 30 in hd. w.i.p.	10½ oz/40 in	11×12	Plain	Dry	500
Hessian bags	47 in × 30 in hd. w.i.p.	11 oz/40 in	11×12	Plain	Dry	500
Hessian bags	40 in × 22 in selv. Ex.	8 oz/40 in	9×10	Plain	Dry	1000
Hessian bags	40 in × 24 in selv. Ex.	8 oz/40 in	9×10	Plain	Dry	1000
Hessian bags	40 in × 22 in selv. Ex.	10 oz/40 in	11×12	Plain	Dry	1000
Hessian bags	40 in × 24 in selv. Ex.	10 oz/40 in	11×12	Plain	Dry	1000
Onion pockets	40 in × 22½ in hd.	12 oz	9×12	3 blue	Tar	1000
Wheat bags	36 in × 22 in hd.	12 oz	11×12	Plain	Dry	1000
Australian bran bags	49 in × 30 in selv.	20 oz	11×12	Plain	Dry	600

[a]Weight in proportion to.

Table 21.2 Plain double warp bags.

Name	Size	Weight (lb)	Porter and shot	Stripe	Sewing	Packing (yards)
Heavy Cee bags	40 in × 28 in hd.	2¼	8×9	Plain or stripe	Dry	400
Light Cee bags	40 in × 28 in hd.	2	8×8	Plain or stripe	Dry	400/500
E bags	40 in × 28 in hd.	1¾	5×8	Plain	Dry	400/500
K bags	40 in × 28 in hd.	1⅞	6×8	Plain or stripe	Dry	500
DW bags flour bags	56 in × 28 in hd.	2½	7×9 8×8 }	Plain or stripe	Dry	400
DW salt bags	45 in × 26 in hd.	1¾	6×8	3 blue	Dry	500

Standardization

Certain categories of standard Calcutta goods, when intended for export, have now necessarily to conform to the requirements of the relevant Indian Standards Institution specification in respect of cloth construction, strength, size, weight, etc., and all bales of such goods must carry a certification mark to that effect.

Failing such compliance/bale marking; shipment is not allowed. Goods at present included in this scheme are shown in Table 21.4. Particulars and copies may be obtained from the Indian Standards Institution, Manak Bhavan, 9 Bahadur Shah Zafar Marg, New Delhi 1, India.

Table 21.3 Twill sacking bags.

Name	Size	Weight (lb)	Porter and shot	Stripe	Sewing	Packing (yards)
A twills	44 in×26½ in hd	2⅜	8×9	3 in blue	Dry	400
Liverpool twills	44 in×26½ in hd	2½	8×8	3 in blue	Dry	250/300
B twills	44 in×26½ in hd	2¼	6×8	3 in blue	Dry	300/400
B twills	44 in×26½ in hd	2	6×7	3 in blue	Dry	300/400
Cuban sugar bags	48 in×29 in hd.	2½	7×9/8×8	2 in blue	Dry	400
Egyptian sugar bags	48 in×28 in hd	2½	6×8	{2 in blue / 2 in magenta}	Dry / Tar	400
Australian cornsacks	41 in×23 in hd		8×9	Plain	Dry	300
NZ cornsacks	46 in×23 in hd	{w.i.p. 2¼ / 41×23}	×9	Plain	Dry	250
Egyptian grainsacks	60 in×30 in hd	5	6×8	2 in magenta	Tar	200
Egyptian grainsacks	60 in×30 in hd	3¼	6×8	2 in magenta	Tar	250

Table 21.4 Indian Standards Institute specifications.

Specification	Name	Weight	Porter and shot
Cloth			
I.S.I. No. 2818/1964	Hessian	10 oz/40 in	11×12
I.S.I. No. 2818/1964	Hessian	7½ oz/40 in	9×9
Sacks			
I.S.I. No. 1943/1964	A twills	26 in×44 in	8×9
I.S.I. No. 2874/1964	Heavy Cees	28 in×40 in	8×9
I.S.I. No. 2875/1964	2¼ lb corn sacks	23 in×41 in	8×9

Sacks and bags of Dundee manufacture for delivery to the UK market are usually made up in uncovered bundles of 50 or less, but may be obtained in baled form if specified.

Lined sacks and bags

Jute sacks and bags may be obtained with a variety of liners. A selection of these together with examples of their applications is set out in Table 21.5.

Where loose liner bags are used, the top of the liner should be folded over and closed before the jute bag or sack is closed. Loose paper liner bags should be not less than 3 in longer than the outer bags to allow for folding over, and not less than 1 in wider to prevent the inner bag bursting. Where air should not be excluded from the contents, as with certain glues, a double jute bag may be preferable.

Table 21.5 Jute bag liners.

Liners	Usage
1. Jute fabrics combined with crêpe or plain paper (united by bitumen or other water-resisting adhesives) generally known as 'paper lined'	For protection against moisture and sifting; this is used for packing such commodities as nitrates and other crystalline chemicals, pigments, fertilizers, etc., provided the contents are not susceptible to contamination by the adhesive
2. Loose crêpe kraft paper liners (waxed if required)	For general protection against sifting where moisture is not detrimental
3. Loose kraft or crêped kraft paper liners (1 waxed, 1 plain)	For use under severe climatic or other conditions where jute fabric and paper united by bitumen or other adhesive is unsuitable
4. Loose kraft paper liners; each liner made from one or more plain kraft plies and one or more union kraft plies, the number and position of the union plies in the liner depending on the protection required	*(a)* Liners with external ply (that ply in contact with the outer package) of union kraft when it is desired to prevent contamination of products by weather or contact with other materials in transit
	(b) Liners with internal ply (that ply in contact with the contents) of union kraft when it is desired to prevent contents from contaminating other products during transit
	(c) Liners with one or more plies of union kraft as the central plies, with the external and internal plies of plain kraft when the liner is used to reduce the permeability of the package
5. Plastics in the form of liners, either loose or bonded to the sack, together with plastics-coated fabrics are nowadays finding increasing use, and it is reasonable to assume that this trend will continue; in general, such sacks are supplied to customers' particular requirements, and it is therefore not practicable here to define standard specifications	

Selection of correct size/type of sack

The selection of the correct type and size of sack for a given purpose is manifestly important, and in most instances the supplier is well qualified to advise on this. In many instances, the particular application of a sack is apparent in the name by which it is normally known, for example Cuban sugar, D.W. Flour, Grain and Chaff, etc. Some typical specifications are given in Table 21.6.

In rare instances, it is necessary to determine mathematically the flat dimensions of a sack to hold a given weight of commodity. This may be achieved by use of formulae devised by H.L. Parsons (BS 1133 Section 9).

Table 21.6   Jute bag and sacks.

Commodity		Typical specification
Bran	Hessian bag:	Capacity: 112 lb
		Sizes: 27×54 in ex. 7½ oz/40 in; 28×56 in ex. 7½ oz/40 in; 30×56 in ex. 7½ oz/40 in
		Sewing: overhead or herakles (dry jute), or union (cotton), hemmed or selvedge at mouth
Chaff	Hessian bag:	Capacity: 8 bushels
		Size: 33×58 in
		Weight of bag: varying from 20 oz to 32 oz
		Sewing: overhead or herakles (dry or tarred jute), or union (cotton), hemmed at mouth
Flour	Hessian bag:	Capacity: 140 lb
		Size: 24×40 in to 26×40 in ex. 10 oz/40 in to 10½ oz/40 in, various specifications; chested, cropped and chested, mangled or cropped and mangled
		Sewing: union (cotton), selvedge at mouth
	or	
	Tarpaulin bag:	Size: 25×44 in
		Weight of bag: 2 lb ex. 14 porter 15/16 shots
		Sewing: splay seam, two sides (dry jute) hemmed at mouth
	or	
	Twilled sack	Size: 25×44 in
		Weight of sack: 2½ lb ex. 10 porter twilled sacking 11/12 shots
		Sewing: Splay seam, two sides (dry jute), hemmed at mouth
Grain	Twilled sack:	Capacity: 4–4½ bushels
		Size: 28×56 in, 10 or 12 porter sacking
		Weight of sack: 4 lb
		Sewing: splay seam or overhead 2 sides (jute or hemp), hemmed at mouth, 3 in vents tarred both sides in the case of splay seam
Nuts and bolts	Pockets made from good second-hand bagging or heavy twill sacking:	
		Sewing: union (flax or heavy cotton) or lockstitch
	Alternative –	
	New twilled sack: ex 8 porter 16 oz)27 in sacking	
		Sewing: herakles, overhead or single lockstitch (jute)
Oats	Hessian bag:	Capacity: 4 bushels
		Size: 27×54 in ex. 9 oz/40 in to 12 oz/40 in hessian
		Sewing: Union (cotton), herakles or overhead (dry jute), hemmed or selvedge at mouth
Potatoes	Hessian bag:	Capacity: 56 lb
		Size: 19½×35 in or 20×34 in ex. 7½ oz/40 in to 10 oz/40 in hessian
		Sewing: union (cotton), herakles or overhead (tarred) jute, hemmed at mouth
		Capacity: 112 lb
		Size: 22×40 in to 24×40 in ex. 7½oz/40 in to 10 oz/40 in hessian
		Sewing: union (cotton), herakles or overhead (tarred jute), selvedge at mouth
	or	
	Twilled sack:	Capacity: 112 lb
		Size 23×41 in, 2¼ lb Australian corn sack
Powders	New laminated bag (paper-lined as described in Table 21.5)	Quality, weight, material, sizes and linings to meet requirements
		Sewing: lockstitch

Methods of closing

Sacks and bags may be closed by hand or machine stitching or by tying with wire ties or twine.

Machine stitching can be accomplished by using either a stationary machine, using the normal conveyer belt feed, or by a portable stitching machine held in the hand. In either case, the line of stitching should be not less than 1 in from the edges of the fabric at the open end of the bag.

For *hand stitching*, the bag should be rolled at the mouth to form two ears, and then oversewn with a packing needle. The filled bags should carry not less than 7 in of free cloth at the mouth, the mouth then being stitched to its full width, rolled down tight on its contents, twisting out an ear at each end. One ear is then securely knotted at its base, the rolled mouth overstitched, finishing with a loop knotted round the base of the other ear.

In *wire tying*, the bag, which should have at least 7 in of free cloth, is bunched at the mouth and tied with a wire tie by means of a suitable tool. The wire tie should be made from not less than 16 SWG wire and carry loops at each end. After tying, the end of the tie should be laid flat along the bag.

To *tie with twine*, the 7 in of free cloth is 'bunched' and the twine, knotted at one end is passed through both thicknesses of cloth by means of a packing needle, wound twice around the bunch, passed again through two thicknesses of cloth and securely knotted.

Bales

The process of baling is normally confined to compressible articles that are not likely to be damaged by compression. The press packing of bales reduces their volume, thereby effecting saving in shipping and warehouse space and costs.

A *bale* is a quantity of supplies, often compressed, generally of cubic shape, and formed in one of the following ways:

1. Unwrapped, and tied with rope or cord, or secured with wire or metal strapping
2. Protected by one or more coverings and stitched on the outer fabric cover
3. Protected by one or more coverings, stitched on the outer fabric cover and strapped

A *truss* is similar to a bale, but the term is generally used to denote a smaller and less highly compressed unit package. Bales and trusses may, within reason, be of any size, dependent upon the nature of the materials to be packed, the forming pressure to be applied, and the handling facilities available at the points of assembly and destination.

The sequence of operations in the press packing of bales is:

1. Where wrappings are used, the lower portion of such wrappings is positioned on the press
2. The materials to be packed are stacked on the wrappings
3. The upper half of the wrappings is positioned on the goods or materials
4. Pressure is applied
5. The outer covering of the wrapper(s) is (are) sewn
6. The ropes, wires or metal straps are positioned and fastened
7. The pressure is released
8. The bale markings are applied

In assembling the goods to be baled note should be taken of the following points. Bulky parts should not be placed one on top of another, but positioned relative to adjoining articles in such a manner as to compensate for any irregularity of form. Metal protuberances, such as buckles attached to the articles intended for baling, must be staggered, otherwise the application of pressure in the process of baling will most probably result in damage to both the metal (or similar part) and the article to which it is attached. Similarly, where the goods to be baled comprise a number of tied parcels, the knots must be at the ends of the parcels. It is essential that all articles to be baled are dry, otherwise mildew is likely to develop during transit or storage.

Covering

In general, bales are covered with jute fabrics, with or without a separate liner or liners of paper or plastic film, with a view to protecting the contents from the effects of moisture, dirt, abrasion, insects, etc. The number and type of such protective wrappings will vary according to the length and hazards of the journey and the storage conditions at the destination.

Bales intended for dispatch to a domestic destination might well be adequately protected by one covering of 9.5 oz/40 in hessian with or without an inner covering of kraft paper. Where overseas or export transport is involved, two coverings of hessian, one layer of waterproof crêped paper, plastic film or plastic-coated kraft, and one layer of plain kraft paper are necessary. In the case of long sea voyages, particularly to tropical countries and for which purpose it is normal to form heavy bales, the outer hessian covering should not be lighter than 10.5 oz/40 in hessian.

It will be apparent that it is not possible to lay down any standard as to the type of fabric wrappings for use in baling, as these will be dependent on the degree of compression and the method of tying employed in forming the bale. In general, however, for heavy bales, where the maximum pressure is applied and steel hoops are used, the inside hessian should be 6.5 or 7 oz/40 in and the outside 10.5 oz/40 in hessian. Small bales formed under relatively

light pressure and strapped with steel may be sufficiently protected by one
layer of 9 oz/40 in hessian, though 10.5 oz/40 in would be preferable.

Forming of the bale

As previously indicated, the positioning of goods on the baling press is of
great importance if damage from pressure is to be avoided. It is of equal
importance when baling bundles or parcels to 'break-joint', otherwise it will
be impossible to form a firm and compact bale, and cutting of the goods by
the steel straps will be almost inevitable. In such instances, the pattern of
piling of the goods should, as far as possible, resemble the positioning of
bricks in a wall, i.e. the units in each course should lie at right angles to those
in the course upon which they are placed, and joints should not lie one above
another. When baling articles of a resilient nature, capable of accepting a
high reduction in volume without damage, it is usual to build these up in a
former or mould (box baling) in order to hold them in shape during the
compression.

Compression

One of the objects of baling being the conservation of space, it is axiomatic
that all bales should be of a maximum density consistent with the protection
of the contents. Pressures employed in baling may vary from 600 lb/in^2 to
4500 lb/in^2 but the amount of compression desirable in a particular instance
is entirely dependent on the nature of the goods being baled. Whilst, for
example, textiles in piece-goods form will withstand a high degree of
compression, stuffed and quilted textiles are preferably not compressed at
all. It is obvious, therefore, that it is impossible to lay down a common
standard for the degree of compression, and that this must be determined by
the type of goods.

Strapping

Rope is not an entirely suitable material for binding of bales in that it
contracts in damp weather and stretches when exposed to a dry atmosphere.
Its use should be confined to lightly pressed bales, the contents of which are
of relatively low value.

Steel or plastic strapping are the most suitable tying materials and the
most commonly employed. Steel strapping should be treated to prevent
rusting. When applying strapping, the outside straps should be positioned
not less than 5 in from each end of the bale, the intermediate straps being
placed equidistant one from another and from the end straps. Stretching
should be achieved by the use of a mechanical stretching tool, and the degree
of tension applied must be sufficient to ensure that the compression of the

bale is fully held. Sealing should be carried out by the use of a sleeve or similar device, designed to be crimped or punched. The loose ends of straps should be either folded under or cut off close to the seal so that no projecting edges are left.

For bales of large size and weight, normally subjected to a high degree of compression, the straps should not be less than 1 in wide and made from hot-rolled hoop iron referred to as 'baling hoops' which are lacquered or otherwise treated against rusting. Such hoops should be fastened by buckles or metal studs.

Marking

Bale marking is carried out using stencils and a waterproof ink or paint. It is important to avoid the use of penetrating fluids. Tags or labels should not be used. Although, as has been implied above, it is not possible to lay down standards for the process of baling, the typical packing specifications for textile goods given in Table 21.7 may prove useful as a guide.

Table 21.7   Typical packing specifications.

Contents		Gross weight (lb)	Approx. size of bale	Covering material	Approx. density ft³/cwt	Strapping or banding
Nature	Yards					
28 in khaki drill	1000	521	2'5"×2'3" ×2'2" = 11.8 ft³		2½	3 hoops 1¼ in wide, 19 gauge
80 in bleached shirt	400	343	1'11"×1'11" ×2'2" = 8.0 ft³		2½	3 hoops 1¼ in wide, 19 gauge
36 in rayon dress goods	1200	436	3'2"×1'10" ×2'6" = 14.5 ft³	(i) Kraft paper inside (ii) Crepe union waterproof paper	3¾	3 hoops 1¼ in wide, 19 gauge
36 in butter muslin	3600	355	3'2"×2'2" ×2'2" = 14.6 ft³	(iii) Double thickness of 9½ oz or 10 oz 40 in hessian outside, boards top and bottom (optional)	4½	3 hoops 1¼ in wide, 19 gauge
46 in printed cotton furnishing	800	342	3'10"×2'0" ×2'4" = 14.0 ft³		4½	3 hoops 1¼ in wide, 19 gauge
24 in terry towelling	1000	462	4'2"×2'2" ×3'0" = 27.1 ft³		6½	3 hoops 1¼ in wide, 19 gauge
Turkish towelling	doz. 46	268	3'4"×2'2" ×2'3" = 16.3 ft³		6¾	3 hoops 1¼ in wide, 19 gauge

MULTI-WALL PAPER SACKS

Introduction

The origin of the multi-wall paper sack dates back to the late nineteenth century when the need to increase the filling and packing speeds for salt demanded a better, flexible container. In 1898, the basic valve-type sack and machinery to fill it was patented. The early paper sack construction consisted of a single wall made of heavy-duty rope stock paper. This type of sack continued to be used until World War I, when a shortage of supplies of hemp and salvage rope forced the sack manufacturers to experiment with kraft mixed with rope fibres, sisal and jute. This research eventually led to sacks made entirely of heavy, inflexible sheets of kraft paper. Further experimentation developed the advantages in strength and flexibility of two walls, or plies of lighter kraft paper combined with corn, potatoe paste, or tapioca.

Although the advantages of more than two plies for sack construction were realized, it was not until 1925 that this construction became feasible through the development of multi-ply sack forming equipment. This equipment permitted the sewing of ends to supplement the pasted satchel-bottom sacks then used, a technique limiting the number of plies combined.

Prior to World War II, the use of multi-wall paper sacks was confined to powdered materials such as lime, cement and basic slag. Today they are used for many different commodities (Figure 21.4). There are few statistics published on the end-uses of paper sacks in the United Kingdom but Table 21.8 gives an analysis of U.S. practice.

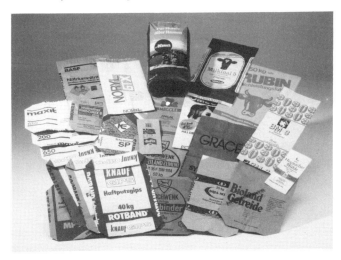

Figure 21.4   The paper sack—a multi-purpose product. Courtesy of Eurosac, European Federation of Multi-wall Paper Sack Manufacturers.

Table 21.8   End-uses of paper sacks (U.S.A.).

	%
Food and agriculture	37
Building materials (cement, etc.)	15
Chemicals (including drugs)	32
Minerals	10
Miscellaneous	6

Paper sacks are made from several thicknesses or plies of paper nested inside one another, so that the total load imposed on the sack during handling and when subjected to strain is distributed over all the plies. The number of plies in any particular sack may vary from 2 to 6 according to its use; and it is this use of several thicknesses of comparatively thin paper, rather than one or two thicker ones, that gives the multi-wall sack its strength and flexibility.

There are two principal types of multi-wall paper sack: open-mouth sacks and valve sacks. Either type may be made with or without gussets, and have either a sewn or pasted closure. However, open-mouth sacks are commonly sewn, and valve sacks are normally pasted. The various styles of sack in standard usage and corresponding UK terminology are illustrated in Figures 21.5 and 21.6. The types of end closures are summarized in Table 21.9.

Open-mouth sacks

Open-mouth sacks are perhaps the more popular class in the United Kingdom. The five types are illustrated in Figure 21.5.

Valved sacks

Valved sacks are extremely important packages and offer the user the opportunity of extremely high packing speeds with a minimum of labour. The two basic styles are illustrated in Figure 21.6. 'Pasted end' and 'pasted and stepped end' sacks are the most economic and popular types but need to be ordered in relatively large quantities. However, sewn valve sacks, either flat or gusseted, are still used for smaller requirements.

Valves

The term *valve* includes both the opening and (where used) the paper or film tube, known as the *valve sleeve* which is inserted and pasted or sewn in the opening that has been left in one corner of the sack, as illustrated in Figure 21.6.

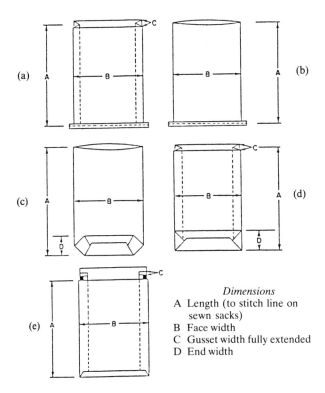

Dimensions
A Length (to stitch line on
   sewn sacks)
B Face width
C Gusset width fully extended
D End width

Figure 21.5 Open-mouth sacks (a) Sewn end—gusseted: the most common type of all normally supplied with a taped and sewn end closure. (b) Sewn end—non-gusseted: usually with a similar end closure. (c) Pasted end—non-gusseted: a flat sack with a glued end which, when filled, forms a rectangular shape. (d) Pasted end—gusseted: similar to (c) but with a gusset; not often used in the United Kingdom; sometimes known as the SOS style (self-opening satchel). (e) Pinch bottom: unlike the other types of open-mouth sacks, which are flush cut at the open end and require mechanical closing, the plies of the pinch bottom sack are staggered so that after folding they may be glued to the opposing edges of the same plies.

Most valved sacks are fitted with an 'internal valve', i.e. where the greater part of the valve sleeve is inside the sack, and on these sacks closure is effected automatically by the contents pressing on the valve sleeve when the sack is inverted in its fall from the packing machine. 'External valves' (or sleeves) can be provided where greater security or moisture protection is desired; after the filling operation the valve sleeve is manually folded down and back into the recess under the sleeve, or alternatively it is heat-sealed if provided with a thermoplastic coating. Such manual operations greatly reduce the potential packing speed which is one of the major advantages of the valved sack system.

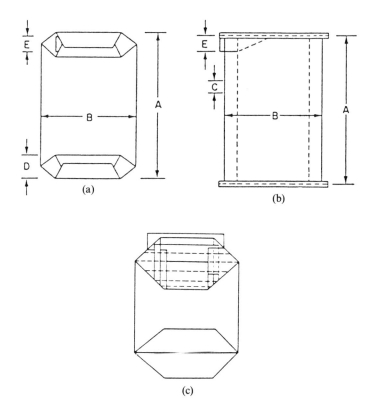

Figure 21.6 Valved sacks. (a) Pasted (block bottom) valved. (b) Sewn valved. (c) Stepped end valved. Dimensions: A, length; B, face width; C, gusset width fully extended; D, end width; E, effective valve width.

Materials

The main materials used in multi-wall paper sacks are:

1. *Kraft paper:* usually denoted in specifications by the letter (K).
2. *Low-stretch crêpe kraft* (LSCK): kraft paper to which crêpe has been applied on the paper-making machine before it is fully dried.
3. *Extensible kraft* (EK): kraft paper treated by a special process to give a controlled degree of stretch. The best known process is the 'Clupak' process.

All these papers can be made 'wet-strength' by adding resins to the paper stock after heating. They are then designated WSK, WSLSCK and WSEK, respectively.

Union kraft (UK) consists of two sheets, combined together with

Table 21.9 Types of end closures

Sewing	The stitches used in paper sack construction are either the chain stitch, or more usually the double locked stitch
Taped and sewn closures	As the name implies the end of the sack is overcapped with a crêpe paper tape and the sewing operation takes place through the tape; the spacing of the stitches is between 12 and 16 per 100 mm, and the sewing line is at least 15 mm from the end of the sack
Sewn and taped closures	Here the sewing operation immediately precedes the application of a capping tape which is secured by wet adhesive, pressure sensitive adhesive, or (if the tape is thermoplastic coated) by heat sealing; the objective is to seal the stitch-holes to prevent sifting of fine powders, or to minimize moisture ingress
Pasted end	The ends of the multi-ply tube are cut evenly across (flush cut) and are lightly pasted to each other; the ends are folded back to form block bottoms and/or valve ends, and are secured with adhesive; frequently an extra reinforcing piece of kraft is glued over the folded end
Pasted and stepped end	The plies forming the ends are arranged in echelon in such a manner that when the folding operation is complete each ply is glued to itself; this is the strongest form of construction for sacks with four or more plies

bitumen, wax or polyethylene. The results are known as bitumen union kraft, wax kraft union and polyethylene union kraft, respectively.

Manufacture

Paper sacks are manufactured on a machine called a *tuber*. The reels of paper, one for each ply in the sack, are arranged on a reel stand at one end of the tuber. Each reel is offset laterally relative to the next so that, when several webs of paper are bent over to form a tube and stuck together, the resulting seams will be staggered (Figure 21.7a). This results in a distribution of the 'side line seams' over the back face of the sack, and gives better strength. When one or more plies in the sack require special protective or barrier properties, the reels from which such plies result are simply placed in the proper position in the reel stand at the end of the tuber (Figure 21.7b and c).

As has already been mentioned, it is very important that the plies are properly nested within each other, and the tuber crew keep careful watch on this during production. Quality inspection also checks this point by examination and test.

The reels of paper are thus converted into a tube, and this is cut into pre-determined lengths as it comes off the tuber. For open-mouth sewn sacks, it is simply a matter of cutting the tube at right angles to its length; but for valve sacks a somewhat different 'chop' is made, which allows the tube to be suitably folded at the ends and so formed into the finished sack.

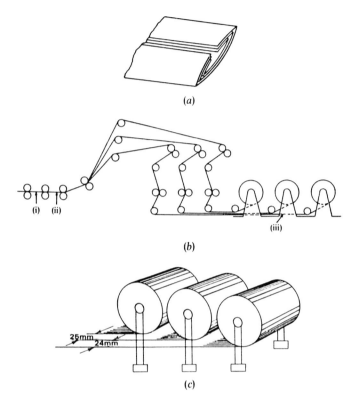

Figure 21.7 (a) Staggering of back seams. (b) Schematic drawing of a tuber: (i) chop; (ii) forming plates; (iii) reel stand. (c) reel stack at end off tuber.

Filling and closing sacks

Open-mouth sacks

Open-mouth sacks may be closed quite simply by 'bunch tying' with wire using a hand-tool; potato sacks are so closed in a field.

Free-flowing products are usually closed with a simple manually operated sack filler. A clamp holds the mouth of the sack open beneath a hopper with a feed gate or flow regulator.

After filling, the usual method of closing open-mouth sacks is by over-sewing through a crêped paper tape. If the sacks are to be used for fine powders, and it is required that they are sift-proof, the closed end may be dipped in wax. Alternatively, the sack is closed by sewing only, and then crêped paper is stuck over the sewing line with a latex adhesive. Open-mouth sacks can also be closed by pasting in the same way as valve sacks.

Valve sacks

Valve sacks are normally closed by folding and bending the ends over, and pasting them. Sometimes the ends are strengthened by cutting the plies in a series of steps so that, when the ends are bent over, the joins will be staggered, somewhat after the same fashion as the side seams formed on the tuber.

Both ends of a valve sack are, of course, closed by the sack manufacturer, and a valve inserted in one end.

To fill a valve sack, it is first placed on the filling spout and the filling operation takes place by one of several methods, depending on the flow characteristics of the product which passes through the filling spout into the sack.

Materials with good flow properties are filled by gravity alone, from an overhead hopper from which the product passes into a chute with a curved end and a filling spout. The sack is clamped on to the spout and supported while a feed gate in the chute is operated. An improved type of gravity packer (Figure 21.8) can, with an easy-flowing product, give high filling rates (8–10 sacks per minute).

Rotary impellers (Figure 21.9) or screws (Figure 21.10) are used for products which do not flow quite so easily. The rotary impeller, for example, handles such products as cement and limestone on multi-head fillers at 20 sacks a minute with one operator. Weighing is done on the spout with an automatic cut-off when the required gross weight is reached. The screw filler can be fitted with an agitator above the screw for sticky products.

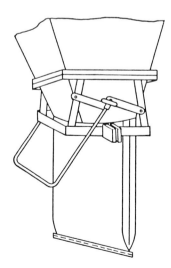

Figure 21.8 Vertical-fall gravity packer.

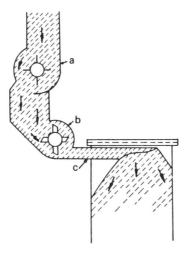

Figure 21.9   Rotary impeller (valve filler) for use with semi-free-flowing materials. (a) Flow regulator. (b) Impeller blades. (c) Filling tube.

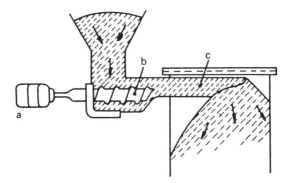

Figure 21.10   Horizontal screw packer (valve filler) for use with semi- and non-free-flowing materials. (a) Motor. (b) Filling screw. (c) Filling tube.

Granular or crystalline products which could be broken up by impellers or screws may be filled by a belt packer (Figure 21.11). The charge is pre-weighed and falls onto a fast-moving endless belt which delivers the product into a horizontal spout and thus into the sack.

Difficult powdered or granular products may require fluidizing to make them flow, and here a pressure packer is used. The product is fluidized by introducing air through the porous base of a storage/pressure chamber, and then pushed into the filling tube, either by the weight of the product in the storage chamber, or by a secondary air supply above the porous base. Weighing is done on the spout with an automatic cut-off when the required weight is reached (Figure 21.12).

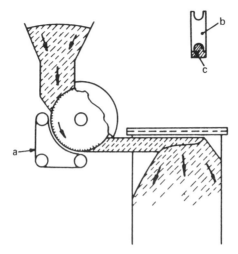

Figure 21.11 Centrifugal belt packer (valve filler) for use with free-flowing and semi-free-flowing granular materials. (a) Continuous belt. (b) Side view of rotating pulley. (c) Belt and pulley groove.

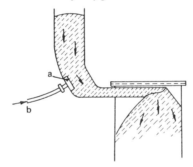

Figure 21.12 Air float packer (valve filler) for use with difficult materials. (a) Air pad. (b) Air.

Storage of sacks

The strength of the paper, and hence the performance of the sacks made from it, varies with the moisture content of the paper. The strength increases with moisture content up to a certain level, and dry conditions tend to lower the strength of the sack.

Now the temperature at certain places on the tuber (because of frictional effects) during manufacture of the sacks tends to dry out the paper and thus reduces its strength. If such sacks were used immediately after manufacture, the breakage rate during handling which is normally negligible, could increase beyond acceptable limits. Accordingly, it is usual to store the sacks until they come to equilibrium with the surrounding atmosphere, and the moisture content of the paper is restored to its previous value. The period of storage required is usually about two weeks, though this can be accelerated by installing a moisture spray in the store or keeping it at a suitable controlled humidity.

Handling of sacks

The filling operation and subsequent handling procedures vary considerably from factory to factory. However, sacks must be able to withstand the more severe rather than the milder conditions, as it is not always known how they may be handled after leaving the factory. Sacks are subjected to a considerable number and variety of hazards. Firstly they are filled from a hopper. After filling, they fall off the filler spout, vertically if they are open-mouthed sacks, and almost horizontally if they are valve sacks. They then proceed through the packing shed to the loading bay or warehouse on a series of conveyers and chutes. Finally, they are either stacked on a pallet or loaded on a lorry or rail wagon, and during such operations they may be dropped more than once. Also, if they are stored in a warehouse, they will probably be stacked to a considerable height.

At a later date the sacks will travel to their final destination by road, rail or sea. During this journey they will be shaken about; they will be handled again during unloading, and then probably re-stored until ready for use. At each loading and unloading they may be dropped intentionally or otherwise, and each and every hazard imposes a strain on the paper from which it never fully recovers. The first drop when falling off the filling spout of the hopper, during which the contents may consolidate, often produces the greatest strain, all subsequent strains being less but of an additive nature. Thus, the total strain imposed on a sack during its lifetime may be considerable.

It is to be expected that paper sacks will fail occasionally, but the percentage of failures must be kept to an absolute minimum, and reduced to zero if possible. Tables 21.10 and 21.11 are an analysis of breakages observed during a survey of damage in a number of factories packing three different products and totalling 330,000 sacks.

Thus, it is very important to check that a particular style of sack will be strong enough to withstand the handling it is likely to receive. For this purpose it is essential that a laboratory assessment be carried out both of the package and the paper from which it was made.

The assessment of paper and sacks

At some stage prior to the manufacture of the sack, the paper from which it is to be made is examined for a number of physical properties. This

Table 21.10 A comparison of sack breakages on a plant and industrial basis[a].

| Type of failure | Animal feed industry | | Fertilizer industry | | | Sugar industry | | | |
	Factory 1 (II)[b]	Factory 2 (II)	Factory 1 (I)	Factory 2 (I)	(III)	Factory 1 (I)	(III)	Factory 2 (I)	(III)
Body burst	14	24	7	14	33	30	21	15	18
Gusset burst	19	21	86	17	33	17	18	17	20
Top sewing	17	6	1	6	13	5	14	21	27
Bottom sewing	4	6	2	29	11	4	10	13	11
Snagging	33	29	1	34	9	22	27	20	20
Torn corner	8	9	3	–	1	18	8	13	4
Unidentified	5	5	–	–	–	4	2	1	–
Total	100	100	100	100	100	100	100	100	100

[a]These tables are taken from a lecture given by P. N. Harvey to the EUROSAC Conference in Lucerne, 1963.
[b] (I) Sacks dispatched direct from production line, or equivalent to direct dispatch.
(II) Sacks stored temporarily in piles of 8 to 10 high prior to dispatch.
(III) Sacks stacked in the full sense of the word prior to dispatch.

Table 21.11 Mean breakage distribution patterns (at manufacturing plants)[a].

| | Methods of dispatch | | | Grand average |
	(I)[b]	(II)	(III)	
Body and gusset bursts	51	39	48	46
Sewing line failures	20	17	29	22
Snagging failures	19	31	19	23
Others	10	13	4	9
	100	100	100	100

[a]These tables are taken from a lecture given by P. N. Harvey to the EUROSAC Conference in Lucerne, 1963.
[b]As in footnote to Table 21.10.

examination may be at the paper mill laboratory, or in the quality control laboratory of the sack factory, or both.

It is obviously important that the paper has strength when under load, and also that it should stretch under these conditions. Thus, the tensile strength and stretch characteristics of the paper are important, and these qualities must not fall below a certain level. Other properties such as the tearing resistance, the porosity of the paper, and its degree of sizing are also measured.

The sacks themselves may be assessed by drop testing. There are a number of ways in which this can be done, but the two most commonly used and which best correlate with field performance involve the repeated dropping of each sack until failure occurs. In one method the sack is dropped from a fixed height until failure occurs. The sack is placed on a trap-door type platform (either flat or standing on one end) and raised to a pre-set height, when the trap opens and the sack falls onto a selected type of floor. The sack is then replaced on the platform, in the same position as before, and the drop is repeated. The procedure is continued until the sack finally fails. The number of drops before failure occurs, and the nature of the failure are recorded. At least ten and preferably more sacks of the same type must be tested before the average number of drops causing failure is calculated. This is known as the *drop number* for the particular style of sack under investigation.

Testing may either be carried out with the sack falling so as to impact one of the large faces, by falling onto a gusset or by being dropped on to its butt. The choice is largely dependent on the mass. 25 kg sacks are usually tested by face drops and 50 kg sacks by butt drops. This reflects the likely impacts they will receive in manual handling during transport.

The disadvantage of the method is that the number of drops to failure is greatly influenced by the nature of the contents and the strength of the sack. Also at realistic drop heights the number of drops can be inconveniently large.

The second method is known generally as the 'Progressive Drop Height Method'. Here if the sack survives its first drop the next drop is made from a greater height by a fixed amount or increment. Successive drops without failure are similarly treated until failure occurs. The number of drops and the height reached before failure give a more convenient assessment of the strength of the sack.

By testing different types of sack, using the same contents and dropping from the same height, a comparison of strength can be made. It is important, however, that sufficient sacks of each type are tested to ensure that the results are representative. Also, testing should be carried out in a conditioned atmosphere, so that variability due to different moisture contents of the papers involved is kept to the minimum.

The sacks should be left in the conditioned atmosphere in which they are to be tested for sufficient time for them to come to equilibrium with it. This usually means overnight before the day on which the test is made. If a conditioned laboratory is not available, the sacks should be tested alternately, i.e. one from the first set, then one from the second set and so on, until one sack from each set has been examined. The second one from each set is then tested, followed by the third one from each set, until all have been examined. In this way, variations in atmospheric conditions will be spread throughout the various sets as evenly as possible.

The drop number of a sack depends not only on the paper from which it is made but also on the nature of the contents. The more free flowing the contents the greater the stress applied to the sack during impact. Whereas an immobile powder behaves almost as if it were a solid, a free flowing powder, particularly when the sack is dropped on to its end (butt drops) flows almost like a liquid to the bottom of the sack producing 'hoop' stresses which stretch and eventually rupture the paper at a much lower height or fewer repeated drops. Thus, free flowing products always need stronger sacks than more cohesive products.

It is therefore very important that the same contents are used for each type of sack to be compared. Moreover, it is advisable to use the product for which the sacks are intended as the contents during testing.

Laboratory testing, if carried out carefully by an experienced operator, can (within limits) determine the suitability of a particular style of sack for packaging any product. However, handling procedures in practice can vary, and a field trial is very useful before full commitment to a particular type is made.

So far only the mechanical strength of the sacks and the corresponding mechanical hazards have been discussed. Sacks, however, may have to withstand other hazards. The most important of these is moisture, either as liquid or in the vapour state. In some instances it does not matter if the contents of the sack get wet. Here it is necessary for the outer ply at least to be made from 'wet strength' paper. This is a paper so treated that it does not disintegrate when wet or lose all its strength under such conditions. Certain products must be protected against both liquid water and water vapour. To achieve this, one or more plies of a bitumen laminated kraft paper (union kraft) or a polyethylene-coated kraft paper are incorporated into the make up of the sacks. The latter is a better barrier, and can also be used as a means of preventing the contents becoming contaminated with paper fibres if it is inserted as the innermost ply, with the coating facing inwards. Polyethylene-coated kraft also behaves better over a wider temperature range, being able to withstand a higher temperature without the melting experienced by bitumen, and not suffering from the latter's fault of cracking at low temperatures.

If a water-vapour barrier is incorporated in the plies of a sack, it is important that moisture is prevented from entering through the ends of the sack. This is particularly important with sewn sacks, as moisture can enter through the sewing line. The moisture enters down the space formed by the two sheets of barrier material in the gaps between the stitch holes, rather than through the holes themselves. However, an improved result can be effected by the technique of sewing and over-capping as previously mentioned.

Other speciality or barrier papers such as wax-laminated paper or silicone-impregnated paper may be used as plies in a sack. Paper may also be sprayed with flame-proofing agents or rodent-repellants. These papers are, of course, more expensive than conventional kraft, and sometimes the special chemical treatment can weaken the paper.

Present and future use of paper sacks

The total number of paper sacks used in the United Kingdom has decreased over the last 10–12 years, and there have been changes in the pattern of usage. A greater use of bulk handling has captured some markets, e.g. cement, previously packed in paper sacks, and is now also making inroads into the market for flour. The introduction of plastic sacks has made large inroads into the packaging of fertilizers in paper sacks and other products requiring high weather resistance such as refuse collection.

Baler bags

Baler bags (Figure 21.13) may be defined as flexible containers for a number of smaller units or consumer packages. As their name implies, they were first used in connection with the packaging of bales of wool, where a suitably styled large bag is required, though not necessarily of great strength. Bales of wool do not have jagged edges or corners which may poke their way through the sides of the bag, and the bulk density is not high. Also, little protection against deformation is required, and thus a lightweight bag forms an excellent container.

However, baler bags are also used for packing a variety of small units such as bottles and cartons. In other words, they not only serve instead of a piece of wrapping paper to provide a stronger package, but they also replace corrugated cases and sometimes wooden boxes.

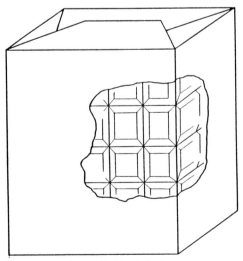

Figure 21.13   Baler bag.

As with any outer containers to carry smaller containers, the style of a baler bag and, in particular, its dimensions must be decided with greater care than if the bag were to contain powder or granules. It must be tailor-made to fit the contents, so that they are held securely. The method of closure will also depend on the contents. Thus, the style, construction and method of closing a baler bag will be determined by the size, weight, nature and number of the smaller units contained in it.

Mailing bags and sacks

Besides baler bags, the use of mailing bags has increased over the last few years. The early types used shredded paper (newsprint) to provide a padding or cushion between two layers of kraft paper. More recently the bags have been improved and made lighter in weight by using either a sheet of expanded polystyrene or a bubble film between the kraft paper layers.

SACKS MADE FROM PLASTIC FILM

Early developments

As early as 1957, some horticultural products were being distributed in France in 25 kg packs made in plastic film. These sacks were manufactured from calendered polyvinyl chloride (PVC) and part of the graphic design was devoted to explaining how the empty sacks could be converted into waterproof aprons as an 'after-use' benefit for the user of the product.

Then in 1960, PVC sacks were produced in Italy by the blow extrusion process. These sacks were made from 0.25 mm thick lay-flat tubing, and used for the packaging of fertilizer in 50-kg units.

The first commercial uses of polyethylene (PE) as a sack material were in 1958 for the packaging of 25 kg quantities of fertilizer and of polyethylene resin. These developments took place in the United States, and the sacks for both applications were made from what was then known as 1000 gauge material (i.e. film of 250 μm thickness). Because of technical problems, however, including those of polymer grade selection, progress was rather slow during the next two or three years.

By the early part of 1961, two large companies in Canada and the United Kingdom were using 1000 gauge (250 μm) polyethylene film sacks for commercial dispatches of fertilizer. The Canadians were packing 38 kg of product per sack, but in the United Kingdom 50-kg units were employed, matching the standard weight offered in multi-ply paper sacks.

From then on, progress in the use of heavy-duty polyethylene sacks was fairly rapid, particularly in the United Kingdom, Canada, the United States and South Africa, and mainly for fertilizers. On the continent of Europe, and especially in Italy, PVC was used to a greater extent than polyethylene, largely due to a price advantage.

In the United States and Canada, polyethylene was always the preferred material, and after some early indecision on the part of some U.K. fertilizer manufacturers, the choice finally settled on polyethylene film which was, and still is less expensive than PVC. It also offers some advantages in the U.K. climatic conditions during severe winters. It is claimed that all problems of filling and sealing, of transit and of climatic protection, permitting the storage of packed fertilizer in the open, are solved by both materials, but the poor low-temperature impact resistance of PVC can prove hazardous if sacks have to be moved from outdoor stacks during very cold weather.

A comparison of PVC and polyethylene films for sacks is given in Table 21.12.

Advantages of plastic sacks for fertilizer

The claim that plastic sacks could be stored in the open during all weathers was the original reason for packing fertilizer in them, even when both polyethylene and PVC sacks were more expensive than the multi-ply paper sacks they began to replace. In addition to the higher cost of the plastic sack itself during these early days of transition from paper, as a plastic sack weighed only two-thirds of the weight of a multi-ply paper sack (even taking into account the heavy gauge of film then used) and since fertilizer was sold in '1 cwt gross' packs, there was a further extra cost to the fertilizer manufacturer in terms of the additional small quantity of product required.

But with an all-year-around production and packaging requirement, because of the overall tonnage demanded and a largely seasonal market for purchases of fertilizer, the reduction in storage costs that could be achieved was very considerable. This advantage also applied to both merchants and farmers, who could relieve the pressure of seasonal demands and also derive benefit from 'early delivery rebates'—a customary off-season price structure system operating within the fertilizer industry.

Even if outdoor storage was not used, the farmer was able to stack his fertilizer packed in polyethylene film sacks in types of barns or sheds that would be quite inadequate for storing this product in paper sacks, as most compound fertilizers are extremely susceptible to deterioration by moisture uptake.

Finally, during the season when the farmer uses the fertilizer, polyethylene film sacks may be taken to various parts of the farm and left out in the open, so that the fertilizer is available where and when it is eventually needed, and the farmer's handling and transport costs are consequently reduced.

Table 21.12 Comparison of PVC and PE films for sacks.

	PVC	PE
Raw material	More expensive than PE in the form required for sack making. The possibility of plasticizer migration must be considered for some products. Many plasticizers that would normally be selected are not free from odour. Formulations which do not give toxicity problems are necessary for animal feeding stuffs.	Less expensive than PVC, particularly in natural (transparent) form. Chemically inert. Comparatively odour free. Coloured sacks can be obtained at extra cost for the necessary polyethylene 'Masterbatch' which obviously slightly narrows the margin between these and PVC sacks.
Processing	Calendering or extrusion (extrusion method produces lay-flat tubing which is obviously easier and cheaper to convert into sacks than the flat sheet film obtained from a calender; PVC lay-flat tubing, however, is more susceptible to 'blocking' problems, which cause the inner surfaces of the sack to stick together, and therefore filling rates may be seriously reduced.)	Extrusion only (standard print-treatment of lay-flat film overcomes printing difficulties with regard to 'keying' of inks.)
Fabrication	High-frequency (HF) welded joints. (Radio frequency (RF) is a synonymous term.)	Heat-sealed joints.
Physical properties	Tensile strength in both machine and transverse direction very good. Slightly higher resistance to puncturing and snagging than polyethylene. Normally adequate moisture-barrier properties in thicknesses suitable for sack applications. Fairly good odour barrier. Poor low-temperature impact resistance. Sacks likely to deform if filled with product at a temperature higher than 50°C, particularly if stresses are applied immediately after filling. Not affected by ultraviolet light.	Tensile strength in both machine and transverse directions good. Very good moisture barrier. Poor odour barrier. Good low-temperature impact resistance, even at temperatures as low as −20°C. Distortion insignificant at temperatures up to approximately 70°C. In parts of the world where ultraviolet rays are more intense than the UK, outdoor storage periods longer than 6 months may cause significant degradation of unpigmented polyethylene sacks. For longer periods black pigmented polyethylene should be employed.

With all these advantages, it is perhaps not surprising that by the end of the 1964 season, the fertilizer industry had changed from paper to polyethylene for nearly all its output of packed fertilizer particularly as by that time the 800 gauge (200 μm) polyethylene film sack was less expensive than the conventional five-ply paper sack it had replaced. Today the most commonly used polyethylene film sack for 50 kg of product is still 200 μm in thickness, although some 175 μm sacks have been introduced. The main difficulty with these thinner sacks is not that they do not provide adequate strength, but that, because the thinner film conforms more closely to the shape of the granules of fertilizer, it becomes more easily abraded during transit, and this can result in failure to give adequate moisture protection or even spillage.

Naturally, even with 200-μm polyethylene film, care must be exercised in handling sacks during filling and conveying operations. Broken metal chutes and projecting bolts or wooden splinters on pallets in need of repair can cause snagging and punctures. Also vehicles must be in reasonable condition, and the decks and walls free from splinters or protruding nails. This is because punctures leading to leakage of product are obviously caused more easily in a single wall of polyethylene sack than in the several plies of a multi-wall paper sack. Polyethylene film sacks have the advantage that, if small accidental holes are present, the high tear strength of the film prevents such punctures from running into larger holes or tears, and hence the spillage of granular contents is greatly reduced.

Another advantage of polyethylene film sacks over multi-wall paper sacks is their greater impact strength. Although it is not possible to simulate exactly the impact stresses encountered in normal transport and stacking by laboratory drop tests on sacks, these clearly indicate the greater strength of the plastic sack.

Valved polyethylene film sacks having all the established features of valved paper sacks, are available for fertilizer manufacturers equipped with valve-sack-filling equipment. Woven plastic sacks are employed for the export of fertilizer to all parts of the world.

Other applications for heavy-duty polyethylene sacks

The good impact strength and tear propagation resistance of polyethylene film permits the economic choice of slightly thinner plastic sacks than those of 200 μm film thickness common for fertilizers, to be used adequately for packaging of products which are not abrasive or free-flowing. The thinner film may be more easily punctured during transit, but these products will not spill out seriously, even if small holes have been caused by mishandling, while the walls of the sack will not readily burst open.

Use of plastic sacks for the retail sale of fertilizers has grown considerably with sales through gardening centres and other retail outlets. Here there is a requirement for quality graphics which do not deteriorate under outdoor conditions. The plastic films used are often pigmented white to present a good image.

In addition to permitting outdoor storage, polyethylene film sacks also provide protection against moisture gain or loss in the product. Furthermore, polyethylene film is a more suitable material for the packaging of certain chemicals which cause paper to deteriorate. It is often economically advantageous to obtain product protection using a polyethylene sack rather than a separate film liner inside a multi-ply paper sack, or one with an inner ply made of polyethylene-coated paper.

Although the largest single use of polyethylene film sacks is for fertilizers, other products packed include horticultural peat, plastic moulding powders and masterbatches, granular and powdered chemicals, certain animal feedstuffs, such as mineral supplements and calf-milk equivalents, and some food products such as dried peas, ingredients for the manufacture of meat pastes and sausages, and confectionery mixes. Other products include pre-packed coal for domestic use, insulation fibres and wadding materials, rubber underlay and other rubber products, and other miscellaneous products such as pipe fittings, cleaning cloths and wood chips.

An interesting application in the animal feedstuffs category, not included in the above list, is the replacement of polyethylene-coated fibreboard containers by less expensive polyethylene sacks (normally 200 μm in thickness) for the packaging of cattle feeding blocks. These are blocks or circular cakes about 480 mm in diameter and 100 mm thick designed for the animal to lick rather than to chew, to obtain a proper balance of vitamins and minerals. Different products containing various ingredients are distinguished by the use of opaque white and other coloured film sacks which are themselves printed in alternative colour combinations.

Such feeding blocks are individually machine-packed into the polyethylene sacks which are then automatically conveyed through a check weigher to a continuous band-sealing machine. As the strength of the vitamins may be affected if the blocks are not consumed within a period of 12 months or so, each sack is automatically date-coded during the heat-sealing operation, to ensure proper stock rotation through the distribution cycle.

The fully sealed polyethylene film sack aids the final hardening process of the feeding block during the 48 hours immediately following the packaging operation, but the blocks are already sufficiently hard to be automatically palletized in the film sacks as the conveyor feeds them from the sealing machine to the palletizing unit. Each pallet load of 1 tonne of feeding blocks in their sacks is then covered with a hood of polyethylene shrink film and passed into the shrink tunnel. The shrink-wrapped pallets are then stored in the open for several months if required (Figure 21.14). But even when the

Figure 21.14   Outdoor storage of animal feeding blocks packed in polythene film sacks.
Courtesy of Bakelite Xylonite Ltd.(BXL).

pallet load is broken into units by a distributor or farmer, and the shrink-wrap cover removed, the individual sacks will protect the feeding blocks for a sufficient period until required by the cattle.

This is another market which traditionally uses an 'early delivery rebate' scheme, and there is an advantage to farmers able to buy seasonal products of this kind during financially favourable periods without the expense of storage under cover. The product manufacturer also has lower storage costs, and a better opportunity to plan production on a consistently even basis throughout the year.

Heavy-duty polyethylene sacks

Properties

    1. Good tensile strength
    2. High tear resistance   Refer to BS 4932 for test methods and minimum
    3. Good impact resistance   values for, various thicknesses of sack film, etc.
    4. Good bursting strength
    5. Chemically inert
    6. Odour-free (film may be specified as not being capable of imparting objectionable odour or taint to foodstuffs, etc.; refer to BS 3755)
    7. Waterproof
    8. Heat-sealable
    9. Printable

Manufacture

Heavy-duty polyethylene film sacks are made from unsupported single-wall low-density polyethylene lay-flat tubing not less than 125 μm in thickness and 380 mm in width. The sacks themselves will have a minimum length of 600 mm, and will be intended to carry not less than 25 kg (refer to Definitions in BS 4932).

Most manufacturers use resins with a density of about 0.923 g/ml with a melt flow index below 1.0 (BS 4932 specifies a maximum density of 0.927 g/ml and a maximum melt flow index (MFI) of 1.40, determining density in accordance with method 509B or C, and MFI in accordance with method 105C of BS 2782).

Polyethylene resins are classified according to their density and melt flow index. The melt index correlates with melt viscosity, which is related both to the processability of a polyethylene resin and to its mechanical properties. A low melt index results from a high molecular weight and, at a given density, the higher the molecular weight the greater is the chemical resistance of the polyethylene. For films of highest impact strength, a resin of the highest possible molecular weight (i.e. the lowest possible melt flow index) would be selected, limited only by processability becoming more difficult as the molecular weight increases (i.e. MFI decreases).

Sack production. Most polythene film sacks are made from tubular blown film of the desired width and thickness, although some sacks are made from flat reeled film forming a longitudinal seam in exactly the same way as paper sacks are 'tubed'.

The molten polythene is extruded upwards from a circular die and expanded outwards by internal air pressure. As will be seen in Figure 21.15 the 'bubble' is held airtight by the two pinch rolls. The cooled film can then be reeled for subsequent processing, or fed directly to the printing and sack making machines. To make the surface of the polythene film receptive to printing inks it is oxidized by an electric discharge method using electrodes positioned as near to the pinch rolls as possible.

The flexographic printing process is most suitable for printing large, bold designs on polythene film, using cheap flexible rubber stereos. Up to four colours in one pass are possible but two colours on both faces of the sack are more useful. The printed tubular film is readily converted into sacks by a single weld across its width; this is usually combined with the cutting operation.

An in-line process such as that described above is linked to the slowest operation. Production methods which are designed to work from reels of film are claimed to be faster because of the higher printing and heat-sealing speeds that can be achieved, but an in-line production line is truly continuous, not intermittent.

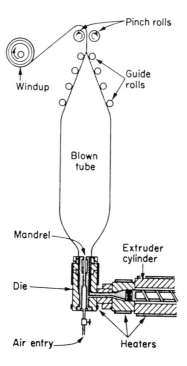

Figure 21.15   Blow extrusion of film.

Styles and construction of film sacks. Fold-in-corner valved sacks (Figure 21.16a) are primarily confined to PVC film sacks. The majority of film sacks in use are of the pillow-shape–open-mouth type (Figure 21.16b), with a heat seal at the base. A typical thickness of sack for 50 kg granular fertilizer is 200 μm (0.008 in). In the gusseted–open-mouth sack (Figure 21.16c), gussets can be formed in the tubular film either before or after the printing operation. The base is heat-sealed through four-two-four thicknesses of film across the face of the sack. Various methods have been used to provide a filling aperture in one face of a fully heat-sealed pillow or gusseted sack. All involve making a slit in the face and the heat-sealing to the sack wall of an inner flap which will automatically cover the slit after the filling operation is completed (patch valved, Figure 21.16d). These types are not widely used. The block bottom valved sack (Figure 21.16e) is essentially the same as the paper sack. The most common type employs special adhesives to secure the ends, which also have an overcapping piece. A second type overcomes the very difficult heat-sealing problem by taking all the joints to one edge only of each end.

Filling and closing. Open-mouth film sacks are packed on similar packing lines to those used for paper sacks. They can be clamped to the filling spout,

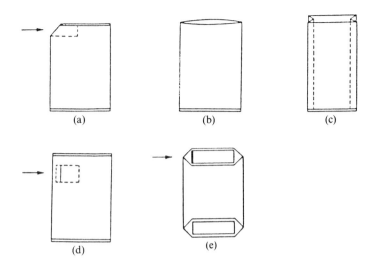

Figure 21.16 Styles of film sack. (a) Fold-in-corner valved. (b) Pillow shape-open mouth. (c) Gusseted-open mouth. (d) Patched valved. (e) Block bottom valved (adhesive overcap).

provided point suspension is avoided, and after the pre-weighed charges have been dropped the sacks are conveyed through the closing operation. This can be by a conventional sewing machine but preferably the sacks should be closed by heat-sealing.

Continuous band heat-sealers are the most widely used closing machines. As shown in plan-view in Figure 21.17a, a pair of PTFE coated steel bands, operating at the same speed as the sack conveyor, grip the mouth of the sack to pass it first over a series of electrically heated blocks and then over a series of water cooled blocks. Under the pressure exerted by the bands, the two inner surfaces of the film sack are fused into a homogeneous mass the width of the band, e.g. 12 mm. Careful attention needs to be paid to the heat-sealing operation, the temperature controls on the heater blocks, the gaps between the heater and cooling blocks, and the avoidance of undue stresses before the seal is completely cooled. Dust on the inner surfaces of the mouth of the sack can interfere with the formation of a satisfactory seal. For such dusty products the surfaces to be sealed can be cleaned mechanically by rotating metal brushes with a vacuum dust extractor, or alternatively a polythene film tape can be applied over the end of the sack and the heat seal made between the tape and the two outer surfaces of the sack. Sealing rates of 20 sacks per minute are common. For low output plants static jaw sealers can be employed.

Heat-sealing filled gusseted sacks is a more difficult operation if weakness of the seal where the thickness drops from four to two is to be avoided, but at slightly lower band speeds satisfactory seals are obtained.

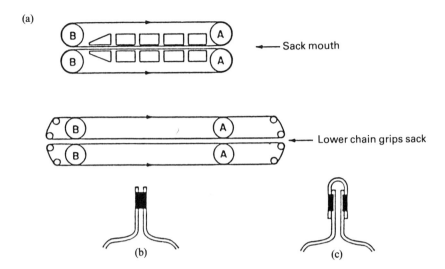

Figure 21.17 Heat-sealing polythene film sacks. (a) Bandsealer. Two separate pairs of bands pass round gears AA BB. Lower chain grips sack whilst upper PTFE coated bands pass mouth of sack between three pairs of heater blocks and two pairs of cooling blocks. (b) Conventional heat seal. (c) Modified heat seal.

Care must be taken with all film sacks to keep entrapped air to a minimum, or instability in transit and in stacking will result. As much air as possible must be excluded before the heat-sealing operation but even then some residual air will require releasing. This is done by making micro-perforations in the sack at the time of manufacture, or applying a maze-seal to the sack base.

Valved sacks are filled in the same manner as are film sacks but due to the impervious nature of the package, care has to be taken to see that air pumped into the sack can escape. Vertical fall packing machines have generally been found suitable.

Types of plastic film sack

Valved plastic film sacks

The earliest valved plastic sacks were made from PVC film by folding in one corner of the top of the sack and, in order to obtain sufficient penetration of the valve aperture into the sack, a projecting 'turret' of film was left on the valve side edge of the top of the lay-flat sleeve. Then both top and bottom seals were applied, trapping the upper edge of the folded-in valve within the top seal of the sack. This type of valve construction was not successful in polyethylene film, because of the less pliable nature of this material compared with the plasticized PVC film formulated for sacks.

Attempts were made to design a polyethylene-film valved sack with an opening for the filling lance or spout on one side of the face of the sack which was covered with a patch of film sealed on three edges only. The main problem with this patch valve sack was that packing rates were reduced because of the difficulty in placing the sacks into position for filling with product at the required speed.

Several other types of plastic valve sack were introduced, including square-ended designs, but as these were all based on semi-automatic production methods, none were commercially successful.

Some time elapsed before the automatic production of a block-bottom valved sack in polyethylene film was developed, providing a plastic equivalent of the valved paper sack. Indeed it was not until late 1969 that the first polyethylene film sacks of this type were manufactured in the United Kingdom, although they had been available from the Continent earlier. Special adhesives are used to secure the ends, which are reinforced with overcapping panels, so that when filled these valved sacks have a neat square-ended profile.

Laminated film sacks

During the past ten years or so, many attempts have been made to overcome the generally limp nature of low-density polyethylene-film sacks, and various blends of high-density polyethylene with other materials have been tried in an endeavour to obtain the stiffness and more paper-like handling characteristics of high-density film whilst retaining the better functional properties of the low-density film. Whilst the tensile strength and the bursting strength of high-density film are very good, the impact strength and the tear resistance of low-density polyethylene film are more favourable for sack applications.

None of these attempts has been successful in producing a single-layer sack film that meets all the requirements, but a possibly better solution to the problem is the manufacture of a two-layer sack film by a lamination process from high-density and low-density polyethylene.

Co-extruded laminated sacks

Co-extruded laminates are produced by linking two (or more) extruders to a single die. In the United States film sacks have been manufactured by this process, using high-density polyethylene to achieve stiffness, and a blend of low-density polyethylene with ethylene vinyl acetate (EVA) to obtain the other properties required. The inclusion of ethylene vinyl acetate in the polythene blend widens the temperature range for heat-sealing.

These and other combinations of compatible materials could be co-extruded to provide the properties required for sack applications, but the higher costs involved in sack manufacture by this process are probably difficult to justify at the present time, in comparison with the existing types of sack that are adequate for the wide range of current packaging requirements. The process could become more important in future.

Cross-ply laminated sacks

Sacks made from a cross-ply laminate of two layers of high-density polyethylene film were first developed in 1968 in the United States and Europe.

The basic concept of cross-laminated films was conceived by a Danish engineer, Ole Benat Rasmussen. The Van Leer organization acquired the rights to this concept and, through a research and development programme, converted it into a workable industrial process. The registered trade name of the material is 'Valéron' and two Van Leer manufacturing plants are now in operation, one in Essen, Belgium, and the other in Houston, Texas. A contract has been signed allowing a Japanese company to produce Valéron cross-laminated high-density polyethylene film under licence in that country.

Manufacture. The material itself consists of two separate layers of high-density polyethylene films which have been produced in such a way that the lines of orientation run at an angle of 45° to the longitudinal direction of the film, and then lamination takes place so that the direction of the orientation of the two webs is opposed, forming a 'cross' (Figure 21.18).

The film produced by this process is an opaque white which provides a good background for printing. The same kind of print treatment as was described for low-density polyethylene film is carried out on both sides, and similar inks are employed, which provide very good resistance to rubbing or

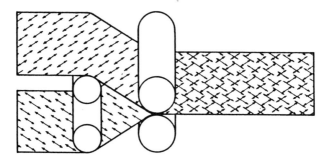

Figure 21.18   Cross-ply laminated high-density polyethylene Valéron® film. Courtesy of Van Leer (U.K.) Ltd.

smearing on sacks printed by the flexographic process in up to four colours. An anti-slip lacquer coating can be applied if required. Special UVI grades of laminated film are available containing a UV absorber component.

It is possible to produce pigmented cross-laminated film sacks if the quantity required is sufficient to justify the special programme required to manufacture the film.

Standard web widths and thicknesses have been introduced to reduce the costs of film production and conversion. A slightly different range is available in the United States to that in Europe, for, although this type of sack has experienced slow development and growth on both sides of the Atlantic, greater interest and activity has occurred in the United States since 1973 where three thicknesses of cross-laminated film are available for the manufacture of sacks and bags (100 μm, 75 μm and 63 μm). In the EEC the first two are manufactured, but only the 100-μm film is used for both valved and open-mouth sacks. However, some European sacks are also supplied in a thicker (120 μm) material. The size range of Valéron sacks in the EEC is: width from 500 mm to 650 mm (in 50-mm increments); length from 460 mm to 1200 mm (up to 1500 mm for sewn open-mouth sacks); block-bottom width from 90 mm to 180 mm (with corresponding size valves).

In the United States sack converting companies have placed considerable emphasis on the supply of smaller sizes of sacks and bags (some with carrier handles) for consumer market packs where more use is made of four-colour printing than is usual for industrial packs.

In both the United States and Europe, the earlier sack-production methods were based on the use of modified paper-sack tubing machines, where a single web of laminated high-density polyethylene sheet was formed into a tube, and then an adhesive back seam applied. This method has now changed to provide an even stronger weld by extruding a copolymer melt for the back seam.

The folding and gluing required to produce the block-bottom and valve sections of the sacks were originally carried out by modified paper-sack bottomer machines. This confirms the paper-like characteristics of this type of laminated film which can be folded, creased, cut, sheared, punched and blanked with conventional tools. Again, however, the method has changed in that block-bottom and valved sacks are now manufactured on the same type of automatic sack-making machines used for low-density polyethylene film valve sacks.

Open-mouth sacks with block bottoms are produced on the same machine, but open-mouth sacks without block bottoms are also available with sewn ends, and these sacks may be of the flat-pillow type or side-gusseted. In the United States pinch-bottom sacks are manufactured, i.e. one face of the sack is longer than the other, and this longer portion is turned over and adhered to the opposing face. Valve sacks with side valves and pinch-bottom closure, top and bottom, can also be made.

Closure. Valved sacks are manufactured with a low-density polyethylene film valve which allows high-speed filling on conventional equipment, the valve closing automatically in the normal way for block-bottom valved sacks.

Because of the cross-lamination of the high-density polyethylene film, the sacks possess a very high tear strength, and this allows the use of conventional stitching machines with a crêpe tape binder and cotton thread for closing open-mouth sacks. The needle should have a conical-shaped head, not one shaped like a pyramid with flat edges.

Heat-sealing of cross-ply laminated sacks is not recommended, because the orientation of the film layers of the laminate results in loss of strength in the seal area.

A more recent development has been the production of micro-perforated Valéron film which, when made into valved sacks, can be used with pneumatic filling equipment for fine powders. For very fine powders, the micro-perforated plastic sacks are provided with a single ply of light-weight kraft paper liner to act as a filter.

Improved filling speeds can be obtained with this type of micro-perforated plastic sack compared with paper sacks, and the strength properties of the cross-laminated film are not significantly weakened by the micro-perforation process.

Properties. High tear resistance/tensile strength: as already stated, the manufacturing process involved provides cross-laminated high-density polyethylene film with very high tear and tear propagation resistance. It also results in an optimal balance of non-directional tensile strength and elongation in the material.

Good impact strength: sacks made from this laminate possess exceptionally good impact strength. Indeed, the normal test methods for low-density polyethylene film are inadequate, and a special testing method has been developed in which a free-falling metal ball is dropped on to a clamped circular specimen of material. The ball must be dropped from heights considerably in excess of that which punches a hole in five plies of paper sack material, or makes an indentation in 200-μm low-density polyethylene film before a 100-μm laminate specimen is damaged. The height from which the ball has to be dropped to damage the specimen is recorded, and this height multiplied by the weight of the ball gives an impact failure-energy (see Figures 21.19, 21.20 and 21.21).

Chemically inert/odour free: as with all polyolefin materials, cross-ply laminated high-density polyethylene film sacks can be used in contact with foodstuffs for both human and animal consumption without risk of contamination.

Puncture/snagging resistance: Valéron sacks are extremely resistant to puncture by sharp objects and possess very good snag resistance.

Figure 21.19   Drop test on 5-ply Kraft Sack Wall, from a height of 1.2 m. Courtesy of Van Leer (U.K.) Ltd.

Figure 21.20   Drop test on low-density polyethylene film (200 μm) form a height of 2.5 m. Courtesy of Van Leer (U.K.) Ltd.

Waterproof/water vapour resistance: the high-density polyethylene film construction ensures that the sacks are waterproof, and sacks made from the film have a low water-vapour transmission rate if good closures are made.

Applications. The manufacturing process for cross-ply laminated sacks results in their being more expensive than normal low-density polyethylene-film heavy-duty sacks, and the cost factor restricted development during the earlier years of availability.

Figure 21.21   Drop test on Valéron® film (100 μm) from a height of 4.0 m. Courtesy of Van Leer (U.K.) Ltd.

The increasing prices of conventional sacks, particularly multi-ply paper sacks with polyethylene film liners, and the successful conclusion of long-term development programmes has now resulted in a growing use of cross-ply laminated high-density polyethylene-film sacks in both the United States and the EEC.

Applications, include the packaging of chemical products for export; plastic polymers, especially where sharp granules are involved; food additives and special animal feedstuffs; herbicides and pesticides; and building products such as special plasters.

SACKS OF OTHER MATERIALS

Cotton sacks are still used in the United Kingdom and in many parts of the world but they are no longer important commercially. Their after-use value as clothing might extend their commercial life for certain traffic, e.g. the despatch of salt to West Africa.

Sacks have been made from man-made fibres, e.g. rayon, nylon and polyester filament yarn; although very strong they have not been developed commercially due to high cost. One man-made fibre, a modified polyvinyl alcohol, has been used in large quntities in Japan as a sack material. Polyvinyl alcohol (PVA) is normally associated with water soluble film, but the modified PVA which in fibre form is strong, very fine with the feel of cotton, rot proof and suited to fertilizer packing. The sacks are made up by sewing, producing open-mouth sacks. The fabric is often extrusion coated with polythene, or laminated to polythene film or crêped kraft paper.

Woven plastic film tape sacks

A major development in the late 1960s was the manufacture of sacks made from slit plastic film by conventional circular or flat loom weaving processes. This originated in Japan, was quickly taken up in the United Kingdom, and is also a growing industry in Europe, South Africa, Australia and North America.

High density polythene or polypropylene resin is turned into film by conventional extrusion casting techniques. Whilst still in the web the film is slit into narrow widths and then highly stretched by a drawing process orienting the film in the longitudinal direction. The individual tape is then usually wound up on to bobbins, although in some plants it passes straight to the weaving process.

The bobbins of tape are used in the weaving process exactly as for more orthodox thread or yarn. Fabric produced from flat looms has, of course, a selvedge at both sides and can be extrusion coated or laminated to polythene or polypropylene film. A plain weave, hessian type, is normally employed.

Manufacture

Polypropylene or high-density polyethylene film is extruded from a slot die in the usual manner (although some processes employ an annular die) and the sheet of film is then multi-slit into narrow widths which are passed through a heating and stretching process which orientates the film tapes in the longitudinal direction. The tapes are then wound up on cores or bobbins for the weaving operation, which can be carried out on flat looms or by a circular loom process.

Fabric made on flat looms has a 'selvedge' at both sides of the woven material where the 'weft' tapes are turned back on themselves. These are the tapes which are carried by the shuttle backwards and forwards across the width of the fabric, and pass alternately over and under the 'warp' tapes which run along the length of the fabric.

With circular-loom weaving, the 'warp' tapes run along the length of the fabric, but the 'weft' tapes are carried by rotating shuttles around the circumference of the woven tube of material, and there is no 'selvedge'.

The film tape is now normally specified by its *tex*, i.e. the weight in grams of 1000 m of tape. In the past, before SI units were adopted, the term *denier* was more commonly used, i.e. the weight in grams of 9000 m of tape. The number of warp and weft tapes per 100 mm of fabric are specified, and sometimes the weight of the fabric in g/m^2.

Although high-density polyethylene (HDPE) has a better low-temperature resistance, polypropylene (PP) has a higher creep resistance. Polypropylene is also 6–8% lighter than high-density polyethylene, and therefore produces a greater yield per unit weight.

In the United Kingdom, most woven plastic film sacks are manufactured from polypropylene tape made into fabric by the circular-loom weaving process, which, by obviating the necessity for a sewn side-seam on the sack, results in a stronger package. This type of sack was first manufactured in the United Kingdom in 1966.

The tube of fabric is cut into the required length by a hot-knife process which seals the raw edge of the woven plastic material. Indeed, the top edge of an open-mouth sack can be left unhemmed unless the customer specifies otherwise. If hemming is required at the top of the sack, this can be provided either as a single hem or a double hem. The bottom is folded over and stitched through all four thicknesses of material, normally with thread made from polypropylene filament yarn, but other sewing threads can be used. The advantage of polypropylene thread is that it matches the properties of the woven sack itself, in not being susceptible to rotting or attack by chemicals, etc.

Before any hemming or stitching operation is carried out, the sleeve of woven sack material can be printed, using special inks.

Liners

The use of a polyethylene film liner (made from low-density polyethylene film) in a woven sack provides a completely waterproof package and, because of the almost transparent nature of unpigmented woven poly-propylene fabric, any printing required can be carried out on the liner. This method not only gives protection to the print, but also enables the woven sack to be left plain, and therefore re-usable for another purpose or product.

Polyethylene film liners can either be inserted loosely, held in position by a solvent-based adhesive at the mouth of the sack, or stitched into the manufacturer's base closure, using the 'skirt' of the polyethylene liner for this purpose (i.e. the area below the base seal line which will be longer than on a normal film bag or liner). Obviously, this type of liner has to be ordered to specification.

Liner gauges vary according to the product packed and the degree of protection required. The normal range of thickness is 25 μm to 125 μm. Perforated film liners are used for some applications.

Closure

After filling with product, the woven polypropylene sack can be closed by machine stitching or by bunch tying with wire ties.

When stitching, it is advisable to arrange that the stitchline passes through four thicknesses of fabric, although a stitchline through two thicknesses is satisfactory for some applications if the top of the sack has been hemmed, or if it is applied at least 50 mm from the top of the sack.

If a polyethylene film liner is used, it can be folded over and the sack stitched sufficiently close to prevent the film unfurling. When the liners are glued to the mouth area of the inside of the sack, this can be right at the top, so that the stitchline of the closure will go through the adhered section; or the adhesive can be applied about 200 mm down from the top of the sack, so that the liner can be heat-sealed to protect any hygroscopic products, and then folded down away from the final stitchline closure of the woven outer sack.

The woven plastic film-tape sack itself cannot be heat-sealed to provide a satisfactory closure, as the orientation of the tapes would be affected, causing a serious reduction of strength in the seal area.

Valved woven plastic sacks are available on the Continent of Europe. These are of the side-entry type, fitted with inner polyethylene film liners.

Properties

Woven polypropylene sacks are much lighter in weight than jute sacks and, weight for weight, polypropylene tape is significantly stronger than conventional sacking materials.

Polypropylene tape is unaffected by water and does not absorb moisture. Woven polypropylene sacks do not rot in damp conditions. With the use of suitable UV-stabilized polymers, improved protection can be given against ultraviolet light degradation. These sacks therefore offer many advantages over traditional sacks for use in tropical areas.

Woven polypropylene sacks are unaffected by most chemicals. Sacks made from polypropylene tapes are clean and do not impart any odour or taint to their contents. There is also an absence of contamination by loose hair-like fibres, a problem often associated with conventional sacking materials.

Applications

The largest packaging application for woven polypropylene sacks in the United Kingdom is for the shipping of British-made fertilizers and other chemical products to various export markets (Figure 21.22). Other main uses are for the packaging of potatoes, vegetable seeds, metallic abrasives, castings and other light-engineering products.

Sugar, coffee, cocoa, and other world crops are often imported into the United Kingdom in woven polypropylene sacks. The packaging of grain is another obvious use. Also flour, particularly in Nigeria, where about 12 000 000 woven sacks are employed every year for this product. In Peru, woven polypropylene sacks are widely used in the fishmeal trade for their export markets.

Typical examples of woven polypropylene sacks used for the packaging of chemicals are as follows:

1. Sack 510 mm wide×900 mm in length, 40 warps×40 wefts, 110 tex, with 25-μm polyethylene film liner for 50 kg of sodium chloride
2. Sack 510 mm wide×820 mm in length in the same specification, with 125-μm polyethylene film liner for 50 kg of sodium hydroxide
3. A typical lighter-weight woven polypropylene fabric would be 40 warps×35 wefts, 100 tex. A heavy-weight fabric could be either 40 warps×40 wefts, 120 tex; or 45 warps×45 wefts, 110 tex. The more intersections per unit area provided by the 45×45 specification produces a tighter weave and therefore a slightly better sift-proof fabric.

Figure 21.22   Woven polypropylene sacks used for exporting fertilizer. Courtesy of Fairbairn Lawson Packaging Ltd.

22
Pallets and unit loads

Introduction

Whenever applicable for distribution and storage it is advantageous to handle goods mechanically rather than manually and to group a number of items into a single larger unit for mechanical handling. Broadly speaking, these Unit Load Concepts have been applied for centuries—ever since man started unitizing items onto a base and using a windlass to lift them. Awareness of the value of aids of this type in modern distribution systems started with the development of pallets and fork-lift trucks in World War II.

A *pallet* has been defined as a flat portable platform constructed to sustain a load and permit handling by mechanical equipment. Many pallets fall within this definition, but there are types which are not flat and/or not properly described as a platform.

A *unit load* has been defined as one or more packages secured to a pallet or skid in such a way that the entire unit may be handled and stored by mechanical equipment. Again many unit loads come within this definition, but there are methods of unitization which do not employ a pallet or skid.

For our purposes therefore the wider definition which is linked with the principal means of handling will be used. This is: a *palletized* or *unit load* is one in which one or more items can be handled as a single unit by fork-lift trucks or similar equipment.

Modes of conveyance may be classified into road, rail, water, air or other (such as pipeline).

Large quantities of products in liquid, powder or granular form may be conveyed by pipeline, bulk tanker or intermediate bulk container (IBC). These are all outside the scope of this chapter, although an IBC is a form of unitized load. This chapter is concerned with liquids, powders and other products in conventional packs, such as cases, drums and sacks; also with other items, such as some engineering products, which may not be packed, but where unitization is a method of preparing them for delivery. The three principal methods of conveyance are Break-bulk; Unitized Loads; and Freight Container. A unitized load may in some instance be an alternative method to freight containers; in other instances it is complementary when unitized loads are transported in freight containers.

Methods of palletization and unitization

There are a large number of methods for unitizing loads. Most, but not all, are based on a pallet. The pallet type used in any given situation depends on a number of interrelated factors. The following are generally the most important and are basic to a consideration of pallet types.

1. The nature of the goods to be handled, and whether the pallet is to be used for one specific purpose or as a general-purpose pallet for a range of packs and products.
2. The distribution system(s) and particularly:
   (a) The number of directions from which fork-lift equipment will need to enter the pallet. Two-way and four-way entry pallets are common. Occasionally an eight-way (entry on the four sides and the four corners diagonally) entry pallet is used.
   (b) Whether the pallets will be handled by pallet trucks. These have small diameter rollers on the fork ends which need to be able to come through the pallet base and bear on the ground.
   (c) General handling and storage methods, including type of fork-lift or similar equipment, slings, spreaders, pallet racking, stack height, etc.
   (d) Whether the pallet 'footprint' is significant in the downward loading applied. This may be important for aircraft floors and for corrugated fibreboard cases on the top of the pallet load below.
3. Cost of the pallet in relation to the expected service life.

Figure 22.1 illustrates by sketches and notes some of the principal types of pallet.

Figure 22.1 Principal types of pallet. (a) A single deck, flat, timber, two-way entry pallet. (b) Four-way entry pallet. (c) Full perimeter base pallet (inverted). The loading area of the base is normally 40% or more of the total, giving an acceptable footprint for many applications. The cruciform base is compatible with pallet trucks. (d) Reversible pallet with similar top and bottom decks. The decks may be of sheet material such as plywood, hardboard or timber. They may be two-way or four-way entry, typically using nine wooden blocks instead of three bearers. (e) Wing pallet. Both decks may project beyond the outer bearers to facilitate the use of lifting slings. In other designs, e.g. to run in live pallet racking systems, only the upper deck projects. (f) Timber and steel stillage with uninterrupted entry for a stillage truck which cannot enter a pallet with centre bearers or bottom deck. (g) Moulded pallet (inverted) made in one piece of plastic or wood fibre bonded with adhesive. Pallets of similar appearance but made with two materials may have a top deck of corrugated fibreboard or hardboard with plastic feet. Both types may be designed to nest when empty. (h) Injection-moulded plastic pallet. (i) Box pallet typically has three or four sides and may have a lid. Vertical sides may be fixed, removable or collapsible. (j) Collapsible pallet. (k) Post pallet, usually made in steel with fixed or detachable posts, plus possibly rails or sides. Used more for storage than for transport. (l) Trough pallet. If the dimensions are compatible with the units (cases or sacks) making up the load, one layer of these can lie in the troughs reducing the volume of the total loaded pallet. (m) Keg pallet. An illustration of one of the many types of special purpose pallets. This one is designed for kegs loaded on their sides.

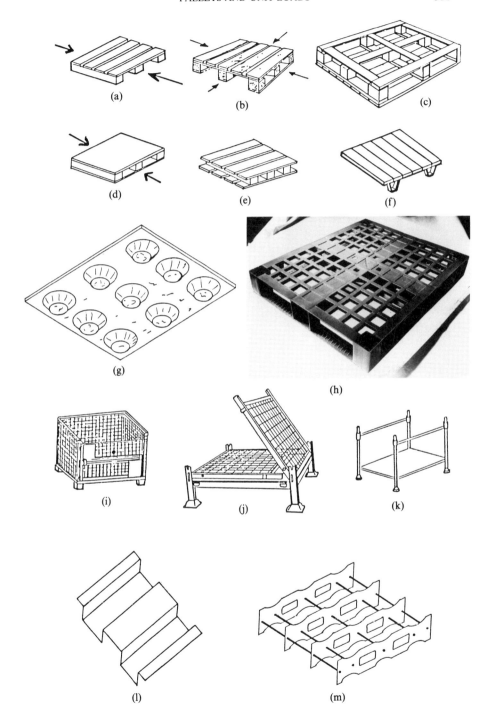

(a)

(b)

(c)

(d)

(e)

(f)

(g)

(h)

(i)

(j)

(k)

(l)

(m)

Figure 22.2 illustrates some methods of unitization where a pallet is not used.

In addition to those shown in Figure 22.2, there are other ways of unitizing loads without conventional pallets. The slip board system uses a thin flexible base board in place of a pallet; this can be gripped by an attachment on a special truck and pulled with the load onto a rigid platform on the truck. The SCULL system employs a 'pallet' of wooden slats and wire made up with the load and which can be dismantled for return and re-use.

Equipment for handling palletized and unitized loads

There is a great variety of mechanical equipment to handle palletized and unitized loads. Two common types are illustrated in Figure 22.3. Other general types include reach trucks and narrow aisle trucks. There is also a variety of special attachments for fork trucks, e.g. attachments so that four pallet loads can be handled by one counterbalanced truck and a single pole attachment in place of the forks to handle large reels.

Wooden pallet construction and specification

The terms used to describe the components of wooden pallets are shown in Figure 22.4. Most wooden pallets are made by specialist manufacturers, although some user companies make their own. Pallet manufacturers generally produce a range of types in various dimensions; timber of relatively low grade is usually used which often has considerable dimensional variation. These two factors make fully automatic manufacture impractical, even on the largest scale. Production rates in excess of 25 pallets per hour are, however, achievable by semi-automatic methods. Pallet components are generally fastened together with nails applied by hopper-fed nailing machines. The two principal methods of nailing are illustrated in Figure 22.4; the clenching is accomplished in the same operation as the nail driving by the use of a steel backing plate. Screws and staples (particularly those with divergent points so that the staple legs splay to increase withdrawal resistance) are sometimes used, but adhesives are rarely practical, partly because of the relatively high moisture content of the timber.

The moisture content of timber is important not only for pallets but for most timber applications including wooden cases. A short account is therefore included here. The moisture content of wood is usually expressed in relation to the oven dry weight and is calculated as

$$\text{Moisture content (\%)} = \frac{\text{Weight of sample} - \text{oven dry weight of sample}}{\text{Oven dry weight of sample}} \times 100$$

If therefore more than half the weight of a freshly felled log consists of water, the moisture content will be greater than 100%. Elm is an example of a hardwood used for pallet blocks; the moisture content green (freshly sawn) is about 135%, and the weight about 1040 kg/m³ (65 lb/ft³); seasoned elm weighs 544 kg/m³ (34 lb/ft³). Pine is a softwood used for pallet deckboards and stringers; the moisture content green is about 85% and the weight 800 kg/m³ (50 lb/ft³); seasoned pine weighs about 500 kg/m³ (32 lb/ft³). Hardwoods come from deciduous trees and may be relatively soft like balsa; softwoods come from coniferous trees and may be relatively soft like some some pines or hard like yew.

After logs are sawn into boards, the latter are usually stacked with spacers (stickers) to season or dry out. Sawn timber can be kiln-dried in a controlled atmosphere to any desired moisture content: timber for furniture is often kiln-dried to 12% moisture content, will then be approximately in equilibrium with the atmosphere of a heated house, and will not dry out further. Shrinking and deforming would occur otherwise. Kiln-drying is relatively costly, and most pallet and packaging timber is air-dried. In the U.K. climate it is difficult to air-dry timber under cover to below 20% moisture content. The moisture content cannot be controlled in the open, since it will depend on the time of year and incidence of rain, as well as on factors such as thickness and length of time in stack.

Wet timber has several limitations:

1. It is relatively heavy.
2. When a palletized load is stored in closed situations such as a freight container, the wood will dry out under warm dry conditions; when the temperature drops, condensation can easily occur and may affect the load. A pallet weighing say 30 kg may contain more than 5 kg of water.
3. It may shrink and warp when it dries out.
4. It may rot. Decay and rot are caused by fungi, and their activity is very dependent on moisture content. Below about 22% moisture, the water is held in the cell walls of the wood, which will not then be attacked by most fungi; above about 22% there is free water in the cell cavities, and the wood is susceptible to fungal attack.

A typical specification for a wooden pallet is given in Table 22.1 and is shown in Figure 22.5.

Pallet performance

There are no generally agreed methods or standards for describing or measuring pallet performance. Structural standards exist in the United Kingdom and United States for wooden pallets and these mainly use some form of compression or flexing test. Pallets made in metal or plastic are also made to conform to dimension and structural strength requirements.

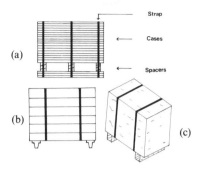

Figure 22.2 Unitization for fork-lift handling without a pallet. (a) A load of heavy-duty corrugated cases folded flat and unitized for fork-lift handling with spacers (typically short lengths of spirally wound fibreboard tubes) and straps. No pallet is used and the lowest cases are placed directly on to the warehouse and vehicle floors. (b) Unitized load of sheet materials requiring feet and strapping only. (c) Unitized load consisting of a single wooden case with skids.

Performance is often judged by actual usage in conditions similar to those anticipated in practice. With the number of pallets in service, and the dangers to life and property should palletized loads collapse in high stacks or in transit, progress should be made in this area. The main factors which need to be taken into account to quantify the performance of pallets like those in Figure 22.1a–h are summarized here, although additional factors may be important in certain situations, e.g. the ease of cleaning a pallet to be used for foodstuffs, and the vertical load resistance of the posts in the pallet shown in Figure 22.1k.

Load capacity

The load capacity of a pallet is a prime consideration. Will the pallet safely transport 0.5, 1 or 5 tonnes? Unfortunately, a simple statement of capacity is meaningless unless other conditions are also specified. These conditions are discussed relative to the type of load, and methods of stacking and lifting.

Loads are classified into three types:

1. Self-supporting
2. Divided
3. Secured

As illustrated in Figure 22.6, a self-supporting load is one which exerts no bending forces on the pallet under the range of lifting and storage conditions. An example is a pallet load of wall boards, each one of approximately the same area as the pallet top deck. A divided load (Figure 22.6b) is one which can exert bending forces during lifting and storage. An

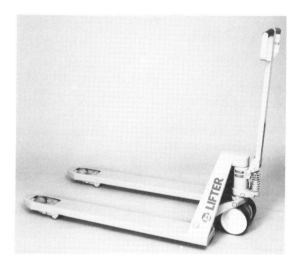

Figure 22.3 Some mechanical equipment for unit load handling. (a) counterbalance fork-lift truck conveying a unitized load two high. Occasionally trucks are used with the forks considerably shorter than the pallets which can result in the latter being unduly stressed. (b) Hand pallet truck. After the forks are pushed beneath the pallet, the two small wheels are moved downwards hydraulically to lift the pallet and allow it to be readily moved. If these wheels rest on the lower deck boards they can be forced off during the lifting operation.

example is a pallet load of small filled cases which are not banded or shrink-wrapped onto the pallet. The effect of lifting a pallet load of this type with a fork truck is illustrated in Figure 22.6d and e. A secured load (Figure 22.6c) is a divided load which has been banded in both directions, shrink-wrapped or similarly secured so that it behaves as a self-supporting load. Three typical stacking conditions are illustrated in Figure 22.6f–h.

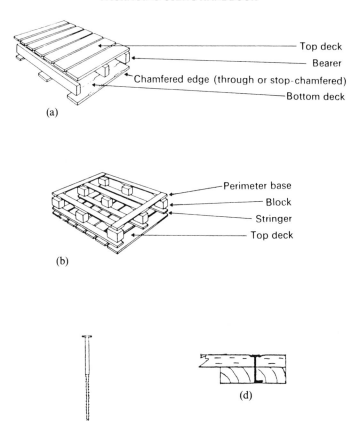

Figure 22.4 The components of wooden pallets. (a) Two-way entry non-reversible (topside view). (b) Four-way entry (underside view). (c) Annular ring nail used typically for nailing through the deck boards into the blocks or bearers. The withdrawal resistance is much higher than that of a plain shank nail and comparable with that of a screw. (d) Section through a top deck nailed to the stringer with clenched nails. Again these have a high withdrawal resistance. Note that in all pallet nailing the nail head should neither project (which may damage the load) nor be over driven (when the board being fastened may pull off the head). It should be flush with the surface as illustrated.

The load capacity can thus be considered under two main sets of conditions:

1. *All loaded pallets are subjected to downward vertical forces taken by the blocks, bearers, feet or other spacers.* Where these are of solid timber, failure is virtually unknown, since short lengths of wood have very high

Table 22.1 Example of a wooden pallet specification.

The timber species, tolerances, nailing pattern, etc., will normally be arranged with the pallet supplier to accord with the particular application.

Type	Four-way entry, perimeter base, 1000×1200 mm
Materials	(a) Boards: the timber species, country of origin, grade or specification, and method of seasoning will be stated (b) Blocks to be a suitable wood
Dimensions	(a) Overall height to be subject to a tolerance of plus 5 mm minus 0 mm, i.e. maximum height 165 mm, minimum height 160 mm (b) Overall length and width to be subject to tolerance of plus 0 mm minus 5 mm (c) The pallet to be square and the pallet deck flat within 5 mm

Timber components	Description	Material	Dimensions (mm)	Quantity
	1. Top deckboards	Sofwood	125×19×1000	7
	2. Base	Softwood	100×19×1000	2
		(chamfered)	100×19×1200	3
	3. Stringers	Softwood	100×22×1200	3
	4. Blocks	Elm	138×100×100	6
			100×100×100	3

Construction	(a) The top deck boards will be evenly spaced with gaps of approx. 21 mm (b) Nails will be positioned not less than 20 mm from the edge of any block or board (c) Nail heads must not project above the boards nor be overdriven more than 4 mm; clenched nails must be embedded in the underside of the stringers (d) The timber components will be fastened securely with the following nails: top deckboards to stringers except over blocks, 2 off 50×3.35 mm plain wire nails clenched; top deckboards and stringers to blocks, 3 off Tilgrip or similar annular ring rails 75×4 mm; bottom deckboard to blocks, 2 off Tilgrip or similar annular ring nails 75×4 mm
Marking	Two diagonally opposite blocks stencilled with user's name
Inspection	Pallets may be subjected to inspection on receipt and those not in accordance with the foregoing specifications rejected

compressive strength. With self-supporting loads, therefore, wooden pallets have virtually unlimited capacity, and the stacking height is restricted by aspects such as stability and the compressive strength of the load, more than by the pallet strength. With spacers of material other than wood, such as plastics or fibreboard, the downward forces may limit both the pallet load capacity and the number of loaded pallets which may be stacked on one another. This is particularly so when the strength of the spacers is reduced by moisture or high temperature.

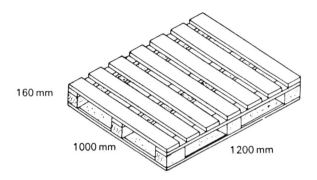

160 mm

1000 mm   1200 mm

Figure 22.5   The pallet of the example described in Table 22.1.

2. *All pallets with divided loads are subject to bending forces when lifted by fork-lift trucks.* When pallet racking or sling handling is employed, such pallets are subjected to additional bending forces. The capacity of most types of pallet for divided loads will thus be limited by the pallet bending strength.

A third condition may be significant with very light timber pallets using a minimum of deck boards. As illustrated in Figure 22.7a, some divided loads may apply bending forces to the deckboards between the bearers, and the capacity may be limited in this way more than when the pallet is lifted.

Other pallet performance factors

The load-bearing capacity is the most important factor in pallet performance. There are a number of other factors, the importance of which will vary greatly with the type of pallet and the material with which it is made. These performance factors include:

1. The resistance to fork tines, which may break, puncture or abrade many parts of the pallet
2. The ability to be put down on uneven ground
3. The resistance to impact when lowered rapidly onto level or uneven ground
4. The resistance to vibration during transit; vibration tends to loosen fastenings and generally fatigue the material
5. The resistance to handling when not loaded, which typically occurs when multi-trip pallets are being returned empty

Pallets often have to perform under conditions of high or low temperature and relative humidity. They are commonly stored outdoors and may be put down in puddles of rainwater. Conditions of use such as these should be borne in mind when selecting and specifying pallets.

Some common forms of damage to pallets, and their causes, are illustrated in Figure 22.7.

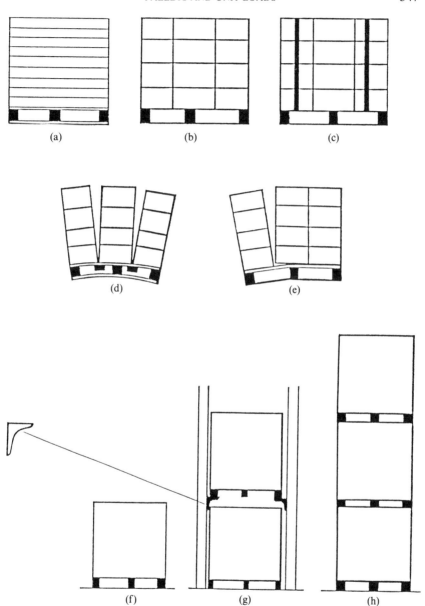

Figure 22.6 Pallet load capacity. (a) Self-supporting load. (b) Divided load. (c) Secured load. (d) Pallet with divided load bending when lifted. The bending moment is accentuated when the tines are close-set. (e) Pallet with divided load bending when lifted either when the tines are offset or at right angles, if the tines are shorter than the pallet. (f) One-high on ground. The area of deckboard supported depends on the design of the pallet base. (g) Effectively one-high in pallet racks. All pallets apart from those in the bottom layer are supported on two bearers only. (h) Three-high on the ground.

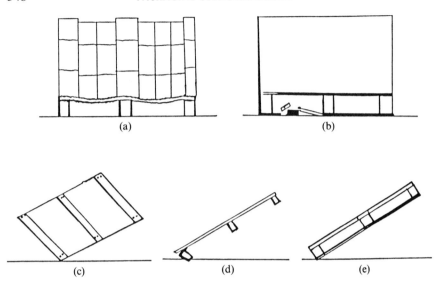

Figure 22.7 Some forms of damage to pallets. (a) A two-way entry pallet with light deck boards bowing under a divided load. (b) Lower deck boards of a four-way entry pallet broken when the loaded pallet is lowered on to uneven ground or on to an obstruction such as a large stone. (c) A wooden pallet changing from a rectangular shape to that of a parallelogram on being dropped empty on a corner. Many wooden pallets are liable to wracking in this way, particularly if they have narrow deck boards providing only a short distance between fastenings. (d) An outside bearer (or its equivalent lower deck board unit in a four-way entry pallet) being knocked away from the top deck when the pallet is dropped on to one edge. (e) A perimeter deck board pallet is less likely to be damaged when dropped on to one edge.

Forming and securing unit loads

With unit loads such as box pallets or skid-based wooden cases, the methods of forming the load are inherent in the design. Internal locating and securing devices, such as checking and bracing engineering products in wooden cases, are often required—but these are outside the scope of this chapter.

Flat pallets can be loaded in various ways, but in general they should be loaded as evenly and squarely as possible. The factors affecting the arrangement of the load will depend largely on its nature; in this chapter it is discussed briefly in relation to pallet loads of small filled corrugated cases only. These may be block-stacked (with each case vertically over the one below) or bonded (by varying the arrangement in alternate layers in various patterns) so that more cases can be loaded per pallet or to restrict shifting. Where the cases themselves, and not the contents, provide most of the stacking resistance required, this will be reduced by bonded stacking. The cases may also be stacked within the pallet deck perimeter or overhang the edge. The latter arrangement should further reduce stack resistance and makes the cases more liable to damage from contact with other pallets, backs of vehicles, etc.

Once the pattern and number of layers have been decided, the method of product-to-pallet transfer is determined. Hand palletizing is the most versatile method but is slow. Recently technological advances permit automated palletization of almost every product, but this is not always economical.

High level palletizers

High level palletizers pick up the product at a level about the height of a full pallet and form the pattern layout on a bed. The pallet is then raised to the bed level, and the product is transferred (by a sweep or rake-off system) to it. Then, the pallet is lowered one layer, and this process continues until the prescribed height is reached.

Low level palletizers

Low level palletizers (also called 'fixed-pallet' palletizers) developed more slowly than high level palletizers, but with the introduction of programmable logic controllers have become highly competitive with high level units in cost and speed. The low level palletizer operates at floor level (see Figure 22.8).

Bulk palletizers

The shift to bulk handling has reduced the demand for case depalletization of empty containers. There are numerous applications for palletizing without benefit of a case; for example, the transfer of empty cans from the can manufacturer to the filler.

In bulk palletization, the product accumulates on a table and is transferred by lifting a layer or clamping the perimeter and sweeping the layer onto a pallet. A corrugated sheet the size of the pallet, is generally used between layers to provide stability.

Robotic palletizers

In the past, 'pick-and place' systems picked up a product at one point and placed it at a second point repeatedly. Robotic palletizers may be thought of as intelligent, discriminating systems capable of picking up several different products and placing them at several different points . They are usually capable of movement in two or three planes, often on rotational coordinates operating from a fixed point at the centre of a circle. They are capable of picking up a variety of products from one point on the circumference of the circle and placing it on one of several pallets located at other points around

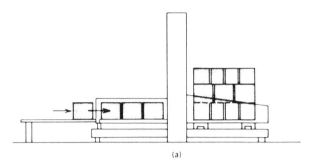

(a)

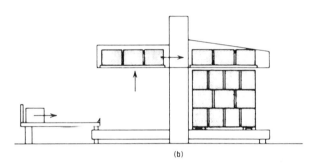

(b)

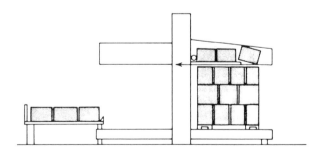

Figure 22.8 Operation of a low-level palletizer. (a) Sealed shipping cases feed in and are oriented to the pre-programmed pattern. When one row is formed, the cases move forward. The next row forms and moves forward, continuing until the layer is complete and the loading plate is filled. (b) The layer is lifted to the height of the existing pallet stack. The filled loading plate moves into position just above the pallet stack. The filled loading plate retracts allowing the cases to settle, row by row on to the top of the pallet stack. The pallet is squared by a bar which assures complete unloading. The loading plate returns to the starting position where another accumulated load is ready.

the circumference. Robotic palletizers are most applicable when a variety of products require different pallet patterns and relatively slow speeds are acceptable.

Load restraint

Flat pallets provide no built-in means of internal load restraint. Despite this, unit loads may be moved on them without restraint, particularly when, as with some goods, e.g. sacks, there is little tendency for the load to shift in the chosen distribution system. When required, there are three main alternative or complementary methods for securing the load: adhesives, strapping and enveloping.

Adhesives

Low-shear adhesives are typically used, which bond cases together on the pallet, but enable them to be pulled loose without significant damage. The adhesives are often sprayed on during palletization.

Strapping

High-tensile steel, plastics or textile strapping is widely used, mainly applied with special hand or automatic tools. Relatively high forces are involved, and edge protectors may be necessary to prevent the straps cutting into the load. Strapping should be in continuous contact with the load and the underside of the pallet deck; with four-way entry pallets it should be parallel to the top deck stringers to avoid damage from forks (Figure 22.9). More detail on strapping is given at the end of this chapter.

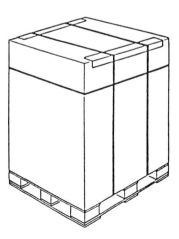

Figure 22.9   Restraint by strapping.

Enveloping

Enveloping may take a number of forms.

Shrink-wrapping: a loose hood of plastic film dropped over the load and the pallet is then shrunk by heat. Shrinking can be performed in a tunnel, in an oven, by lowering a rectangular heating element over the load or by portable guns. Features of shrink-wrapping include:

1. Relatively low forces over the whole load area, edge protectors unnecessary
2. Effective with a wide range of loads, regular and irregular
3. Weather protection provided
4. Shrink-wrap below the pallet not affected by forks
5. Nature and condition of load can be observed through the film

Stretch-wrapping is similar to shrink-wrapping in result. An elastic film is used, applied with special equipment.

The application of a *fibreboard snood* over the load, usually fastened to the pallet by nailing or strapping.

Vacuum packing may be accomplished with a heat-sealed plastic envelope or a heavy-duty returnable envelope. The purpose is more usually to provide a high degree of climatic protection rather than to secure the load.

In addition to securing the load on the pallet, the unit loads themselves must be properly secured on or into the vehicles or other means of conveyance. So far as road vehicles are cornered, guidance is given in the Code of Practice 'Safety of Loads on Vehicles' published by the Department of the Environment and available from HMSO. This deals with a variety of loads and suggests the acceleration *g* levels against which their fastening must restrain them. If palletized loads are stacked more than one high, each layer must be secured separately; sheeting is not generally adequate as a means of securing loads.

Pallet and unit load dimensions

The choice of pallet and unit load dimensions for any purpose is related to:

1. The dimensions, stacking pattern, etc., of the components of the load to be carried. For a returnable general-purpose pallet, the range of loads likely to be carried has to be considered.
2. The limitations of the associated equipment, including fork-lift trucks, road vehicles, rail wagons, ships' holds, freight containers, etc.
3. The dimensions of the existing pallets being used by the same company or on the same distribution system.

The following dimensions for through-transit pallets are generally accepted: A, 800×1200 mm; B, 1000×1200 mm; C, 1200×1200 mm; D, 1200×1800 mm. Sizes A, B and D are also recommended by ISO.

It should be noted, however, that the internal width of an ISO Freight Container is 2299 mm and none of the recommended pallet dimensions is modular to this. This generally means that to load ISO pallets onto a Freight Container requires the use of the more expensive four-way entry pallets. Modification to the ISO standard has now been accepted and the first modular dimensions of 600 mm and 400 mm has now been joined by a module of 550 mm permitting two one-way pallets to be loaded side by side into a freight container.

A most important dimension is the free height for entry of the forks, which should be a minimum of 96 mm. Most pallets are made to this standard, but skids on wooden cases are often made more shallow; they may therefore not be handled by fork trucks as intended, or be damaged by forks being forced under the case.

Pallet utilization

A common way of classifying pallets is into single-trip (or expendable) and multi-trip (or returnable). To take two extreme situations, an expendable pallet will typically be used on an export journey where there is no prospect of it being returned economically. The single-trip pallet will normally be selected to give acceptable performance at minimum purchase price. In contrast, a closed-circuit distribution system, wholly within the pallet owner's control, will normally employ returnable pallets. The capital cost of the pallet can be balanced against the service life and repair cost (if a recovery and repair system is instituted) and a pallet with the lowest purchase price may not necessarily be the most economical choice. Higher first cost may be more than justified by longer life and lower maintenance.

On many distribution systems, however, pallets and their loads are delivered to a customer. It usually is not practicable to depalletize immediately, so that the delivery vehicle can collect and return the pallet. Various arrangements may be tried, an empty pallet from a previous delivery can be collected in exchange for a full one, or the pallets can be invoiced, and a credit given for return. Arrangements of this type work better in theory than in practice, and high pallet loss rates are common. The supplying company may therefore consider the use of an expendable pallet. The cost equation needed to make a comparison is straightforward so far as the expendable pallet is concerned, but requires an actual average service life figure to complete the returnable cost side. The difference between an expendable and returnable pallet is therefore one of utilization rather than of type.

Performance requirements are the same for expendable and returnable pallets, except that the latter have to meet them repeatedly. The requirements are relatively severe, specially for heavy divided loads—which any general-purpose pallet may have to meet. For wide acceptance as an alternative, an expendable pallet has to cost less than the returnable by an amount which renders its use economic. From the nature of materials, it is difficult to achieve the bending and crushing properties needed within the tight dimensional and cost limitations.

As indicated earlier in this chapter, there are a wide variety of methods for palletization and unitization, some being specially suitable for certain types of loads. While, therefore, a general-purpose expendable pallet may be unattainable at acceptable cost, it may be quite practical to design a low-cost pallet for specific loads and distribution systems.

Many returnable pallets are purchased and owned by individual users. An alternative system is the pallet pool. An example is the European Pallet Pool operated by the railways. This operates on one size of pallet, 800×1200 mm and two types, with or without sides. To participate, a company purchases the number of pallets they expect to use. For every loaded pallet delivered to the railway, they receive an equal number of empty pallets in return.

Economic factors

The economic advantages of unitization are both *direct* (in lower distribution costs) and *indirect*, e.g. in pack cost and damage rate reductions.

The *direct advantages* accrue not only to the consignor, but also to the transport agencies and often to the consignee as well. The consignor saves on handling and storage costs in his own warehouse, as well as on shipment charges. Unitization enables transport agencies to achieve larger and faster throughputs on very flexible systems. To encourage unitization, shipping companies, for example, may exclude the pallet in calculating the shipping volume, and may offer lower rates.

The Unit Load Concept based on the theory that cargoes should be packed so that they can be moved and handled by mechanical equipment at all links of the transport chain is now well established.

The consignee may receive a direct benefit from unitization in a similar way to the consignor, saving on handling and storage costs. Some companies have instituted systems with all their suppliers so that all goods are sent unitized on pallets of agreed dimensions.

Prior to World War II, virtually all freight was shipped break-bulk. The most publicized postwar change has been the introduction of containerization. This, however, demands major capital investment in ships, Containers, handling depots, etc., and Container lines still connect only a limited number of the world's ports. It is often assumed that containerization offers the ultimate in economy, but this is not necessarily so in comparison with unitization. Both certainly offer considerable economies compared with break-bulk. It should be noted, too, that a high proportion of Containers are filled with unit loads.

The *indirect advantages* of unitization can also be considerable. They stem from the fact that large unit loads handled mechanically are subject to lower transit hazards, especially in regard to impact. One pack weighing say, 10 kg and exported individually may have to withstand drops from heights of 900 mm on to edges and corners. Fifty such packs unitized onto a pallet may have to withstand impacts equivalent to less than 300 mm effectively onto the base only, and protected to some extent by the pallet. The unitized packs may therefore require much less protective packaging than those which are to travel individually. As in containerization, the limitations are not in the major portion of the export journey, but distribution to the final destination if pallets are broken down near the port of entry.

Few statistics are available covering the various types of pallet in service. It is estimated that wooden pallets hold the bulk of the world market, much of which is for 1000×1200 mm four-way entry, open-deck, perimeter-base, nailed wooden pallets. In the future, four trends can be expected.

1. The advantages and ease of unitization will result in more pallets and similar aids being used.
2. Greater recognition will be given to the vital part played by the pallet in modern distribution systems, and to the amount of capital which may be deployed in stocks of pallets This will lead to more careful handling, including modifications of fork-lift trucks to minimize the risk of damage.
3. The timber pallet is likely to remain dominant, but more attention will be paid to a number of alternatives to timber.
4. Further development of pallet pools will occur.

The Unit Load Concept is simple and fascinating in its ramifications. It seems to be one of those rare concepts with major applications giving advantages to all concerned and disadvantages to none. Essentially, it is based on the humble but invaluable pallet.

Strapping (steel or non-metallic)

Steel or non-metallic strapping or wire can be used to perform one or more of the following packaging functions:

1. Bundling: grouping and holding together several articles into a larger handling unit
2. Palletizing: securing one or more articles to a pallet
3. Unitizing: bundling or palletizing with provision for pick-up by mechanical equipment

4. Reinforcing: strengthening a shipping unit to withstand the hazards of transportation and handling
5. Closing: securing the lid or top of a container, as with a telescope box or an interlocking flange container
6. Baling: holding material together under compression, especially resilient items, to save space

Steel was originally by far the most predominant strapping used, but non-metallic strapping, especially rayon, nylon and polypropylene are now increasingly used for almost every purpose. Steel has certain advantages. The ability of steel strapping to hold is not affected by age, heat or cold, sunlight or dampness, dirt or oil and steel is not attacked by mould or insects. The non-metallic strapping materials, however, are more easily disposed of (e.g. by incineration), are more resilient and so may stay tight on a shrinking package, are light in weight and are non-staining.

Strapping was first applied by nailing it on to wooden containers. In 1913 tools capable of tensioning and sealing the strapping were introduced, and this greatly speeded the process and made possible its application to a wider variety of shipments, including those enclosed in paperboard, wrapped in paper and unwrapped. The increasing use of mechanical lift trucks and other handling equipment has contributed to a very substantial growth in the use of strapping and similar binders in recent years.

Metal strappings are available in a variety of cross-sections (Figure 22.10). Each of these has its own advantages and disadvantages. Wire can be twisted together at the ends to eliminate the cost of separate seal. It can be draped diagonally across the face of the package if desired, but it tends to cut into the package at the corners, to stand away from the package, is harder to bend at the corners (especially in heavier wires) and is more difficult than flat strapping to pull tight around the package. Flat strapping can generally be applied faster with manual tools than can wire, it conforms better to the shape of the package, offers less friction for more effective tensional

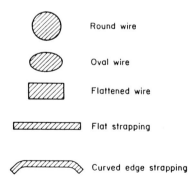

Figure 22.10   Cross-sections of metal strapping materials.

reinforcement and is firm and straight, and so can easily be passed through a pallet or fed through chutes or tracks. Oval or flattened wires have characteristics intermediate between flat strapping and round wire.

Steel strappings and wires are available in a variety of finishes. A galvanized finish provides long-term protection against rust, roughly proportional to the amount of zinc deposited.

Steel strapping is also available in a choice of painted finishes. Some of the best painted finishes rival galvanized strapping in rust resistance. Both blue oxide and black oxide finishes have been popular in the past, while bright (untreated) metal finish is still manufactured, but rusts too rapidly for most packaging purposes.

Strapping for use in feed-wheel type tensioning tools should have a finish coated with the proper type of wax or other lubricant to withstand the extreme pressures exerted by these tools. This lubrication also minimizes the loss of tension in pulling the strapping around the corners of a package. However, lubrication makes it more difficult to obtain a securely sealed joint; for this reason heavy-duty strapping that will be subject to heavy shock loads is often furnished 'dry' (unlubricated).

Strength

Strapping is bought mainly for strength—strength to reinforce a package, strength to hold a bundle together, strength to hold compression in a bale, strength to hold a load in place. What happens when there is not enough strength? The package breaks open, the bundle comes apart, the bale expands, the load shifts and the purpose is not fulfilled. That is not economy. The selection of the right material for the purpose is essential.

Adequate elongation

The *elongation* is a measure of the amount that strapping can stretch. It is a measure of its ability to absorb impact shock by 'rolling with the punch'. All hot-rolled medium-carbon strapping and cold-rolled, heat-treated medium-carbon strappings should have an elongation of from 5 to 16%. On the other hand, elongation is an undesirable characteristic in those applications for which cold-rolled strapping is used. Too much elongation in such situations allows a strap to become too loose to reinforce packages adequately. The elongation of cold-rolled strapping is only between 1 and 3%, and this is ideal for package reinforcement.

Flat steel strapping is produced in four basic types with important differences in their physical characteristics:

1. *Hot-rolled medium-carbon strapping:* has considerable elongation and is rather ductile, permitting heavy impact shocks to be absorbed by yielding (Table 22.2).

Table 22.2 Tensile strength and weight of heat-treated and hot-rolled strappings.

Width mm (in)	Thickness		Average tensile strength (N)	Weight (m/kg)
	mm (in) hot-rolled	mm (in) heat-treated		
13 (½)		0.50 (0.02)	6450	19.6
13 (½)		0.50 (0.02)	8050	15.8
15 (⅝)		0.57 (0.023)	8900	13.6
18 (¾)		0.64 (0.025)	11650	10.5
18 (¾)		0.79 (0.031)	13790	8.5
31 (1¼)		0.79 (0.031)	24240	5.1
31 (1¼)	0.89 (0.035)		22680	4.4
31 (1¼)		1.12 (0.044)	33850	3.6
31 (1¼)	1.40 (0.050)		31150	3.0
50 (2)	1.40 (0.050)		48900	2.0

2. *Cold-rolled low-carbon strapping:* has a high yield point and low elongation. The cold-rolling process permits an improved finish and closer control of tolerances, and this is the type of strapping most used in packaging (Table 22.3).

3. *Cold-rolled heat-treated medium-carbon strapping:* combines the improved finish and closer control of the cold-rolling process with the greater ductility of the hot-rolled product, together with a substantial

Table 22.3 Tensile strength and weight of cold-rolled low-carbon strappings.

Strap size		Average tensile strength (N)	Weight (m/kg)
mm	inches		
7.5×0.30	⁵⁄₁₆×0.012	1950	52.6
9×0.25	⅜×0.010	2000	52.6
9×0.38	⅜×0.015	2890	35.1
9×0.50	⅜×0.020	3690	26.3
13×0.25	½×0.010	2670	39.5
13×0.38	½×0.015	3825	26.3
13×0.50	½×0.020	4890	19.7
13×0.57	½×0.023	5515	17.2
15×0.38	⅝×0.015	4800	21.0
15×0.44	⅝×0.018	5600	17.6
15×0.50	⅝×0.020	6140	15.8
15×0.57	⅝×0.023	6890	13.7
18×0.38	¾×0.015	5740	17.6
18×0.50	¾×0.020	7340	13.1
18×0.57	¾×0.023	8270	11.5
18×0.70	¾×0.028	10000	9.4
18×0.89	¾×0.035	11560	7.5

gain in strength. It is a premium grade, justifying its extra cost through improved performance in many demanding applications.

4. *Cold-rolled annealed strapping:* a soft strapping capable of repeated bending and easily nailed through. It is widely used as a nailed-on reinforcement on cases for bottles.

Hand tools

Tensioners. Tensioners are used to tighten a loop of strapping around the object to be strapped, to apply tension on two overlapping strap ends. Most tensioners have a base which goes under the strapping (Figure 22.11) and which rests on a flat or nearly flat surface of the object. If this surface is narrow or if its girth is small, the freedom of the tensioner to move with take-up may be very limited. Some tension will be lost when the base is taken out from under the strapping. To avoid such loss, a push-type tensioner which has no base under the strap is required. (Figure 22.12) .

Since it operates by pushing against the seal (by tightening a slip knot) the push type tensioner requires that seals be threaded on to the strapping, and the end of the strapping bent back under the seal. Push-type tensioners are preferred for small or irregular bundles and for narrow packages.

The majority of strapping applications utilize conventional tensioners with a base under the strapping. These are usually faster to use than the push

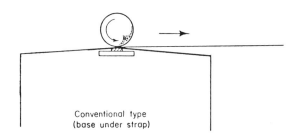

Conventional type
(base under strap)

Figure 22.11 Steel strap hand tensioner with base under strap.

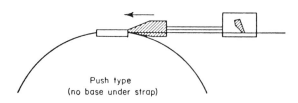

Push type
(no base under strap)

Figure 22.12 Push type tensioner (no base under strap).

type, and permit somewhat higher tensions to be attained, but they require a wide enough surface to permit the tool to move with take-up.

Sealers. Sealers are used to join the overlapping strap ends with a metal seal. They must do this reliably, always producing a strong secure joint. A notch-type sealer cuts into the outer edges of the strapping and seal, and turns the resulting tabs down (regular notch) or up (reverse notch). This type is generally used on waxed or coated strapping.

Combination tools. The functions of both tensioner and sealer, and often cutter as well, may be provided in a single tool. It eliminates reaching for and then putting down these separate tools. A seal-feed combination tool carries a stack of seals in a magazine, and eliminates reaching for a seal and manually placing it on the strapping. For most efficient use on a production line, a combination tool should be mounted close to its operating position.

Power strapping machines

A power strapping machine can bring a variety of benefits. It increases an operator's productivity, so that he can perform additional functions such as inspection, stacking, labelling, marking and the like, as well as speeding the strapping operation. It improves the reliability and uniformity of the strapping job. There are three stages of automation in this equipment.

Semi-automatic machines leave the following functions to the operator: he/she positions the package in the machine, pushes the strap feed forward, takes the end of the strapping from the upper chute, passes it over and down in front of the package and inserts it into a slot in the lower chute, pushes the cycling control, and removes the package. These machines save the cost and complication of the more automatic control systems, are extremely flexible as to package shapes and sizes, and are used where little would be gained by freeing the operator for other tasks.

Automatic machines require only three functions of the operator: positioning the package in the machine, pressing the foot switch, and removing the package. An operator with an automatic machine often applies more than a thousand straps per hour, in addition to performing other functions such as labelling and marking. Such speeds are, of course, limited to packages or bundles, light and small enough to be moved in and out of the machine at that rate.

Operator-less machines make use of a powered conveyor to bring the package in and out of the machine, a system of automatic controls to provide proper spacing and sequence, and a pusher cylinder or skewed conveyor rollers to position the package against the machine head. Operator-less machines have been installed which include the following position controls. A photoelectric cell or feeler switch senses entry of the package. Straps are put on at pre-determined distances from the front of the package. The feeler switches locate cleats on a crate or box; a strap is applied over each cleat. The carton or wrapper carries conductive or magnetic ink marks at each strap location. The head is sensitive to these marks, but to no other marking on the package.

Automatic machines have also been developed for specific industrial requirements, e.g. to secure steel coils, bricks, tubes, paper reels, compressed loads of corrugated containers and bagged products.

Non-metallic strapping

Non-metallic strapping in nylon, polypropylene or rayon cord was introduced to overcome certain shortcomings found in all forms of steel strapping and wire. Table 22.4 gives some average figures for the various types of material used. In choosing a strapping material, a number of factors should be considered: price of strap, strength, elongation, elastic recovery, preferred method of application, danger of rusting and of damage to the packaged goods, safety hazards, disposability problems, etc. The first question is whether the strap must restrict or accommodate movement of the package. Basically there are three types of package: those which tend to expand, those which do not change in size and those which tend to shrink. There are mainly three types of strapping in use today: steel, nylon and polypropylene. Figure 22.13 illustrates how different types satisfy the service and package requirements.

Table 22.4 Comparison of strapping characteristics.

	Types of strapping		
	Steel strapping	Nylon and plastics strapping	Typical rayon cord strapping
Tensile strength (kN/m^2)	700 000– 1 000 000	400 000– 450 000	220 000– 240 000
Tension transmitted around a 90°C corner (%)	72	72	56
East of feeding and threading	Excellent	Good	Poor
Smoothness	Excellent	Excellent	Fair
Stability under long-term load	Excellent	Good	Fair
Weight (density kg/m^3)	7750	1100	1400
Resistance to abrasion	Excellent	Excellent	Poor
Resistance to shearing (as on a sharp corner)	Good	Good	Fair
Ease of disposal	Good	Excellent	Excellent
Can be incinerated	No	Yes	Yes

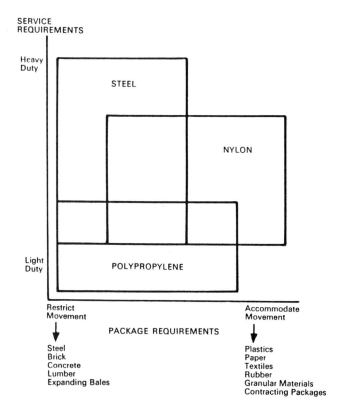

Figure 22.13 Fields of application for principal strapping materials.

Elongation and elastic return

When non-metallic strapping is placed under tension, it will elongate between 5% and 7.5% as the tension is increased to load limit, according to the type of non-metallic strapping used. Elongation expresses how much the strap stretches when it is tensioned. Normally only the elongation at break is given, i.e. how far the strap will stretch before it breaks. This information is useful, but it is also very important to know how much the strap will elongate at any given tension. In other words to know the curve in Figure 22.14 which compares steel strap, nylon strap and a typical polypropylene strap.

The first part of the curve (shaded) indicates which elongation of a non-metallic strap can be obtained in practice during a typical application. Generally there is a greater elongation with nylon and plastic strapping than with the rayon type. When the tension is released, non-metallic strapping in most instances will return to its original length. The benefit of such elasticity is that, in the course of time, the tension on certain packages may be reduced

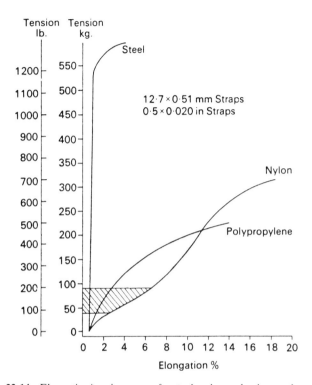

Figure 22.14 Elongation/tension curves for steel, nylon and polypropylene strap.

because of fatigue (fibreboard cases), drying out (wooden boxes) or settling (bales).

Transmission of tension

As non-metallic strapping is brought under tension, it can be seen to glide smoothly and easily around the package. The low coefficient of friction prevents binding at the corners. The difference in strap tension between top, sides and bottom of the package is thereby reduced to a minimum.

Package conformation

The flexibility of non-metallic strapping enables it to 'hug' the package and remain in position despite some irregularity in package shape. It can be successfully applied to some irregular shapes where a non-flexible binder, in order to remain in place, may require tension sufficient to damage the package. Good conformation also makes the strapping less subject to snagging.

Moisture resistance

Moisture has no effect on non-metallic strapping for all practical purposes. Although shippers normally take precautions against damage to their packaged products, conditions cannot always be controlled. Where nylon or polypropylene strapping is used, high humidity conditions will not affect performance, but humidity can cause rayon strapping to change in length and lose tensile strength. Moreover, when rayon strapping becomes wet, it may disintegrate.

Types of tools available

It is true to say that most non-metallic strapping can now be used with the same type of equipment as for steel strapping. In addition, machines which form the joint in non-metallic banding by a welding process have also been developed. The heat for these is provided by electricity, or by friction and impact welding, and a completely non-metallic binding with no metallic seal is formed. Light binders, i.e. automatic machines using strap sections of about 6 mm are now available with specially adapted models for packing wet fish and bundling newspapers. However, developments are so rapid in non-metallic strapping and tools that it is advisable to consult the suppliers. It is also suggested that comparisons should be made between each type of product.

23

Package cushioning systems

The term *packaging cushioning systems* covers a wide variety of techniques and materials for protecting goods from the effects of impacts and vibration in transit. They range from the traditional application of straw and woodwool in the packaging of glassware, pottery and ceramics, to the highly sophisticated shock isolation systems, incorporating springs and hydraulic shock absorbers, used in packages for aero-engines and space satellites. Goods can be protected against impact in three ways.

1. By spreading the forces on impact over a large area so that the force per unit area is reduced and no part of the product is subjected to concentrated forces
2. By supporting the product at its stronger points so that the forces on impact are directed to those points and not to the weaker ones
3. By reducing the forces and hence the deceleration on impact by using materials which compress on impact and cushion the product by absorbing the energy of the impact and converting the short duration/high intensity forces to lower intensity/longer duration forces which are less damaging

Methods 1 and 2 do not require a compressible material or system to reduce damage to the exterior of a product but will not necessarily prevent internal damage for example in electronic equipment or a complex mechanical assembly. In practice, cushioning systems combine method 3 with load spreading or directing forces through the area of contact and the placing of the system in contact with the goods.

Cushioning systems fall into three main groups: substantially elastic material, less elastic materials and load spreaders. Most of the materials and devices used fall into one of these categories (Figure 23.1). However, most bulk cushioning materials also act as load spreaders due to the often quite large area in contact with the packaged article and most load spreaders are often more or less compressible so that they too reduce the forces on impact to some extent.

Some load spreaders are loose fill materials often called space fillers, used as the name implies to fill the space around an irregularly shaped article placed in a regularly shaped container. They permit the controlled movement which is essential to reduce the shock transmitted to the contents

Package Cushioning Systems

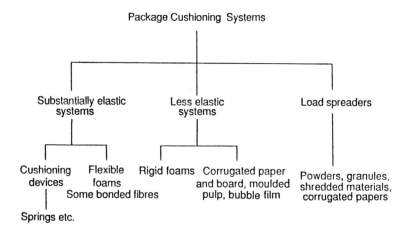

Figure 23.1   Cushioning systems.

when the package receives an impact, but the level of shock protection is not easily predicted because of the non-homogeneous nature of these materials, and the difficulty of ensuring a uniform density throughout the space filled. This is not important in the bulk packaging of glassware and pottery, where the main function of a space filler is to prevent the pieces from making contact with one another, or with the walls of the container, but it is important in packaging more shock-sensitive products. For this reason, space fillers are normally used for fairly robust or low-value articles.

Less elastic systems cover a wide range of materials from corrugated boards and rigid foams which crush on impact and provide one drop protection to those such as expanded polystyrene which provide protection by a combination of crushing of the polymer matrix and mainly elastic compression of the air entrapped in the closed cell structure of the material. The level of shock protection is predictable although it deteriorates with repeated impacts due to crushing of the material's structure.

There are two distinct types of elastic system: *bulk cushioning materials* and *cushioning devices*. Expanded plastics and bonded foams are typical of the first type. Their performance is predictable, and simple engineering design techniques can be applied to the package design, provided suitable performance data are available. Such data can often be obtained from the material manufacturers. The term *cushioning device* includes a wide variety of rubber and steel springs. In general, resilient systems are more expensive, and their use in consumer packaging is restricted to higher value fragile goods, although they are sometimes employed in re-usable packages where the high cost can be justified. They are also widely used in the packaging of stores and equipment for the Defence Services where the long life required of the package (normally 5 years) justifies the cost.

Nature of the product to be protected

The effect that any impact will have on an article will depend on its fragility or robustness—the ease or the difficulty with which it can be damaged. With products such as milk, fishpaste, beer, custard powder, fruit juices and frozen vegetables, it is the primary package into which they are filled (the glass bottle or jar, the paperboard carton, the metal can or the plastic container) that receives the shock of impact and it is these that must be protected not the product itself. The cost of such primary packagings is low enough for transit testing in the laboratory to check their impact resistance without measurement of their fragility.

With more complex equipment such as domestic appliances, televisions and computers, after ascertaining the basic physical data (weight, dimensions, surface finish, centre of gravity and centre of symmetry) we must measure or estimate its fragility.

The nature of the damage sustained in impacts will obviously depend on the materials from which an article is made and also on its structure. A typewriter, a word processor, or a domestic refrigerator, may have some parts broken, some parts deformed or distorted or it may fail electrically due to shocks from impacts.

Internal damage is dependent not on how the impact forces are applied to the product, but on their magnitude since the internal damage arises from the deceleration occurring during the impact, which is directly proportional to the applied force. The peak deceleration the product receives is proportional to the peak force applied during impact and is usually measured in multiples of the normal gravitational acceleration ($g = 9.81$ m/s).

The fragility of a product is then defined in terms of the fragility factor or impact load factor in multiples of g. Thus an item with a fragility factor of 50 could withstand an acceleration of 50 g. This does not mean that it could withstand such acceleration indefinitely, but for the short periods of time (not more than a few hundreths of a second) which occur on impact.

It is theoretically possible, if we know enough about the structure of an article, the properties of the materials of which it is made and the masses of its component parts, to calculate its fragility. But for all but the very simplest articles, the calculations are so complex that the method is quite impractical.

Test equipment for determining product shock fragility consists of a massive table (on to which the article to be tested can be firmly secured) which can be released from the desired height to fall freely in a guided manner on to a shock programmer (Figure 23.2). This is a controllable and variable type of cushion which pre-determines the distances which the table moves before being brought to rest (the stopping distance) and hence determines the deceleration (shock) it receives. The shock is transmitted through the supports securing the table to the article under test. A rebound brake mechanism prevents any bouncing or secondary impacts. Using this equipment over a range of shock impulses, the fragility in any particular direction of impact can be determined (see ASTM D3332) for details.

Properties of cushioning materials

The properties required of a cushioning material depend on the nature of the article to be protected. As we have seen the range is very wide. The properties listed here do not apply in all instances. In general, only some of them will be relevant.

Dynamic performance

Dynamic performance describes the first and most important property—the ability to provide shock protection. The performance data are obtained on a drop test machine in which a flat bottomed carriage falls from selected heights down guides on to test pieces of cushion of different thicknesses and measuring the maximum deceleration (peak g_n) experienced by the carriage at each impact. The mass of the carriage is varied to obtain a range of static stresses (i.e. weight of carriage divided by area of cushion) (Figure 23.3). The data are converted to peak deceleration (g_n)/static stress curves. Since most materials suffer some fatigue performance curves are usually based on the decelerations recorded at the third impact. A typical set of curves is shown in Figure 23.4. The parameter *static stress* is used because it is readily calculated from the design conditions.

Loose fill materials are tested similarly, but instead of a drop test machine, a dropping box is used. A layer of the required thickness of loose fill is placed in a rigid open topped box and a weighted block fitted with an accelerometer is placed on top. The box is then filled with more loose fill and a lid fixed in place. The box is then dropped from the specified height and the peak acceleration recorded. Weights are added or removed from block to cover the required range of static stress. Since loose fill materials always settle during transport, and allow the packed product to move, preliminary tests are made to determine the excess fill needed to give the required thickness to minimise movement.

The term *cushion factor* is commonly used to describe the efficiency of a cushioning material and reference figures are given for individual materials in Table 23.1. The cushion factor, C, for any drop condition (drop height, h, thickness of cushion, t and static stress) is given by

Figure 23.2 The fragility tester.

$$C = \frac{g_n \times t}{h}$$

where g_n is the peak acceleration. The figures quoted in the table are for the cushion factor at or near the minimum value of the g-static stress curves. A low cushion factor denotes high efficiency, e.g. the thickness of a polyurethane foam with a cushion factor of 2.2 required for a given level of shock protection is only half that required if rubberized hair with a cushion factor of 4.5 is used. Thus an appreciably smaller package size is achieved with the polyurethane.

Fatigue and thickness loss

Few cushioning materials are completely resilient; their structure is damaged by impact and this results in a progressive increase in the maximum deceleration recorded at each successive drop. This is known as *fatigue*. In addition, some materials do not recover to their original thickness after impact. This is sometimes undesirable, as it permits the cushioned article to rattle.

In general, space fillers and non-resilient cushioning materials suffer considerably both from fatigue and from thickness loss. With resilient

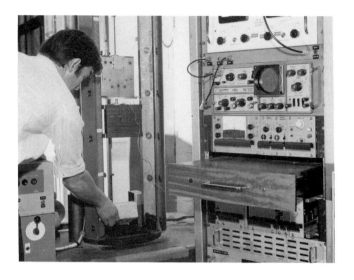

Figure 23.3   The drop test machine

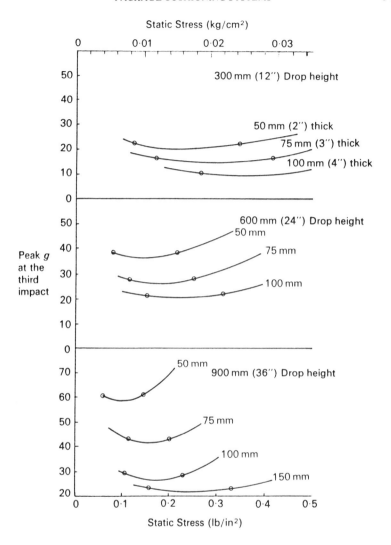

Figure 23.4 Deceleration/static stress curves for chipfoam of density 80 kg/m³ (6 lb/ft³).

materials, thickness loss is usually small and fatigue resistance is good, though materials vary considerably.

Creep

Most cushioning materials suffer a progressive loss of thickness under a static load, such as is applied by the article packaged while in storage. This is known as *creep*. Creep is measured by applying a static load to a test

Table 23.1   Properties of bulk cushioning materials.

Material	Cushion factor	Density (lb/ft³)	Thickness recovery	Fatigue resistance	Creep resistance	Temperature limits (°C) −	Temperature limits (°C) +	Moisture content (by weight) %	Water absorption	Corrosive effect	Mould resistance	Dusting
Wood wool	4.0–5.0	up to 5	Poor	Good	Fair	10	45	12–20	High	High when wet	Fair	Fair
Expanded polystyrene	3.0–3.5ᵃ	1.0–2.0	Poor	Poor	Good	30	70	0.2ᵇ	Low	Low	Good	Low
Rubberized hair	4.0–4.5	4 & 6	Fair	Fair	Fair	30	60	6–8	High	Low	Good	Low
Orientated rubberized hair	2.3–2.8	4 & 6	Good	Fair	Fair		60	6–8	High	Low	Good	Low
Crimped rubberized hair	3.0–3.3	5	Good	Good	Good	30	60	6–8	High	Low	Good	Low
Expanded rubber	3.5–9.5	11,18 & 28	Good	Good	Fair	0	60	up to 3	Low	Low	Good	Low
Polyurethane foam	2.0–3.0	1½–6	Good	Good	Good	20	60	10–18	High	Low	Good	Low
Bonded polyurethane chipfoam	3.0–3.5	3–15	Good	Good	Fair	–	–	–	High	Low	Good	Low
Expanded polyethylene	3.0–3.3	2.5	Good	Poor	Good	20	60	7–8	Low	Nil	Very good	Nil
Expanded EVA	3.0–3.5	2.8	Good	Fair	Good	–	–	–	Low	Nil	Very good	Nil
Bubbled polyethylene	4.0–5.0	–	Good	Good	Good	–	–	Very low	Low	Nil	Very good	Nil

ᵃ1st drop.
ᵇBy volume.

specimen of the cushion and measuring the loss of thickness at suitable intervals of time.

Thickness loss due to creep usually decreases roughly logarithmically with time. Plotting the results on a log scale gives a straight line, from which it is possible to extrapolate to estimate the thickness loss over a longer time. A typical graph of creep characteristics is shown in Figure 23.5. Such information is useful in comparing different materials but, if the data are used to calculate thickness loss over a long period, it should be borne in mind that the tests of short duration are carried out under constant climatic conditions. For cellular flexible intervals there is usually a % loss at which the creep rate increases rapidly as the cells buckle and caution is needed in predicting long term behaviour from very short tests.

Moisture content

Many materials, and especially those derived from plants or animals, tend to be hygroscopic, i.e. if dry materials are exposed in a damp atmosphere, their moisture content will increase until it reaches equilibrium with the surrounding atmosphere. All the figures in Table 23.1 were obtained after equilibrating at 35°C and 95% relative humidity. In general, a high moisture content has little effect on dynamic performance, even at low temperatures, but it does encourage mould growth.

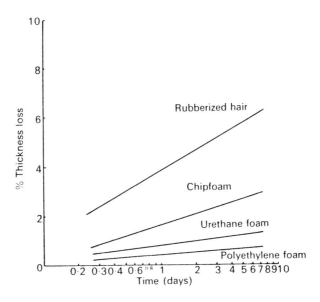

Figure 23.5 Measurement of creep, i.e. thickness loss with time.

Water absorption

Water absorption is usually measured by an immersion test. Clearly, materials of open structure, such as polyurethane foam, will have a high water absorption under these conditions and, if there is any likelihood of a package being exposed to liquid water, it would be wise not to use such materials. Instead a closed-cell material, with a low moisture absorption should be used (e.g. expanded polyethylene).

Corrosive effects

Corrosion will occur on metals if an electrolyte is present. In a packaging context, the electrolyte is usually contaminated water. Corrosion can be prevented in packages by controlling the internal relative humidity, and by ensuring that packaging materials are free from contaminants and the product has been cleaned and is free from finger prints. Relative humidity may be controlled by enclosing the product in a water vapour barrier, often with a suitable desiccant included, and in special instances carrying out the packing operation in a controlled environment. Contaminants are controlled by prescribing acceptable levels in the packaging materials.

The standard test for 'freedom from corrosive impurities' is to boil a sample of the material in distilled water and determine the pH of the solution. Briefly, pH is a measure of the acidity or alkalinity, and refers to the concentration of hydrogen ions in the solution. Pure water is slightly ionized, and with a pH of 7 is considered neutral. All solutions with a pH below 7 are acid, and those with a pH above 7 are alkaline. Most cushioning materials are required to have a pH not less than 5 or not greater than 9.

With some materials including those using rubber, corrosion may also be caused by the presence of reducible sulphur. This is checked by a silver staining test in which a small piece of the material is placed on a larger piece of silver foil in a covered dish and held at 70°C in an oven for 30 min. Non-silver staining grades of rubberized hair and expanded rubber are obtainable.

Mould growth

Mould grows rapidly on damp materials, and it is, therefore, desirable to select a cushioning mterial which is resistant to fungal growth, especially if storage in warm damp climates is expected.

Dusting

When packing products such as optical instruments or those containing delicate mechanisms, it is important that the cushioning material used does not contain or generate dust. This is a difficult property to measure precisely, and a rough order of merit is given in Table 23.1.

Space fillers

Powders and granules

Cork is the outer bark of the holm oak, an evergreen species, *Quercus suber*, which grows mainly in Spain. It is tough, light and elastic, hence its use in granulated form as a space-filling cushioning system. No cushioning data have been published.

Kieselguhr is a fine white siliceous powder containing the remains of diatoms. It is highly absorbent and non-flammable, hence its main use as a space filler around tins or carboys of corrosive or other dangerous liquids.

Vermiculite is the soft granular material formed by heating thin flat flakes of crude vermiculite (a mineral that resembles mica) at temperatures up to 1100°C. Each flake grows to many times its original size by heat expansion of both free and combined moisture in the flake. After heating it is called *exfoliated vermiculite*. It is normally used as a fireproof space filler and has good heat insulation properties. It is useful for keeping vaccines cool for several hours, and protecting aircraft flight recorders during a crash.

Wrapping materials

Cellulose wadding is made from a crêped web or sheet of lightweight paper of open formation, made of cellulosic fibres and comprises two, three or more plies. It is available in rolls or as sheets.

Single-faced corrugated board consists of a corrugated sheet faced on one side with a flat liner. Both fluting and liner are usually made from low-grade pulps. It is often rolled to form cushioning pads.

Air bubble film is also used for wrapping and consists of two layers of PE film between which regularly spaced air bubbles are trapped. The film may be coated to reduce the loss of air from the bubbles.

Shredded materials

Two forms of shredded paper are available, plain and waxed. The waxed form is to be preferred if damp conditions are anticipated.

Regenerated cellulose film is also shredded and often used as an alternative to shredded paper. The widths of the shredded materials may vary from 1.5 to 6 mm depending on usage.

Less resilient materials

Wood wool

Wood wool has been widely used as a space filler for packaging glassware, pottery, ceramics, etc. Its use has declined considerably in the last decade. Wood wool is a mass of tangled strands manufactured from sound well-seasoned softwood of low resin content. Cushion performance depends on the size and thickness of the strands, the packing density and moisture content, which is normally 12–20% by weight. Strands are normally between 1.5 and 2 mm wide and about 0.25 mm thick. Packing densities vary from 24 to 144 kg/m^3. The use of waterproof lining in the outer case will help to keep moisture content low during transit. For engineering products, a water-vapour resistant barrier is advisable, since damp wood wool can cause corrosion.

When using loose material, care is required to ensure that the correct weight of wood wool is applied to each face, and that it is uniformly distributed. This can often be achieved more readily by using pre-fabricated pads or bolsters.

Wood wool can also be bonded into conforming moulds for squaring off irregular shapes. A range of densities is available. Wood wool is primarily a space filler but, used with care, it qualifies as a non-resilient cushion and some performance curves have been published.

Expanded polystyrene

This is a closed-cell material, made from beads of polymerized styrene, which are expanded by pentane and air, and moulded in a steam heating process. It is available as slab stock in sheets of various thicknesses. The standard density is a nominal 1 lb/ft^3 but other densities are available. Performance curves are published for 1 lb/ft^3 slab stock material

Expandable polystyrene beads can be moulded very cheaply in a range of densities and colours to make fitments for products such as cameras, TV sets, radios, etc. Although relatively rigid, such material is often used in moulded form with ribs on the outside of the moulding. Under impact these ribs compress thus enabling the material to provide a degree of cushioning to lighter products than would otherwise be possible. Expanded polystyrene is also widely used in small shaped pieces which can be fed from hoppers as a free flow material to provide a combination of space filling, load spreading and cushioning. Packs are slightly overfilled so that, on closing, the compressive forces lock the shaped pieces into place and reduce the tendency for the product to move under vibration during transport. It is also available in strands for use as a space-filling material, although used in this way its performance is claimed to be comparable with that of some resilient slab materials.

Its main advantages are low cost, light weight and good load-bearing capacity; the main disadvantages are flammability (though fire-retardant grades are available) and poor resistance to attack by solvent vapours, including hydrocarbons.

Moulded fittings

Moulded fittings can be used to protect edges and corners and to square off irregular shapes. Using a suitable bonding resin, paper pulp and wood-wool mouldings are available in a range of densities.

Corrugated fibreboard

Corrugated fibreboard consists of a fluted sheet of paper faced on both sides with a flat liner sheet, often of kraft paper. This is known as double-faced or single wallboard. Double and triple wallboards are also made. The main use is in the fabrication of containers, but they are also frequently used to make cushioning pads and fittings.

Resilient materials

Rubberized hair

Rubberized hair is made from animal fibres, usually consisting of about 80% hog's hair and 20% horse's mane, horse tail or cow tail bonded with rubber latex and heat-vulcanized. Plain rubberized hair is most commonly used. In the plain variety, the hair lies parallel to the surface. By cutting the sheet into strips, and bonding the strips together at 90° to their original position, the hairs are re-orientated so that they are at 90° to the plane of the sheet. In this position they are loaded as struts. This improves efficiency, reducing the cushion factor from about 4.5 to 2.5.

Crimped rubberized hair

In this process curled hair is first carded into a delicate uniform layer of low density and then gathered by a system of needles, by which it can be condensed to any desired degree. In this way, not only is the original curl of the hair preserved, but in addition a crimped structure develops as a result of the needling action. Resilience, fatigue resistance and cushion factor are all improved.

Polyurethane foams (flexible)

Polyurethane foams are of open-cell construction and are formed by the polymerization and simultaneous expansion of an isocyanate and a hydroxyl compound. There are two types of foam, polyesters and polyethers. Polyesters tend to be unstable and are degraded by exposure to high humidities and temperatures. Polyethers are, therefore, preferred for most applications, although even these break down if exposed for long periods to sunlight. By altering the chemical constituents, an extremely wide range of foam stiffness and hence cushioning properties can be produced.

Polyurethane foams are amongst the most efficient of cushioning materials, with cushion factors in the range 2–2.5. At impact, the air contained in the open-cell structure is expelled in a very short time, thus doing work additional to that done by the material itself. Hence, it is important when using these materials in a closed container to leave some free space. Furthermore, the dynamic stiffness due to the trapped air is much higher than the static stiffness, so creep under load limits its application more than its dynamic performance.

The material is available in sheets in a range of densities from 12 to 112 kg/m^3. It can be readily and cheaply machined to make conforming shapes.

Polyurethane foam can also be moulded in the conventional way, although mould costs are high. Alternatively, the foam can be moulded in situ. The article to be cushioned is covered by a thin LDPE film, is suspended in its container, and the two liquid chemical components are mixed and the mixture dispensed into the space around the packaged article. Alternatively pre-moulded cushions can be made.

Bonded polyurethane chipfoam

Bonded polyurethane chipfoam consists of small pieces or flakes of polyurethane foam of the polyether type, bonded with a binder and cured to form a homogeneous network of interconnecting cells. It is available in a range of densities between 50 and 250 kg/m^3. A binder of a different colour is used for each density to aid identification. It is supplied in the form of sheet, roll, strips, corner blocks and mouldings.

Expanded polyethylene

Polyethylene is expanded about thirty times from the solid state to form a lightweight closed-cell material. It is available in two forms, cross-linked and uncross-linked. The cross-linked form has superior thickness loss and creep characteristics, especially at elevated temperatures. As no chemical additives are used, expanded polyethylene is free from sulphur staining compounds or other ingredients liable to tarnish silver, e.g. cellulose or lacquers. It can be sawn or cut with a sharp knife or a hot wire. It can also be shaped by moulding, hot hobbing or cold forming.

Expanded polyethylene is a cushioning material of moderate to high efficiency (cushion factor about 3) and light weight. Ethylene vinyl acetate (EVA) foam is a copolymer of polyethylene and vinyl acetate. It is more rubber-like in its behaviour, but is less efficient than expanded polyethylene, having a cushion factor of about 4.

Air bubble film

This is made from two sheets of polyethylene film, one of which is embossed to entrap air bubbles. Several grades are available with bubbles of different sizes. Small bubbles are recommended for protecting small parts, and large bubbles for heavier items. The material can also be used for wrapping, or in rolls or pads as cushions. Used as a heat-sealed envelope, it also provides a barrier to water vapour. It is supplied in rolls 1.25 m wide and up to 240 m long.

Corner pieces

A very simple method of packing articles which are generally rectilinear in shape is to use corner pieces. These are easily applied by unskilled labour and are effective and inexpensive. However, care is needed to ensure that the total loaded area on each face gives a static stress in the optimum range for the material employed.

Corner pieces can be made in most of the materials described here (Figure 23.6). Expanded polystyrene and bonded chipfoam mouldings are commonly used and expanded polyethylene lends itself to the fabrication of corner blocks from sheet. Other designs are also available, including corner pieces blow-moulded from polyethylene sheet (Figure 23.5).

Selection of bulk cushions

To select a package cushion we need to know: about the product—its overall dimensions, its weight and its fragility factor; about possible cushioning materials—their peak g_n/static stress curve and their cost. Finally we need to know the height from which the package is likely to be dropped in transit and the likely number of drops. Trials with instrumented packages have confirmed the general correctness of the traditional assumption that the

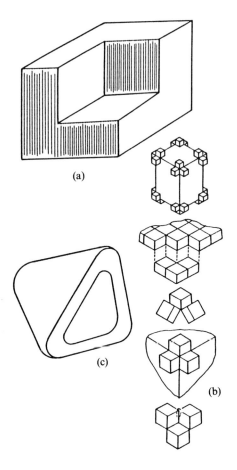

(a)

(c)

(b)

Figure 23.6   Corner pieces. (a) Expanded polystyrene mouldings. (b) Fabrication of corner blocks from sheets of expanded polyethylene. (c) Blow-moulded from polyethylene sheet.

lighter the package the greater the height from which it is likely to be dropped. The measured drop heights agree reasonably well with those derived from observation and judgment, and show that drop height is inversely proportional to the logarithm of gross weight (Figure 23.7).

The size and weight of an item are easily obtainable, but its fragility factor is more difficult to determine. The best way is to carry out a fragility test by dropping the article, fitted with an accelerometer, from increasing heights until damage occurs.

Experienced engineers can often make reasonable estimates of the fragility of closely related equipment by inspection, identifying the critical part, but there is a tendency to err on the safe side. Changes in construction or materials are less easy to estimate.

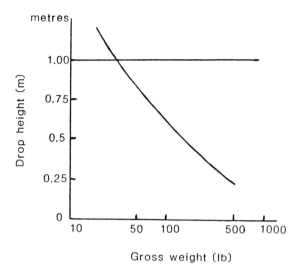

Figure 23.7 Drop height is inversely proportional to the logarithm of gross weight (approximately).

Many fragile articles, particularly those of electronic or complex mechanical structure, have one or two parts or components which are of greater fragility than the rest. Often these can be removed and tested separately and this may then permit redesign to strengthen the item. In many instances, particularly with larger assemblies, impacts will be limited to only one orientation, the base, thus reducing the amount of testing needed. Moreover there are many situations where the fragility test is used in product development, it being much less costly to produce a more robust item, through elimination of the weaknesses by fragility tests, than to package to protect a fragile item.

Peak g_n/static stress curves for many materials are now widely available. The following simple example is given to illustrate the basic design method. It assumes that the article is either rectilinear or can be made so by the addition of suitable fittings.

Consider an article contained within the volume $0.5 \times 0.4 \times 0.3$ m^3 weighing 20 kg and having a fragility factor of 40. We must calculate the static stress on each face as follows:

Sides $\dfrac{20}{0.5 \times 0.3} = 133 \text{ kg/m}^2$

Ends $\dfrac{20}{0.4 \times 0.3} = 167 \text{ kg/m}^2$

Top and bottom $\dfrac{20}{0.5 \times 0.4} = 100 \text{ kg/m}^2$

Next we must specify the drop height against which we wish to protect the item. This requires an estimate of the gross weight of the finished package. Assuming, for the purpose of this example, that gross weight will be twice the cargo weight (i.e. 40 kg) we see from Figure 23.7 that the probable drop height will be about 0.9 m).

Referring now to Figure 23.4 we see that, for a 0.9 m drop, between 75 and 100 mm (say 90 mm) of 6 lb/ft^3 chipfoam will be required. About 40 g_n will be achieved at static stresses between 75 and 150 kg/m^2 (0.1 and 0.2 lb/in^2), which spans the actual loading range.

If the article weighed 10 kg then the actual static stresses would have been halved. In such instances, the area of the cushions must also be halved in order to achieve the static stress at which the cushion is most efficient (i.e gives the lowest peak g_n). If the article weighed 40 kg it would be necessary to use a stiffer material—possibly a high-density chipfoam or a polyethylene foam.

In practice, the choice amongst possible cushioning is guided by the following considerations:

1. The area of cushion needed must not be greater than the exterior dimensions of the article to be protected, governed by the optimum static stress. A low static stress at the minimum of the relevant g_n/static stress curve will result in a large area of cushion. Conversely a high static stress may result in the cross-section of the cushion being small relative to the thickness, with a consequent risk of buckling.
2. The minimum pack dimensions will be given by the cushion with the lowest cushion factor (since the thickness in any application is proportional to the cushion factor of the material).
3. The total weight of the cushioning is proportional to the cushion factor times density divided by static stress.

Some typical fragility factors are listed in Table 23.2.

Other factors

Most cushioning materials are damaged by repeated impact and this shows itself as an increase in peak deceleration with repeated drops. This effect is usually referred to as cushion fatigue. However, the figures usually quoted are based on successive drops on a drop test machine, normally at 1-min intervals and, in practice, drops from the maximum height for which the packages were designed are rare.

The container itself is also resilient and therefore contributes to shock protection, although its contribution is difficult to estimate and is usually ignored. For these reasons, the curves used for design should be based on the peak deceleration recorded on the first drop.

Loss of thickness due to creep under static stress will also occur but, within the working range of static stress, it is not significant on most materials.

Table 23.2 Typical product fragility factors.

Classification	Fragility factor	Types of product
Highly fragile	15–25g	Precision instruments with sensitive mechanical bearings
Very fragile	20–40g	Electro-mechanical measuring instruments
Fragile	40–60g	Electro-mechanical equipment, e.g. typewriters, cash registers, calculating machines, etc.
Moderately fragile	60–85g	Television receivers, optical projectors
Fairly robust	85–100g	Domestic appliances (washing machines, refrigerators, cookers)

Flexible cellular materials are an exception and static stress should be limited with these materials to prevent excessive creep.

The effects of temperature and humidity should also be considered. Many cushioning materials, especially the polymeric materials, become stiffer at low temperatures, so that their cushioning effect is reduced. High temperatures have less effect on dynamic performance but may cause accelerated ageing effects if prolonged.

Polymeric materials are not affected by humidity but some loose fill and paper-based materials may be degraded.

Cushioning devices

Bulk cushioning materials cannot provide the answer to all cushioning problems, and it is sometimes necessary or advantageous to use cushioning devices. This is particularly so when a product is already provided with mounting or attachment points, which can be used to secure it to a shock mounted platform. For cargoes up to say 120 kg weight, the benefits are probably most evident with articles of irregular shape, since there is no need to 'square off' to provide faces for the cushioning material, and this results in a simpler, smaller, and often lighter package. With larger products such as aero-engines, where size and weight are such that the volume of bulk cushioning material required, results in a considerable increase in package size and weight, the cost of the cushioning material may well exceed the total cost of shock mounts and mounting platform. The greatest advantage of shock mounts for this type of product is that the cargo is secured only to the base of the case and, if required, the top, sides and ends can be removed in one piece, sewing-machine fashion, to give instant access to the contents.

In designing a cushioning system with shock mounts, several problems arise which do not occur with bulk cushioning materials. The design procedure is more complex since, in most packages, the system must provide roughly equal protection against shocks from all directions. Shock mounts have less natural damping than bulk cushioning materials, and care is needed to ensure that the spring/mass system is not likely to be excited at resonance in transit. On larger packages, hydraulic dampers are sometimes used to control resonance, and these have the double advantage that they greatly increase the efficiency of the shock isolators. With smaller packages space and cost do not allow this, but experience shows that satisfactory systems can be designed without dampers. In addition, special containers must be designed with local reinforcement, to take the high concentrated loads.

For all these reasons it is advisable, except perhaps for the simplest designs, to seek the advice either of the mount manufacturer or of a specialist firm of packaging engineers.

Bonded-rubber shock mounts

Shear mountings

A shear mount is a block of rubber bonded between two parallel steel plates, which incorporates holes or studs for attachment. Occasional shock deflections up to twice the rubber thickness are possible in the shear plane. In compression, the deflections available are too small to be useful and various geometrical arrangements can be adopted to give the necessary all-round protection. Mountings are available with shear stiffnesses in the range 2000–30 000 kg/m (100–1400 lb/in), and capable of shear deflections up to 60 mm. There are several manufacturers of shear mounts, but other types of shock mount are often porprietary designs.

Conical mounts

Conical mounts are designed to overcome the limitation of the shear mount, by providing flexibility in three planes. Some examples are illustrated in Figure 23.8. Generally, the axial stiffness of these mounts is much lower than the lateral stiffness. This is because they are designed to be mounted beneath heavy equipment where, in simple drops, they are loaded axially in compression, In side impacts, the fact that the centre of gravity is well above the plane of the mountings means that a rolling movement occurs, and the mountings are subjected to both axial and lateral loads. Mountings are available to support static loads in the range 170–800 kg.

Metalastik buckling mounts

The buckling mount is a development of the conical mount. Stiffness is similar in tension, compression and lateral shear (Figure 23.9). Two sizes are available. The smaller gives up to 55 mm deflection in all directions and

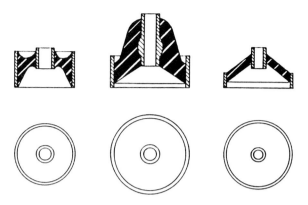

Figure 23.8   Conical mounts.

takes static loads between 30 and 70 kg, depending upon rubber hardness. The larger gives up to 100 mm deflection in all directions, and takes loads from 45 to 100 kg according to hardness.

Tubular mountings (BTR)

The tubular mounting (Figure 23.10) is an alternative to the conical mounting and is used in a similar way, i.e. mounted beneath the load. It is available in a range of hardness to take loads from 200 to 2200 kg per mount.

Delta mountings (Silentbloc)

The Delta mounting (Figure 23.11) represents a novel approach to shock mount design. It is available in a range of sizes. The smallest is 92 mm high

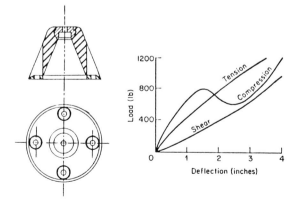

Figure 23.9   Buckling mount.

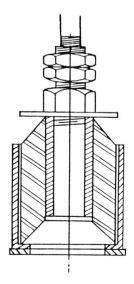

Figure 23.10   Tubular mount (BTR).

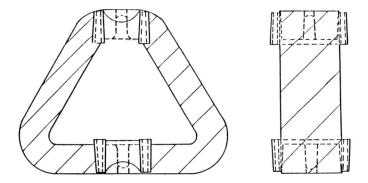

Figure 23.11   Silentbloc Delta mounting.

and gives maximum deflection of about 35 mm. The largest is 550 mm high with a maximum deflection of 225–250 mm.

Steel spring systems

Coil springs

When deflections in excess of about 280 mm are needed, bulk cushions and the conventional cushioning devices are seldom practical or economic. Deflections of this order are sometimes necessary for products of low

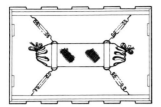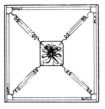

Figure 23.12   The use of coil springs.

fragility and light weight. Many years ago a tension spring system was evolved to deal with this problem, and it was adopted by the Ministry of Defence as a standard method of packaging electronic valves. A design procedure was published by Mindlin in 1945. A typical package is illustrated in Figure 23.12. Damping is low in the system and may sometimes be a problem.

Leaflote springs

This is a laminated leaf spring system suitable for heavy fragile loads requiring large deflections. The springs are curved to form a semi-circle, and the ends are curled over so that they slide easily over the surface of a steel plate secured inside the container. The semicircular shape gives a very high initial stiffness (Figure 23.13) and hence high efficiency. The chief disadvantage of the system is its great weight.

'Pnucush' mounts (Wilmot packaging)

The Pnucush mount is an alternative to the bonded rubber conical mount. It is especially useful in meeting extreme low-temperature requirements. Each

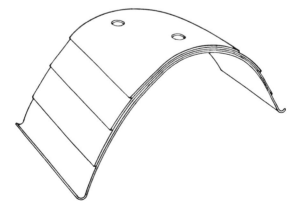

Figure 23.13   Leaflote springs: laminated steel plates of semicircular form giving a very high initial stiffness and hence high efficiency.

mount comprises a number of individual torsion springs secured between
two metal end plates (Figure 23.14). They are designed to provide deflection
by both axial compression and lateral shear, thus giving protection from
both drops and horizontal impacts. Mountings are available to cover the
weight range 14–140 kg per mount and to give impact load factors up to 30.

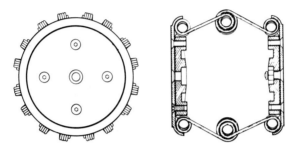

Figure 23.14   The Pnucush shock mount.

Bibliography

R. Adams and K. Harmon, *Reading Between the Lines, An Introduction to Bar Coding Technology*, North American Technology Inc, Peterborough NH, USA (1984).

Anon, CAD/CAM System aids folding carton design, *Boxbed Containers* **94**(9) Apr 1987, p. 42.

ASTM D 999–75, *Standard Methods for the Vibration Testing of Shipping Containers*.

ASTM D1596, *Standard Test Method for Shock Absorbing Characteristics of Packaging Cushioning Materials*, ASTM Philadelphia (1984).

ASTM D3332, *Standard Test Method for Mechanical Shock Fragility of Products ASTM Philadelphia* (1984).

ASTM D4169 *Standard Practice for Performance Testing Shipping Containers and Systems*, American Society for Testing and Materials, Philadelphia (1984).

B. Attwood, *Multi-ply Web Forming*, in Proceedings of TAPPI Annual Meeting, Atlanta, Ga., TAPPI, Atlanta Ga. (1980) pp. 229–241.

M. Bakker, *The Competitive Position of the Steel Can*, in Technology Forecast, Westport, Conn, USA (1984).

M. Bakker (ed.), *Wiley Encyclopedia of Packaging Technology*, J. Wiley & Sons, New York (1986).

C. J. Benning, *Plastic Films for Packaging*, Technomic Publishing Co Inc, Lancaster Pa (1983).

E. Bojkov, *Getranke-verpackung und Umwelt*, Springer-Verlag, Wien and New York (1989).

J. Boulanger, C. Reny, M. Veaux and J. L. Victor, *Emballage sous toutes ce facettes*, Laboratoire National d'Essai and Emballages Magazine, Paris, France, (1989).

I. Boustead and G. Hancock, *Energy and Packaging*, Halstead Pubs, John Wiley & Sons Inc., New York (1981).

I. Boustead and H. Lidgren, *Problems in Packaging, The Environmental Issue*, John Wiley and Sons Inc., New York (1981).

J. H. Briston and L. L. Katan, *Plastics in Contact with Food*, Food Trade Press Ltd, London (1974).

J. H. Briston, L. L. Katan and P. Godwin, *Plastics Films*, Longman, New York (1983).

J. H. Briston, *Plastics in Packaging, Conversion Processes*, Institute of Packaging, Melton Mowbray, Leics. (1983).

J. H. Briston, *Plastics Films*, Iliffe Books, London (1974).

J. H. Briston, *Plastics in Packaging, Properties and Applications*, Institute of Packaging, Melton Mowbray, Leics. (1983).

A. L. Brody, *Flexible Packaging of Foods*, CRC, Cleveland, Ohio (1970).

J. A. Cairns, C. R. Oswin and F. A. Paine, *Packaging for Climatic Protection*, Newnes-Butterworths, London (1974).

A. Caunt, *Factors to be Considered in Machine Selection*, Pira Seminar S29, 13 April 1978.

Checklist for a Packaging Brief, Siebert Head Ltd, London.

Childproof Packaging/Required Tests, DIN 55 559, FRG Standard (1980).

Child Resistant Packaging, ASTM Standard, ASTM, Philadelphia, USA.

A. Cowan, *Quality Control for the Manager*, Pergamon Press, London (1964).

N. Crosby, *Food Packaging Materials*, Applied Science, UK (1981).

D. Dean, *Plastics in Pharmaceutical Packaging*, Pira Reviews of Packaging, Pira, Leatherhead, Surrey (1990).

D. A. Dean, *Packaging of Pharmaceuticals, Packages and Closures*, Institute of Packaging, Melton Mowbray, Leics. (1983).

Drawn and Ironed Aluminium Cans, Aluminium Co. of America, Pittsburg (1975).

M. Fairley, *Bar Coding*, Label and Labelling Data and Consultation Services Ltd, Potters Bar, London (1984).

M. Fairley and R. Brown, *Thermal Labelling*, Label and Labelling Data and Consultation Services Ltd, Potters Bar, London (1984).

M. Fairley, *Label Packaging Processes and Technology*, Label and Labelling Data and Consultation Services Ltd, Potters Bar, London (1984).

M. Fairley, *Labelling Operations*, Pira, Leatherhead, Surrey, (1989).

N. Farmer, *Caps and Closures for Glass and Plastic Containers*, Pira Reviews of Packaging, Pira, Leatherhead, Surrey (1990).

S. Farrell *(Compiler)*, *International Packaging Sources*, Pira Information Centre, Leatherhead (1987).

P. van Gieson, *ASTM History in Child Resistant Packaging*, ASTM Standard News 26 April (1983).

R. Goddard, *Multi Packs*, Pira Reviews of Packaging, Leatherhead, Surrey (1989).

R. Goddard, *Packaging Materials*, Pira, Leatherhead, Surrey (1990).

R. R. Goddard and F. A. Paine, *Optimising the Packaging Costs*, 2nd IAPRI Conference, Munich 1976.

G. A. Gordon, *Use of Shock Testing in Developing Transport Packages*, Paper delivered to the IAPRI Symposium, Vienna, Sept. 1984.

R. C. Griffin and S. Sacharow, *Principles of Package Development*, AVI Publishing Co. Inc. (1980).

Guide to Compilation of Performance Test Schedules for Complete Filled Transport Packages, Part 1, General Principles, Part 2, Quantitative Data, ISO 4180/1 and 2 (–1980) and BS 6082/81.

Guide to Compilation of Performance Test Schedules for Complete Filled Transport Packages, Part 1, General principles, Part 2, Quantitative data, ISO 4180/1–1980.

Guide to Tinplate, Tech Publication No. 622, International Tin Research Institute, Middlesex (1984).

S. G. Guins, *Notes on Package Design*, Private communication of teaching notes for School of Packaging, MSU, East Lansing, USA (1979).

C. M. Harris and C. E. Crede, *Shock and Vibration Handbook*, McGraw Hill, New York (1976).

S. M. Herschdorfer (Ed), *Quality Control in the Food Industry*, Vol 4, Academic Press Ltd, London (1987) pp. 343–382.

A. C. Hersom and E. D. Hulland, *Canned Foods, An Introduction to Their Microbiology*, Chemical Publication Co, New York (1964).

A. Herzka and T. Pickthall, *Pressurized Packaging* (2nd Edition), Butterworth, London (1981).

H. J. Hohmann, Filling Characteristics and Dosing Accuracy of Volumetric Dosing Devices in VFFS Machines, *Pkg. Tech. and Science* **1**(3) (1988) pp. 123–137.

H. J. Hohmann, Folding ratio and erection of folding boxes under production conditions, *Pkg. Tech. and Science* **2**(2) (1989) pp. 109, 119.

International Article Numbering Assoc., Rue des Colonies, Brussels, Belgium.

M. A. Johnsen, *The Aerosol Handbook* (2nd Edition), Wayne E. Dorland Co. Ltd, Mendham, NJ (1982).

G. Jonson and J. Balkedal, Principles of calculating packaging costs, *Meddelande No. 26*, Packforsk, Stockholm, Dec. 1973.

K. Q. Kellicut and E. E. Landt, *Basic Design Data for Use of Fibreboard in Shipping Containers*, Report No. D1911, Forest Products Laboratory, US Dept of Agriculture (1955).

J. Knappe, *New Developments in Glass Packaging*, Pira Reviews of Packaging, Pira, Leatherhead, Surrey (1989).

C. Koppelmann, On the cost structure when using packaging, *Verpack. Rdsch*, **25**(5) May 1974, pp. 416–418, 420.

Yo Kusuda, *Cost Reduction in Japanese Retail and Transport Packaging*, Paper delivered at Intl. Packaging Cong., Paris, 16–17 Nov. 1982.

Le Pack, BSN Emballage, Information and Professions Service, EM, Paris (1989).

E. A. Leonard, *Economics of Packaging*, Morgan Grampian Inc., New York (1975).

E. A. Leonard, *How to Improve Packaging Costs*, AMACOM, New York (1981).

E. A. Leonard, *Managing the Packaging Side of the Business*, AMACOM, New York (1987).

E. A. Leonard, *Packaging Specifications, Purchasing and Quality Control*, Morgan-Grampian Publishing Co., New York (1976).

E. A. Leonard, *Specifications and Quality Assurance*, in Wiley Encyclopedia of Packaging Technology, (ed) M. Bakker, John Wiley and Sons, New York (1986), pp. 588–601.

Logistical Packaging (LP) System, Honshu Paper Co., Ltd, Japan (1982).

R. P. Long, *Package Printing*, Graphic Magazine Inc., New York (1964).

F. Lox, *Verantwoorde Verpakking*, Stichting Leefmilieu vzw, Antwerp (1983).

A. C. McKinnon, *Physical Distribution Systems*, Routledge, London and New York (1989).

C. J. Mackson, *Impact of New Technologies on Food Packaging and Presentation Industries*, MSU School of Packaging Report for US Congress, Office of Technical Assistance (1985).

R. C. Massey and D. Taylor (Eds), *Aluminium in Food and the Environment*, Special Publication No. 73, Royal Society of Chemistry, Thomas Graham House, Cambridge, UK (1989).

Methods of Protection Against Shock, BS 1133, Section 12.

R. D. Mindlin, *Bell System Tech. J.*, **24** (1945), p. 353; Bell System Monograph B 1369.

J. M. Montresor, H. P. Mostyn and F. A. Paine, *Packaging Evaluation*, Newnes Butterworths, London (1974) pp. 26–47.

B. E. Moody, *Packaging in Glass*, Hutchinson & Co. Ltd, London (1963).

C. R. Oswin, *Plastic Films and Packaging*, Elsevier Applied Science Publishers, London (1975).

C. R. Oswin and L. Preston, *Protective Wrappings*, Camm Publications Ltd, (1980).

Packaging Code, BS 1133, Sections 1–22, British Standards Institution, London.

Package Design, Facing Up to the Market, A report based on training lectures by Jean Volkaert delivered at the Nestle Training Centre, higher educational institutions and the International Trade Centre, Interntl. Trade FORUM, Dec. (1969).

Packaging Equipment Check List, *1984 Packaging Encyclopeida*, Cahners Publishing Co., Des Plaines, USA, p. 29.

Packaging in Plastic and Laminate Tubes, Washington Tech Center, American Can Co., Washington.

Packagings Resistant to Opening by Children, BS 6652, BSI, London (1989).

Packaging and Resources, INCPEN (Industry Council for Packaging in the Environment), Premier House, London (1989).

Packaging Standards, American Society for Testing Materials, ASTM, Philadelphia, USA.

F. A. Paine (Ed), *Food Packaging Technology International*, 1989, Cornhill Publications Ltd, London (1989).

F. A. Paine (Ed), *Food Packaging Technology International 1990*, Cornhill Publications Ltd, London (1990).

F. A. Paine (Ed), *Fundamentals of Packaging*, UK Inst of Packaging, (reprinted 1985).

F. A. Paine (Ed), *Modern Processing Packaging and Distribution Systems for Food*, Blackie & Sons, Glasgow.

F. A. Paine (Ed), *Packaging and the Law*, Newnes-Butterworths, London (1973).

F. A. Paine, *Packaging Design and Performance*, Pira, Leatherhead, Surrey (1990).

F. A. Paine, *Tamper Evident Packaging*, Pira Reviews of Packaging, Pira, Leatherhead, Surrey (1989).

F. A. Paine and H. Y. Paine, *Handbook of Food Packaging*, Blackie & Son, Glasgow.

S. J. Palling, *Developments in Food Packaging*, Applied Science Publishers, London (1980).

Pocket Pal, International Paper Co., New York (1983).

Ports of the World, Benn Publications Ltd, London (Annual).

Preshipment Test Procedures, National Safe Transit Association (1982).

L. Preston, *Flexible Packaging Materials – An Update*, Pira Reviews of Packaging, Pira, Leatherhead, Surrey (1990).

The Printing Ink Manual, (3rd Edition), Northwood Books, London (1979).

M. Pyke, *Food Science and Technology*, 4th edition, John Murray Ltd, London (1980).

G. Raimondi, Reduction of packaging costs, *Tech. Imball.*, **5**(11) Nov 1974, pp. 103–106.

H. F. Rance (Ed), *Handbook of Paper Science*, Elsevier Scientific Publications Co., New York (1982).

Recommendations of the United Nations Committee of Experts on the Packaging of Dangerous Goods;

Sea Transport, *International Maritime Dangerous Goods Code*, International Maritime Organization (IMO);

Air, *Technical instructions for the safe transport of Dangerous Goods by Air*, International Civil Aviation Organisation;

Rail, *Internl. Regs. concerning the carriage of Dangerous Goods by Rail (RID), Annex 1 Convention on Intntl. Carriage by Rail*;

Road, *European Agreement concerning the Intntl. Carriage of Dangerous Goods by Road (ADR)*;

RID and ADR; English texts are published by H.M.S.O., London.

S. Sacharow, *A Guide to Packaging Machines*, Harcourt Brace Jovanovich, New York (1980).

S. Sacharow, *A Packaging Primer*, Books for Industry, New York (1978).

S. Sacharow, *Handbook of Packaging Materials*, AVI Publishing Co. Inc., Westport, USA (1976).

D. Satas (Ed), *Handbook of Pressure Sensitive Adhesives*, Van Nostrand Reinhold Co. (1982).

D. A. Seiler, *Preservation of Bakery Products*, FMBRA Bulletin No. 4 (1983).

S. Selke, *Biodegradation and Packaging*, Pira Reviews of Packaging, Pira, Leatherhead, Surrey (1990).

J. Stepek *et al.*, *Polymers as Materials for Packaging*, Ellis Horwood, Chichester (1987).

C. Swinbank, Coordination of National, Regional & International Standardisation in Packaging, 1978, Report ITC/Conf/P/4, ITC (UNCTAD/GATT) Geneva.

C. Swinbank, *Intermediate Bulk Containers*, Institute of Packaging, Melton Mowbray, Leics. (1983).

C. Swinbank, *Packaging of Chemicals*, Newnes-Butterworth, London (1973).

J. S. Thorne (Ed), *Developments in Food Preservation*, Applied Science, UK (1981).

I. Turtle, *Plastics Packaging in Pack Heat Processed Foods*, Pira Reviews of Packaging, Pira, Leatherhead, Surrey (1990).

L. R. Whittington, *Whittington's Directory of Plastics*, 2nd edition, Technomic Publishing Co. Inc., Lancaster Pa., USA (1978).

A. H. Woollen (Ed), *Food Industries Manual*, 20th edition, Leonard Hill (1969).

Wood Crate Design Manual, Agriculture Handbook No. 252, US Dept. of Agriculture, Forest Service (1964).

P. G. Wright, P. R. McKinlay and E. Y. N. Shaw, *Corrugated Fibreboard Boxes*, Australian Paper Makers (Amcor Group Ltd), Camberwell, Australia (1988).

Index